DIFFERENTIAL AND INTEGRAL INEQUALITIES
Theory and Applications

Volume II
FUNCTIONAL, PARTIAL, ABSTRACT, AND COMPLEX DIFFERENTIAL EQUATIONS

This is Volume 55 in
MATHEMATICS IN SCIENCE AND ENGINEERING
A series of monographs and textbooks
Edited by RICHARD BELLMAN, *University of Southern California*

A complete list of the books in this series appears at the end of this volume.

DIFFERENTIAL AND INTEGRAL INEQUALITIES
Theory and Applications

Volume II
FUNCTIONAL, PARTIAL, ABSTRACT, AND COMPLEX DIFFERENTIAL EQUATIONS

V. LAKSHMIKANTHAM and S. LEELA

University of Rhode Island
Kingston, Rhode Island

ACADEMIC PRESS   New York and London   1969

Copyright © 1969, by Academic Press, Inc.
ALL RIGHTS RESERVED.
NO PART OF THIS BOOK MAY BE REPRODUCED IN ANY FORM,
BY PHOTOSTAT, MICROFILM, RETRIEVAL SYSTEM, OR ANY
OTHER MEANS, WITHOUT WRITTEN PERMISSION FROM
THE PUBLISHERS.

ACADEMIC PRESS, INC.
111 Fifth Avenue, New York, New York 10003

United Kingdom Edition published by
ACADEMIC PRESS, INC. (LONDON) LTD.
Berkeley Square House, London W1X 6BA

LIBRARY OF CONGRESS CATALOG CARD NUMBER: 68-8425
AMS 1968 SUBJECT CLASSIFICATIONS 3401, 3501

PRINTED IN THE UNITED STATES OF AMERICA

Preface

The first volume of Differential and Integral Inequalities: Theory and Applications published in 1969 deals with ordinary differential equations and Volterra integral equations. It consists of five chapters and includes a systematic and fairly elaborate development of the theory and application of differential and integral inequalities.

This second volume is a continuation of the trend and is devoted to differential equations with delay or functional differential equations, partial differential equations of first order, parabolic and hyperbolic types respectively, differential equations in abstract spaces including nonlinear evolution equations and complex differential equations. To cut down the length of the volume many parallel results are omitted as exercises.

We extend our appreciation to Mrs. Rosalind Shumate, Mrs. June Chandronet, and Miss Sally Taylor for their excellent typing of the entire manuscript.

V. LAKSHMIKANTHAM
S. LEELA

Kingston, Rhode Island
August, 1969

Contents

PREFACE   v

FUNCTIONAL DIFFERENTIAL EQUATIONS

Chapter 6.
- 6.0. Introduction   3
- 6.1. Existence   4
- 6.2. Approximate Solutions and Uniqueness   9
- 6.3. Upper Bounds   13
- 6.4. Dependence on Initial Values and Parameters   18
- 6.5. Stability Criteria   21
- 6.6. Asymptotic Behavior   24
- 6.7. A Topological Principle   29
- 6.8. Systems with Repulsive Forces   32
- 6.9. Functional Differential Inequalities   34
- 6.10. Notes   42

Chapter 7.
- 7.0. Introduction   43
- 7.1. Stability Criteria   43
- 7.2. Converse Theorems   49
- 7.3. Autonomous Systems   58
- 7.4. Perturbed Systems   62
- 7.5. Extreme Stability   66
- 7.6. Almost Periodic Systems   72
- 7.7. Notes   80

Chapter 8.
- 8.0. Introduction   81
- 8.1. Basic Comparison Theorems   81
- 8.2. Stability Criteria   87
- 8.3. Perturbed Systems   97
- 8.4. An Estimate of Time Lag   100
- 8.5. Eventual Stability   101
- 8.6. Asymptotic Behavior   105
- 8.7. Notes   110

PARTIAL DIFFERENTIAL EQUATIONS

Chapter 9.
- 9.0. Introduction — 113
- 9.1. Partial Differential Inequalities of First Order — 113
- 9.2. Comparison Theorems — 118
- 9.3. Upper Bounds — 127
- 9.4. Approximate Solutions and Uniqueness — 134
- 9.5. Systems of Partial Differential Inequalities of First Order — 136
- 9.6. Lyapunov-Like Function — 144
- 9.7. Notes — 148

Chapter 10.
- 10.0. Introduction — 149
- 10.1. Parabolic Differential Inequaliies in Bounded Domains — 149
- 10.2. Comparison Theorems — 155
- 10.3. Bounds, Under and Over Functions — 163
- 10.4. Approximate Solutions and Uniqueness — 170
- 10.5. Stability of Steady-State Solutions — 174
- 10.6. Systems of Parabolic Differential Inequalities in Bounded Domains — 181
- 10.7. Lyapunov-Like Functions — 186
- 10.8. Stability and Boundedness — 190
- 10.9. Conditional Stability and Boundedness — 200
- 10.10. Parabolic Differential Inequalities in Unbounded Domains — 205
- 10.11. Uniqueness — 210
- 10.12. Exterior Boundary-Value Problem and Uniqueness — 213
- 10.13. Notes — 219

Chapter 11.
- 11.0. Introduction — 221
- 11.1. Hyperbolic Differential Inequalities — 221
- 11.2. Uniqueness Criteria — 223
- 11.3. Upper Bounds and Error Estimates — 229
- 11.4. Notes — 233

DIFFERENTIAL EQUATIONS IN ABSTRACT SPACES

Chapter 12.
- 12.0. Introduction — 237
- 12.1. Existence — 237
- 12.2. Nonlocal Existence — 241
- 12.3. Uniqueness — 243
- 12.4. Continuous Dependence and the Method of Averaging — 247
- 12.5. Existence (continued) — 249
- 12.6. Approximate Solutions and Uniqueness — 254
- 12.7. Chaplygin's Method — 258
- 12.8. Asymptotic Behavior — 264
- 12.9. Lyapunov Function and Comparison Theorems — 267
- 12.10. Stability and Boundedness — 269
- 12.11. Notes — 272

COMPLEX DIFFERENTIAL EQUATIONS

Chapter 13.
- 13.0. Introduction — 275
- 13.1. Existence, Approximate Solutions, and Uniqueness — 275
- 13.2. Singularity-Free Regions and Growth Estimates — 279
- 13.3. Componentwise Bounds — 284
- 13.4. Lyapunov-like Functions and Comparison Theorems — 286
- 13.5. Notes — 288

Bibliography — 289

Author Index — 315

Subject Index — 318

DIFFERENTIAL AND INTEGRAL INEQUALITIES
Theory and Applications

Volume II
FUNCTIONAL, PARTIAL, ABSTRACT, AND COMPLEX DIFFERENTIAL EQUATIONS

FUNCTIONAL DIFFERENTIAL EQUATIONS

Chapter 6

6.0. Introduction

The future state of a physical system depends, in many circumstances, not only on the present state but also on its past history. Functional differential equations provide a mathematical model for such physical systems in which the rate of change of the system may depend on the influence of its hereditary effects. The simplest type of such a system is a differential-difference equation

$$x'(t) = f(t, x(t), x(t - \tau)),$$

where $\tau > 0$ is a constant. Obviously, for $\tau = 0$, this reduces to an ordinary differential equation. More general systems may be described by the following equation:

$$x'(t) = f(t, x_t),$$

where f is a suitable functional. The symbol x_t may be defined in several ways. For example, if x is a function defined on some interval $[t_0 - \tau, t_0 + a)$, $a > 0$, then, for each $t \in [t_0, t_0 + a)$,

(i) x_t is the graph of x on $[t - \tau, t]$ shifted to the interval $[-\tau, 0]$;
(ii) x_t is the graph of x on $[t_0 - \tau, t]$.

In case (ii), f is a functional of Volterra type which is determined by t and the values of $x(s)$, $t_0 - \tau \leqslant s \leqslant t$. Systems of this form are called delay-differential systems. In what follows, we shall, however, consider the functional differential equations in which the symbol x_t has the meaning described by case (i) and study some qualitative problems by means of the theory of differential inequalities.

In the present chapter, we consider existence, uniqueness, continuation, and continuous dependence of solutions and obtain *a priori* bounds and error estimates. Asymptotic behavior and stability criteria are included. An extension of topological principle to functional differential equations, together with some applications, is discussed. Finally, we develop the theory of functional differential inequalities, introduce the notion of maximal and minimal solutions, prove comparison theorems in this setup, and give some interesting applications.

6.1. Existence

Given any $\tau > 0$, let $\mathscr{C}^n = C[[-\tau, 0], R^n]$ denote the space of continuous functions with domain $[-\tau, 0]$ and range in R^n. For any element $\phi \in \mathscr{C}^n$, define the norm

$$\|\phi\|_0 = \max_{-\tau \leqslant s \leqslant 0} \|\phi(s)\|.$$

Suppose that $x \in C[[-\tau, \infty), R^n]$. For any $t \geqslant 0$, we shall let x_t denote a translation of the restriction of x to the interval $[t - \tau, t]$; more specifically, x_t is an element of $\mathscr{C}^n$ defined by

$$x_t(s) = x(t + s), \qquad -\tau \leqslant s \leqslant 0.$$

In other words, the graph of x_t is the graph of x on $[t - \tau, t]$ shifted to the interval $[-\tau, 0]$. Let $\rho > 0$ be a given constant, and let

$$C_\rho = [\phi \in \mathscr{C}^n : \|\phi\|_0 < \rho].$$

With this notation, we may write a functional differential system in the form

$$x'(t) = f(t, x_t). \tag{6.1.1}$$

DEFINITION 6.1.1. A function $x(t_0, \phi_0)$ is said to be a *solution of* (6.1.1) *with the given initial function* $\phi_0 \in C_\rho$ at $t = t_0 \geqslant 0$, if there exists a number $A > 0$ such that

(i) $x(t_0, \phi_0)$ is defined and continuous on $[t_0 - \tau, t_0 + A]$ and $x_t(t_0, \phi_0) \in C_\rho$ for $t_0 \leqslant t \leqslant t_0 + A$;
(ii) $x_{t_0}(t_0, \phi_0) = \phi_0$;
(iii) the derivative of $x(t_0, \phi_0)$ at t, $x'(t_0, \phi_0)(t)$ exists for $t \in [t_0, t_0 + A)$ and satisfies the system (6.1.1) for $t \in [t_0, t_0 + A)$.

We now state the following well-known result.

6.1. EXISTENCE

SCHAUDER'S FIXED POINT THEOREM. A continuous mapping of a compact convex subset of a Banach space into itself has at least one fixed point.

The following local existence theorem will now be proved.

THEOREM 6.1.1. Let $f \in C[J \times C_\rho, R^n]$. Then, given an initial function $\phi_0 \in C_\rho$ at $t = t_0 \geqslant 0$, there exists an $\alpha > 0$ such that there is a solution $x(t_0, \phi_0)$ of (6.1.1) existing on $[t_0 - \tau, t_0 + \alpha)$.

Proof. Let $a > 0$ and define $y \in C[[t_0 - \tau, t_0 + a], R^n]$ as follows:

$$y(t) = \begin{cases} \phi_0(t - t_0), & t_0 - \tau \leqslant t \leqslant t_0, \\ \phi_0(0), & t_0 \leqslant t \leqslant t_0 + a \end{cases}.$$

Then $f(t, y_t)$ is a continuous function of t on $[t_0, t_0 + a]$ and hence $\|f(t, y_t)\| \leqslant M_1$. We shall show that there exists a constant

$$b \in (0, \rho - \|\phi_0(0)\|)$$

such that

$$\|f(t, \psi) - f(t, y_t)\| < 1$$

whenever $t \in [t_0, t_0 + a]$, $\psi \in C_\rho$ and $\|\psi - y_t\|_0 \leqslant b$. Suppose that this is not true. Then for each $k = 1, 2,...$, there would exist $t_k \in [t_0, t_0 + a]$ and $\psi_k \in C_\rho$ such that $\|\psi_k - y_{t_k}\|_0 < 1/k$ and yet

$$\|f(t_k, \psi_k) - f(t_k, y_{t_k})\| \geqslant 1.$$

We now choose a subsequence $\{t_{k_p}\}$ such that $\lim_{p \to \infty} t_{k_p} = t_1$ exists, and we have a contradiction to the continuity of f at (t_1, y_{t_1}).

It now follows that $\|f(t, \psi)\| \leqslant M = M_1 + 1$ whenever $t \in [t_0, t_0 + a]$, $\psi \in C_\rho$ and $\|\psi - y_t\|_0 \leqslant b$. Choose $\alpha = \min(a, b/M)$.

Let B denote the space of continuous functions from $[t_0 - \tau, t_0 + \alpha]$ into R^n. For an element $x \in B$, define the norm

$$\|x\|_0 = \max_{t_0 - \tau \leqslant u \leqslant t_0 + \alpha} \|x(u)\|.$$

With respect to this norm, B is a Banach space. Let $S \subset B$ be defined as follows:

$$S = \left\{ x \in B \quad \begin{array}{l} \text{(i)} \quad x(u) = \phi_0(u - t_0), \quad t_0 - \tau \leqslant u \leqslant t_0, \\ \text{(ii)} \quad \|x(t_1) - x(t_2)\| \leqslant M|t_1 - t_2|, \quad t_1, t_2 \in [t_0, t_0 + \alpha] \end{array} \right\}.$$

Since the members of S are uniformly bounded and equicontinuous on $[t_0 - \tau, t_0 + \alpha]$, the compactness of S follows. A straightforward computation shows that S is convex.

We now define a mapping on S as follows. For an element $x \in S$, let

(i) $T(x_{t_0}) = \phi_0$;
(ii) $T(x(t)) = \phi_0(0) + \int_{t_0}^t f(s, x_s)\, ds,\ t_0 \leqslant t \leqslant t_0 + \alpha$.

For every $x \in S$ and $t \in [t_0, t_0 + \alpha]$,

$$\| x(t) - \phi_0(0) \| \leqslant M | t - t_0 | \leqslant M\alpha \leqslant b.$$

Thus $x_t \in C_\rho$ and $\| x_t - y_t \|_0 \leqslant b$. Hence $f(s, x_s)$ is a continuous function of s and $\| f(s, x_s) \| \leqslant M$ for $t_0 \leqslant s \leqslant t_0 + \alpha$. It therefore follows that the mapping T is well-defined on S. It is readily verified that T maps S into itself and is continuous. An application of Schauder's fixed point theorem now yields the existence of at least one function $x \in S$ such that

(i) $T(x_{t_0}) = x_{t_0}$;
(ii) $T(x(t)) = x(t),\ t_0 \leqslant t \leqslant t_0 + \alpha$,

which implies that

(i) $x_{t_0} = \phi_0$;
(ii) $x(t) = \phi_0(0) + \int_{t_0}^t f(s, x_s)\, ds,\ t_0 \leqslant t \leqslant t_0 + \alpha$.

Since $x \in S$, the integrand in the foregoing equation is a continuous function of s. Thus, for $t_0 \leqslant t < t_0 + \alpha$, we can differentiate to obtain

$$x'(t) = f(t, x_t), \qquad t_0 \leqslant t < t_0 + \alpha.$$

It follows that $x(t_0, \phi_0)$ is a solution of (6.1.1), defined on $[t_0 - \tau, t_0 + \alpha)$, and the proof is complete.

We consider next the existence of solutions of (6.1.1) for all $t \geqslant t_0$. The following lemma is needed before we proceed in that direction.

LEMMA 6.1.1. *Let $m \in C[[t_0 - \tau, \infty), R_+]$, and satisfy the inequality*

$$D\_m(t) \leqslant g(t, | m_t |_0), \qquad t > t_0,$$

where $g \in C[J \times R_+, R_+]$. Assume that $r(t) = r(t, t_0, u_0)$ is the maximal solution of the scalar differential equation

$$u' = g(t, u), \qquad u(t_0) = u_0 \geqslant 0 \tag{6.1.2}$$

existing for $t \geqslant t_0$. Then,

$$m(t) \leqslant r(t), \qquad t \geqslant t_0,$$

provided $| m_{t_0} |_0 \leqslant u_0$.

6.1. EXISTENCE

Proof. To prove the stated inequality, it is enough to prove that

$$m(t) < u(t, t_0, u_0, \epsilon), \qquad t \geq t_0,$$

where $u(t, t_0, u_0, \epsilon)$ is any solution of

$$u' = g(t, u) + \epsilon, \qquad u(t_0) = u_0 + \epsilon,$$

$\epsilon > 0$ being an arbitrarily small quantity, since

$$\lim_{\epsilon \to 0} u(t, t_0, u_0, \epsilon) = r(t, t_0, u_0).$$

The proof of this fact follows closely the proof of Theorem 1.2.1. We assume that the set

$$Z = [t \in [t_0, \infty): m(t) \geq u(t, t_0, u_0, \epsilon)]$$

is nonempty and define $t_1 = \inf Z$. It results from $|m_{t_0}|_0 < u_0 + \epsilon$ that $t_1 > t_0$. Moreover,

$$m(t_1) = u(t_1, t_0, u_0, \epsilon),$$

and

$$m(t) < u(t, t_0, u_0, \epsilon), \qquad t_0 \leq t < t_1.$$

Hence,

$$D_- m(t_1) \geq u'(t_1, t_0, u_0, \epsilon) = g(t_1, u(t_1, t_0, u_0, \epsilon)) + \epsilon. \qquad (6.1.3)$$

Since $g(t, u) \geq 0$, $u(t, t_0, u_0, \epsilon)$ is nondecreasing in t, and this implies, from the preceding considerations, that

$$|m_{t_1}|_0 = u(t_1, t_0, u_0, \epsilon) = m(t_1). \qquad (6.1.4)$$

Thus, we are led to the inequality

$$D_- m(t_1) \leq g(t_1, |m_{t_1}|_0) = g(t_1, u(t_1, t_0, u_0, \epsilon)),$$

which is incompatible with (6.1.3). The set Z is therefore empty, and the lemma follows.

With obvious changes, we can prove

COROLLARY 6.1.1. *Let* $m \in C[[t_0 - \tau, \infty), R_+]$, *and, for* $t > t_0$,

$$D_- m(t) \geq -g(t, \min_{t-\tau \leq s \leq t} m_t(s)),$$

where $g \in C[J \times R_+, R_+]$. Let $\rho(t) = \rho(t, t_0, u_0) \geq 0$ be the minimal solution of

$$u' = -g(t, u), \quad u(t_0) = u_0 > 0$$

existing for $t \geq t_0$. Then $u_0 \leq \min_{t_0 - \tau \leq s \leq t_0} m_{t_0}(s)$ implies

$$\rho(t) \leq m(t), \quad t \geq t_0.$$

The following variation of Lemma 6.1.1, whose assumptions are weaker, is also useful for later applications.

LEMMA 6.1.2. *Let $m \in C[[t_0 - \tau, \infty), R_+]$, and let, for every $t_1 > t_0$ for which $|m_{t_1}|_0 = m(t_1)$, the inequality*

$$D_- m(t_1) \leq g(t_1, m(t_1))$$

be verified, where $g \in C[J \times R_+, R_+]$. Then, the conclusion of Lemma 6.1.1 is valid.

The proof follows from Lemma 6.1.1 in view of the relation (6.1.4).

THEOREM 6.1.2. *Let $f \in C[J \times \mathscr{C}^n, R^n]$, and, for $(t, \phi) \in J \times \mathscr{C}^n$,*

$$\|f(t, \phi)\| \leq g(t, \|\phi\|_0), \tag{6.1.5}$$

where $g \in C[J \times R_+, R_+]$ and is nondecreasing in u for each $t \in J$. Assume that the solutions $u(t) = u(t, t_0, u_0)$ of (6.1.2) exist for all $t \geq t_0$. Then, the largest interval of existence of any solution $x(t_0, \phi_0)$ of (6.1.1) is $[t_0, \infty)$.

Proof. Let $x(t_0, \phi_0)$ be a solution of (6.1.1) existing on some interval $[t_0, \beta)$, where $t_0 < \beta < \infty$. Assume that the value of β cannot be increased. Define, for $t \in [t_0, \beta)$, $m(t) = \|x(t_0, \phi_0)(t)\|$, so that $m_t = \|x_t(t_0, \phi_0)\|$. Using the assumption (6.1.5), it is easy to obtain, for $t > t_0$, the inequality

$$D_- m(t) \leq g(t, |m_t|_0).$$

Choosing $|m_{t_0}|_0 = \|\phi_0\|_0 \leq u_0$, we obtain, by Lemma 6.1.1,

$$\|x(t_0, \phi_0)(t)\| \leq r(t, t_0, u_0), \quad t_0 \leq t < \beta. \tag{6.1.6}$$

Since the function $g(t, u) \geq 0$, $r(t, t_0, u_0)$ is nondecreasing, and, hence, it follows from (6.1.6) that

$$\|x_t(t_0, \phi_0)\|_0 \leq r(t, t_0, u_0), \quad t_0 \leq t < \beta. \tag{6.1.7}$$

For any t_1, t_2 such that $t_0 < t_1 < t_2 < \beta$, we have

$$\| x(t_0, \phi_0)(t_1) - x(t_0, \phi_0)(t_2) \| \leq \int_{t_1}^{t_2} g(s, \| x_s(t_0, \phi_0)\|_0) \, ds,$$

which, in view of (6.1.7) and the monotonicity of $g(t, u)$ in u, implies

$$\| x(t_0, \phi_0)(t_1) - x(t_0, \phi_0)(t_2) \| \leq \int_{t_1}^{t_2} g(s, r(s, t_0, u_0)) \, ds$$
$$= r(t_2, t_0, u_0) - r(t_1, t_0, u_0).$$

Letting $t_1, t_2 \to \beta^-$, the foregoing relation shows that $\lim_{t \to \beta^-} x(t_0, \phi_0)(t)$ exists, because of Cauchy's criterion for convergence. We now define $x(t_0, \phi_0)(\beta) = \lim_{t \to \beta^-} x(t_0, \phi_0)(t)$ and consider $\psi_0 = x_\beta(t_0, \phi_0)$ as the new initial function at $t = \beta$. An application of Theorem 6.1.1 shows that there exists a solution $x(\beta, \psi_0)$ of (6.1.1) on $[\beta, \beta + \alpha]$, $\alpha > 0$. This means that the solution $x(t_0, \phi_0)$ can be continued to the right of β, which is contrary to the assumption that the value of β cannot be increased. Hence, the stated result follows.

COROLLARY 6.1.2. *The conclusion of Theorem 6.1.2 remains true even when the condition (6.1.5) is assumed to hold only for $t \in J$ and ϕ satisfying $\| \phi \|_0 = \| \phi(0) \|$.*

6.2. Approximate solutions and uniqueness

DEFINITION 6.2.1. A function $x(t_0, \phi_0, \epsilon)$ is said to be an ϵ-approximate solution of (6.1.1) for $t \geq t_0$ with the initial function $\phi_0 \in C_\rho$ at $t = t_0$ if

(i) $x(t_0, \phi_0, \epsilon)$ is defined and continuous on $[t_0 - \tau, \infty)$ and $x_t(t_0, \phi_0, \epsilon) \in C_\rho$;

(ii) $x_{t_0}(t_0, \phi_0, \epsilon) = \phi_0$;

(iii) $x(t_0, \phi_0, \epsilon)$ is differentiable on the interval $[t_0, \infty)$, except for an at most countable set S and satisfies

$$\| x'(t_0, \phi_0, \epsilon)(t) - f(t, x_t(t_0, \phi_0, \epsilon)) \| \leq \epsilon \qquad (6.2.1)$$

for $t \in [t_0, \infty) - S$.

In case $\epsilon = 0$, it is to be understood that S is empty and $x(t_0, \phi_0)$ is a solution of (6.1.1).

We shall now give some comparison theorems on ϵ-approximate solutions of (6.1.1).

THEOREM 6.2.1. Let $f \in C[J \times C_\rho, R^n]$, and, for $(t, \phi), (t, \psi) \in J \times C_\rho$,

$$\|f(t, \phi) - f(t, \psi)\| \leq g(t, \|\phi - \psi\|_0), \qquad (6.2.2)$$

where $g \in C[J \times [0, 2\rho), R_+]$. Assume that $r(t) = r(t, t_0, u_0)$ is the maximal solution of

$$u' = g(t, u) + \epsilon_1 + \epsilon_2, \qquad u(t_0) = u_0 \geq 0$$

existing for $t \geq t_0$. Let $x(t_0, \phi_0, \epsilon_1)$, $y(t_0, \psi_0, \epsilon_2)$ be ϵ_1, ϵ_2-approximate solutions of (6.1.1) such that

$$\|\phi_0 - \psi_0\|_0 \leq u_0. \qquad (6.2.3)$$

Then,

$$\|x(t_0, \phi_0, \epsilon_1)(t) - y(t_0, \psi_0, \epsilon_2)(t)\| \leq r(t, t_0, u_0), \qquad t \geq t_0. \qquad (6.2.4)$$

Proof. Consider the function

$$m(t) = \|x(t_0, \phi_0, \epsilon_1)(t) - y(t_0, \psi_0, \epsilon_2)(t)\|,$$

so that

$$m_t = \|x_t(t_0, \phi_0, \epsilon_1) - y_t(t_0, \psi_0, \epsilon_2)\|$$

for $t \geq t_0$. Then, we have, for $t > t_0$,

$$D_-m(t) \leq \|x'(t_0, \phi_0, \epsilon_1)(t) - y'(t_0, \psi_0, \epsilon_2)(t)\|$$
$$\leq \|x'(t_0, \phi_0, \epsilon_1)(t) - f(t, x_t(t_0, \phi_0, \epsilon_1))\|$$
$$+ \|y'(t_0, \psi_0, \epsilon_2)(t) - f(t, y_t(t_0, \psi_0, \epsilon_2))\|$$
$$+ \|f(t, x_t(t_0, \phi_0, \epsilon_1)) - f(t, y_t(t_0, \psi_0, \epsilon_2))\|.$$

Now, making use of the assumption (6.2.2) and the fact that $x(t_0, \phi_0, \epsilon_1)$, $y(t_0, \psi_0, \epsilon_2)$ are ϵ_1, ϵ_2-approximate solutions of (6.1.1), respectively, it follows, from the preceding inequality, that

$$D_-m(t) \leq g(t, |m_t|_0) + \epsilon_1 + \epsilon_2.$$

Choose u_0 such that (6.2.3) holds. Then, on the basis of Lemmas 1.2.2 and 6.1.1, the estimate (6.2.4) results immediately.

COROLLARY 6.2.1. The function $g(t, u) = Lu$, $L > 0$, is admissible in Theorem 6.2.1.

In fact, Corollary 6.2.1 implies the well-known inequality for approximate solutions

$$\|x(t_0, \phi_0, \epsilon_1)(t) - y(t_0, \psi_0, \epsilon_2)(t)\|$$
$$\leq \|\phi_0 - \psi_0\|_0 e^{L(t-t_0)} + \frac{\epsilon_1 + \epsilon_2}{L}[e^{L(t-t_0)} - 1], \qquad t \geq t_0. \qquad (6.2.5)$$

6.2. APPROXIMATE SOLUTIONS AND UNIQUENESS

It is possible to weaken the assumptions of Theorem 6.2.1 in that we need not assume the condition (6.2.2) for all $\phi, \psi \in C_\rho$. To do this, we require the subset $C_1 \subset C_\rho$ defined by

$$C_1 = [\phi, \psi \in C_\rho : \|\phi - \psi\|_0 = \|\phi(0) - \psi(0)\|].$$

THEOREM 6.2.2. Let the assumptions of Theorem 6.2.1 hold except that the condition (6.2.2) is replaced by

$$\|f(t, \phi) - f(t, \psi)\| \leqslant g(t, \|\phi(0) - \psi(0)\|) \qquad (6.2.6)$$

for $t \in J$, $\phi, \psi \in C_1$. Then, (6.2.3) implies the estimate (6.2.4).

Proof. Suppose that, for some $t_1 > t_0$, $|m_{t_1}|_0 = m(t_1)$, where $m(t)$, m_t are the same functions as in Theorem 6.2.1. Setting $\phi = x_{t_1}(t_0, \phi_0, \epsilon_1)$, $\psi = y_{t_1}(t_0, \psi_0, \epsilon_2)$, we see that $\phi, \psi \in C_1$ and $m(t_1) = \|\phi(0) - \psi(0)\|$. Hence, using (6.2.6), we get

$$D_-m(t_1) \leqslant g(t_1, m(t_1)) + \epsilon_1 + \epsilon_2,$$

as previously. The assumptions of Lemma 6.1.2 are verified, and therefore the conclusion follows.

A uniqueness result of Perron type may now be proved.

THEOREM 6.2.3. Assume that

(i) the function $g(t, u)$ is continuous, nonnegative for $t_0 \leqslant t \leqslant t_0 + a$, $0 \leqslant u \leqslant 2b$, and, for every $t_0 < t_1 < t_0 + a$, $u(t) \equiv 0$ is the only differentiable function on $t_0 \leqslant t < t_1$ which satisfies

$$u' = g(t, u), \qquad u(t_0) = 0 \qquad (6.2.7)$$

for $t_0 \leqslant t < t_1$;

(ii) $f \in C[R_0, R^n]$, where

$$R_0 : t_0 \leqslant t \leqslant t_0 + a, \qquad \|\phi - \phi_0(0)\|_0 \leqslant b,$$

and, for $(t, \phi), (t, \psi) \in R_0$ and $\phi, \psi \in C_1$,

$$\|f(t, \phi) - f(t, \psi)\| \leqslant g(t, \|\phi(0) - \psi(0)\|). \qquad (6.2.8)$$

Then, the functional differential system (6.1.1) admits at most one solution on $t_0 \leqslant t \leqslant t_0 + a$ such that $x_{t_0} = \phi_0$.

Proof. Suppose that there exist two solutions $x(t_0, \phi_0)$, $y(t_0, \phi_0)$ of (6.1.1) with the same initial function ϕ_0 at $t = t_0$. An argument similar to that of Theorem 6.2.2 shows that

$$\| x(t_0, \phi_0)(t) - y(t_0, \phi_0)(t) \| \leqslant r(t, t_0, 0), \qquad t_0 \leqslant t < t_1,$$

where $r(t, t_0, 0)$ is the maximal solution of (6.2.7). By assumption, however, we have $r(t, t_0, 0) \equiv 0$, and hence

$$x(t_0, \phi_0)(t) = y(t_0, \phi_0)(t), \qquad t_0 \leqslant t < t_1$$

for every t_1 such that $t_0 < t_1 < t_0 + a$. The uniqueness of solutions is therefore proved.

COROLLARY 6.2.2. *The function $g(t, u) = Lu$, $L > 0$ is admissible in Theorem 6.2.3.*

COROLLARY 6.2.3. *The function $g(t, u) = g(u)$, where $g(0) = 0$, $g(u) > 0$ for $u > 0$, and*

$$\int_{+0} \frac{ds}{g(s)} = +\infty,$$

is another admissible candidate in Theorem 6.2.3.

The next uniqueness result is analogous to Kamke's uniqueness theorem.

THEOREM 6.2.4. *Suppose that*

(i) *the function $g(t, u)$ is continuous, nonnegative for $t_0 < t \leqslant t_0 + a$, $0 \leqslant u \leqslant 2b$, and, for every t_1, $t_0 < t_1 < t_0 + a$, $u(t) \equiv 0$ is the only function differentiable on $t_0 < t < t_1$ and continuous on $t_0 \leqslant t < t_1$, for which*

$$u_+'(t_0) = \lim_{t \to t_0^+} \frac{u(t) - u(t_0)}{t - t_0} \quad \text{exists},$$

$$u'(t) = g(t, u(t)), \qquad t_0 < t < t_1,$$

and

$$u(t_0) = u_+'(t_0) = 0;$$

(ii) *the hypothesis (ii) of Theorem 6.2.3 is satisfied except that the condition (6.2.8) holds for (t, ϕ), $(t, \psi) \in R_0$, $\phi, \psi \in C_1$, and $t \neq t_0$.*

Then, the conclusion of Theorem 6.2.3 is true.

Proof. We define the function

$$g_f(t, u) = \sup_{\|\phi - \psi\|_0 = u} \|f(t, \phi) - f(t, \psi)\|$$

for $t_0 \leqslant t \leqslant t_0 + a$, $0 \leqslant u \leqslant 2b$. Since $f(t, \phi)$ is continuous on R_0, it follows that $g_f(t, u)$ is a continuous function on $t_0 \leqslant t \leqslant t_0 + a$ and $0 \leqslant u \leqslant 2b$. Moreover, because of the condition (6.2.8),

$$g_f(t, u) \leqslant g(t, u)$$

for $t_0 < t \leqslant t_0 + a$, $0 \leqslant u \leqslant 2b$. Theorem 2.2.3 is therefore applicable with $g_1(t, u) = g_f(t, u)$, and hence the conclusion follows from Theorem 6.2.3.

6.3. Upper bounds

We give, in this section, some *a priori* bounds for the solutions $x(t_0, \phi_0)$ of the system (6.1.1).

THEOREM 6.3.1. *Let* $f \in C[J \times C_\rho, R^n]$, *and let* $x(t_0, \phi_0)$ *be any solution of (6.1.1) such that* $\|\phi_0\|_0 \leqslant u_0$. *Suppose that*

$$\|f(t, \phi)\| \leqslant g(t, \|\phi(0)\|) \qquad (6.3.1)$$

for $t \in J$ *and* $\phi \in \mathscr{C}^n$ *such that*

$$\|\phi\|_0 = \|\phi(0)\|, \qquad (6.3.2)$$

where $g \in C[J \times [0, \rho), R_+]$. *Then,*

$$\|x(t_0, \phi_0)(t)\| \leqslant r(t, t_0, u_0), \qquad (6.3.3)$$

on the common interval of existence of $x(t_0, \phi_0)$ *and* $r(t, t_0, u_0)$, *where* $r(t, t_0, u_0)$ *is the maximal solution of (6.1.2).*

On the basis of the proof of Theorem 6.2.2, this theorem can be demonstrated.

THEOREM 6.3.2. *Let* $f \in C[J \times C_\rho, R^n]$, $g \in C[J \times [0, \rho), R_+]$, *and*

$$\|f(t, \phi)\| \leqslant g(t, \|\phi\|_0)$$

for $(t, \phi) \in J \times \Omega$, *where* Ω *is the set defined by*

$$\Omega = [\phi \in C_\rho : r(t, t_0, u_0) \leqslant \|\phi\|_0 \leqslant r(t, t_0, u_0) + \epsilon_0, t \geqslant t_0],$$

for a certain $\epsilon_0 > 0$, $r(t, t_0, u_0)$ being the maximal solution of (6.1.2) existing for $t \geq t_0$. Then, if $x(t_0, \phi_0)$ is any solution of (6.1.1) such that $\|\phi_0\|_0 = u_0$, we have

$$\|x(t_0, \phi_0)(t)\| \leq r(t, t_0, u_0) \tag{6.3.4}$$

as far as $x(t_0, \phi_0)$ exists to the right of t_0.

Proof. Let $[t_0, \tau]$ be a given compact interval. Then, by Lemma 1.3.1, the maximal solution $r(t, \epsilon) = r(t, t_0, u_0, \epsilon)$ of

$$u' = g(t, u) + \epsilon, \qquad u(t_0) = u_0 + \epsilon$$

exists on $[t_0, \tau]$, for all sufficiently small $\epsilon > 0$ and

$$\lim_{\epsilon \to 0} r(t, \epsilon) \equiv r(t)$$

uniformly on $[t_0, \tau]$. In view of this, there exists an $\epsilon_0 > 0$ such that

$$r(t, \epsilon) < r(t) + \epsilon_0, \qquad t \in [t_0, \tau].$$

Furthermore, we have, by Theorem 1.2.1,

$$r(t) < r(t, \epsilon), \qquad t \in [t_0, \tau],$$

which implies that

$$r(t) < r(t, \epsilon) < r(t) + \epsilon_0, \qquad t \in [t_0, \tau]. \tag{6.3.5}$$

To prove (6.3.4), it is enough to prove that

$$\|x(t_0, \phi_0)(t)\| < r(t, \epsilon), \qquad t \in [t_0, \tau].$$

Assuming the contrary and proceeding as in the proof of Lemma 6.1.1 we arrive at a $t_1 > t_0$ such that

(i) $\|x(t_0, \phi_0)(t_1)\| = r(t_1, \epsilon)$;
(ii) $\|x(t_0, \phi_0)(t)\| < r(t, \epsilon)$, $t_0 \leq t < t_1$;
(iii) $\|x_{t_1}(t_0, \phi_0)\|_0 = r(t_1, \epsilon) = \|x(t_0, \phi_0)(t_1)\|$.

Setting now $\phi = x_{t_1}(t_0, \phi_0)$, we see that $\|\phi\|_0 = r(t_1, \epsilon)$, and so, because of (6.3.5), it follows that, at $t = t_1$, $\phi \in \Omega$. Thus, we obtain, defining $m(t) = \|x(t_0, \phi_0)(t)\|$,

$$D_- m(t_1) \leq g(t_1, \|\phi\|_0)$$
$$= g(t_1, r(t_1, \epsilon)),$$

6.3. UPPER BOUNDS

which contradicts the relation

$$D_-m(t_1) \geq r'(t_1, \epsilon) = g(t_1, r(t_1, \epsilon)) + \epsilon,$$

resulting from cases (i) and (ii) above. The proof is therefore complete.

The following theorem gives a more useful estimate.

THEOREM 6.3.3. Let $f \in C[J \times C_\rho, R^n]$, $g \in C[J \times [0, \rho), R_+]$, and

$$A(t) \liminf_{h \to 0^-} h^{-1}[\|\phi(0) + hf(t, \phi)\| - \|\phi(0)\|] + \|\phi(0)\| D_-A(t)$$
$$\leq g(t, \|\phi(0)\| A(t)), \tag{6.3.6}$$

for $t > t_0$ and $\phi \in C_\rho$ satisfying

$$\|\phi\|_0 \mid A_t \mid_0 = \|\phi(0)\| A(t), \tag{6.3.7}$$

where $A(t) \geq 0$ is continuous on $[-\tau, \infty)$. Let $x(t_0, \phi_0)$ be any solution of (6.1.1) such that

$$\|\phi_0\|_0 \mid A_{t_0} \mid_0 \leq u_0$$

and $r(t, t_0, u_0)$ be the maximal solution of (6.1.2) existing to the right of t_0. Then,

$$\|x(t_0, \phi_0)(t)\| A(t) \leq r(t, t_0, u_0),$$

as far as $x(t_0, \phi_0)$ exists to the right of t_0.

Proof. Consider the function

$$m(t) = \|x(t_0, \phi_0)(t)\| A(t),$$

so that

$$m_t = \|x_t(t_0, \phi_0)\| A_t.$$

By hypothesis, we have

$$\mid m_{t_0} \mid_0 = \|\phi_0\|_0 \mid A_{t_0} \mid_0 \leq u_0.$$

Let $u(t, \epsilon) = u(t, t_0, u_0, \epsilon)$ be any solution of

$$u' = g(t, u) + \epsilon, \qquad u(t_0) = u_0 \tag{6.3.8}$$

for $\epsilon > 0$ sufficiently small. Since

$$\lim_{\epsilon \to 0} u(t, \epsilon) = r(t, t_0, u_0)$$

and $u(t, \epsilon)$ exists as far as $r(t, t_0, u_0)$ exists, it is enough to show that

$$m(t) \leqslant u(t, \epsilon), \qquad t \geqslant t_0. \tag{6.3.9}$$

If this inequality is not true, let t_1 be the greatest lower bound of numbers $t > t_0$ for which (6.3.9) is false. The continuity of the functions $m(t)$ and $u(t, \epsilon)$ implies that

(i) $m(t) \leqslant u(t, \epsilon)$, $t_0 \leqslant t \leqslant t_1$;
(ii) $m(t_1) = u(t_1, \epsilon)$, $t = t_1$.

By the relations (i) and (ii), we have

$$D_- m(t_1) \geqslant u'(t_1, \epsilon) = g(t_1, u(t_1, \epsilon)) + \epsilon. \tag{6.3.10}$$

Since $g(t, u) + \epsilon$ is positive, the solutions $u(t, \epsilon)$ are monotonic nondecreasing in t, and hence, by relation (ii),

$$| m_{t_1} |_0 = m(t_1) = u(t_1, \epsilon).$$

Setting $\phi = x_{t_1}(t_0, \phi_0)$, it follows that

$$\| \phi \|_0 | A_{t_1} |_0 = \| \phi(0) \| A(t_1).$$

Thus, at $t = t_1$, (6.3.7) holds true with this ϕ. Hence, using (6.3.6), there results, using the standard computations, the inequality

$$D_- m(t_1) \leqslant g(t_1, m(t_1)),$$

which contradicts (6.3.10). It therefore follows that (6.3.9) is true, and this proves the stated result.

As a typical result, we shall prove next a theorem for componentwise bounds.

THEOREM 6.3.4. Let $f \in C[J \times C_\rho, R^n]$, and let

$$| f_i(t, \phi) | \leqslant g_i(t, | \phi_1 |_0, ..., | \phi_{i-1} |_0, | \phi_i(0) |, | \phi_{i+1} |_0, ..., | \phi_n |_0) \tag{6.3.11}$$

for each $i = 1, 2, ..., n$, $t \in J$, and $\phi \in C_\rho$ satisfying

$$| \phi_i |_0 = | \phi_i(0) | \qquad (i = 1, 2, ..., n),$$

where $g \in C[J \times R_+^n, R_+^n]$, $g(t, u)$ is quasi-monotone nondecreasing in u for each $t \in J$. Then, if $x(t_0, \phi_0)$ is any solution of (6.1.1) with the initial function $\phi_0 = (\phi_{10}, ..., \phi_{n0})$, such that $| \phi_{i0} |_0 \leqslant u_{i0}$, we have

$$| x_i(t_0, \phi_0)(t) | \leqslant r_i(t, t_0, u_0) \qquad (i = 1, 2, ..., n),$$

as far as $x(t_0, \phi_0)$ exists to the right of t_0, where $r(t, t_0, u_0)$ is the maximal solution of the ordinary differential system

$$u' = g(t, u), \quad u(t_0) = u_0$$

existing for $t \geq t_0$.

Proof. Define the vector function

$$m(t) = |x(t_0, \phi_0)(t)|,$$

so that

$$m_t = |x_t(t_0, \phi_0)|.$$

Then, since $|\phi_{i0}|_0 \leq u_{i0}$, we have

$$|m^i_{t_0}|_0 \leq u_{i0}.$$

As before, it is enough to show that

$$m_i(t) < u_i(t, \epsilon), \quad t \geq t_0, \quad i = 1, 2, \ldots, n, \tag{6.3.12}$$

where $u(t, \epsilon) = (u_1(t, \epsilon), \ldots, u_n(t, \epsilon))$ is any solution of

$$u' = g(t, u) + \epsilon, \quad u(t_0) = u_0 + \epsilon,$$

for sufficiently small $\epsilon > 0$. If (6.3.12) is false, let

$$Z = \bigcup_{i=1}^{n} [t \in J : m_i(t) \geq u_i(t, \epsilon)]$$

be nonempty and $t_1 = \inf Z$. Arguing as in Theorem 1.5.1, there exists an index j such that

(i) $m_j(t_1) = u_j(t_1, \epsilon)$,
(ii) $m_j(t) < u_j(t, \epsilon), t_0 \leq t < t_1$,
(iii) $m_i(t_1) \leq u_i(t_1, \epsilon), i \neq j$,

and

$$D_- m_j(t_1) \geq u_j'(t_1, \epsilon) = g_j(t_1, u(t_1, \epsilon)) + \epsilon. \tag{6.3.13}$$

Since $u(t, \epsilon)$ is nondecreasing in t, it follows from (i), (ii), and (iii) that

$$|m^j_{t_1}|_0 = m_j(t_1) = u_j(t_1, \epsilon), \tag{6.3.14}$$

$$|m^i_{t_1}|_0 \leq u_i(t_1, \epsilon), \quad i \neq j. \tag{6.3.15}$$

Setting $\phi = x_{t_1}(t_0, \phi_0)$, it results that $|\phi_j|_0 = |\phi_j(0)|$, and hence, using (6.3.11), we arrive at

$$D_-m_j(t_1) \leqslant g_j(t_1, |\phi_1|_0, ..., |\phi_{j-1}|_0, |\phi_j(0)|, |\phi_{j+1}|_0, ..., |\phi_n|_0).$$

The quasi-monotone property of $g(t, u)$ in u and the inequalities (6.3.15) yield

$$D_-m_j(t_1) \leqslant g_j(t_1, u_1(t_1, \epsilon), ..., u_{j-1}(t_1, \epsilon), \phi_j(0), u_{j+1}(t_1, \epsilon) \cdots u_n(t_1, \epsilon))$$
$$\leqslant g_j(t_1, u(t_1, \epsilon)),$$

because of the definition of ϕ and (6.3.14). This inequality is incompatible with (6.3.13), and hence the set Z is empty, which in turn proves the stated componentwise estimates. The theorem is proved.

6.4. Dependence on initial values and parameters

We shall first prove the following lemma, which will be used subsequently.

LEMMA 6.4.1. Let $f \in C[J \times C_\rho, R^n]$, and let, for $t \in J, \phi \in C_\rho$,

$$G(t, r) = \max_{\|\phi\|_0 \leqslant r} \|f(t, \phi)\|.$$

Suppose that $r^*(t, t_0, 0)$ is the maximal solution of

$$u' = G(t, u)$$

through $(t_0, 0)$. Then, if $x(t_0, \phi_0)$ is any solution of (6.1.1) with the initial function ϕ_0 at $t = t_0$, we have

$$\| x_t(t_0, \phi_0) - \phi_0 \|_0 \leqslant r^*(t, t_0, 0), \qquad (6.4.1)$$

on the common interval of existence of $x(t_0, \phi_0)$ and $r^*(t, t_0, 0)$.

Proof. Consider the function

$$m(t) = \| x(t_0, \phi_0)(t) - \phi_0(0)\|,$$

so that

$$m_t = \| x_t(t_0, \phi_0) - \phi_0 \|.$$

6.4. DEPENDENCE ON INITIAL VALUES AND PARAMETERS

Then, $m(t_0) = 0$ and $m_{t_0} \equiv 0$. Moreover,

$$\begin{aligned}
m_+'(t) &\leq \| x'(t_0, \phi_0)(t) \| \\
&= \| f(t, x_t(t_0, \phi_0)) \| \\
&\leq \max_{\| \phi \|_0 \leq m(t)} \| f(t, \phi) \| \\
&= G(t, m(t)),
\end{aligned}$$

and hence

$$m(t) \leq r^*(t, t_0, 0)$$

by Theorem 1.4.1. The function G being nonnegative, $r^*(t, t_0, 0)$ is nondecreasing in t, and therefore there results the desired inequality (6.4.1).

THEOREM 6.4.1. Let $f \in C[J \times C_\rho, R^n]$, and, for $t \in J$, $\phi, \psi \in C_1$,

$$\| f(t, \phi) - f(t, \psi) \| \leq g(t, \| \phi(0) - \psi(0) \|),$$

where $g \in C[J \times [0, 2\rho), R_+]$. Assume that $u(t) \equiv 0$ is the only solution of the scalar differential equation (6.1.2) through $(t_0, 0)$. Then, if the solutions $u(t, t_0, u_0)$ of (6.1.2) through every point (t_0, u_0) exist for $t \geq t_0$ and are continuous with respect to the initial values (t_0, u_0), the solutions $x(t_0, \phi_0)$ of (6.1.1) are unique and continuous with respect to the initial values (t_0, ϕ_0).

Proof. The uniqueness of solutions of (6.1.1) is a consequence of Theorem 6.2.3, and hence we need only to prove the continuity with respect to initial values. Let $x(t_0, \phi_0)$, $x(t_0, \psi_0)$ be the solutions of (6.1.1) with the initial functions ϕ_0, ψ_0 at $t = t_0$, respectively, existing in some interval $t_0 \leq t \leq t_0 + a$. Then, an application of Theorem 6.2.2 yields that

$$\| x(t_0, \phi_0)(t) - x(t_0, \psi_0)(t) \| \leq r(t, t_0, \| \phi_0 - \psi_0 \|_0), \quad t_0 \leq t \leq t_0 + a,$$
(6.4.2)

where $r(t, t_0, \| \phi_0 - \psi_0 \|_0)$ is the maximal solution of (6.1.2) through the point $(t_0, \| \phi_0 - \psi_0 \|_0)$. By assumption, it follows that, given $\epsilon > 0$, there exists a $\delta > 0$ such that

$$r(t, t_0, \| \phi_0 - \psi_0 \|_0) < \epsilon, \quad t_0 \leq t \leq t_0 + a,$$

provided $\| \phi_0 - \psi_0 \|_0 < \delta$. This, in view of (6.4.2), assures the continuity of the solutions $x(t_0, \phi_0)$ of (6.1.1) with respect to initial functions ϕ_0.

We now prove the continuity of solutions with respect to the initial time t_0. Let $t_1 > t_0$ and $x(t_0, \phi_0)$, $x(t_1, \phi_0)$ be the solutions of (6.1.1) through (t_0, ϕ_0) and (t_1, ϕ_0), respectively, existing in some interval to the right. Define

$$m(t) = \| x(t_0, \phi_0)(t) - x(t_1, \phi_0)(t)\|, \quad t_1 \leqslant t \leqslant t_1 + a.$$

Since, by Lemma 6.4.1, we have

$$m_{t_1} = \| x_{t_1}(t_0, \phi_0) - \phi_0 \| \leqslant r^*(t_1, t_0, 0),$$

we obtain

$$m(t) \leqslant \tilde{r}(t), \quad t_1 \leqslant t \leqslant t_1 + a, \tag{6.4.3}$$

where $\tilde{r}(t) = \tilde{r}(t, t_1, r^*(t_1, t_0, 0))$ is the maximal solution of (6.1.2) through $(t_1, r^*(t_1, t_0, 0))$. Now,

$$\lim_{t_1 \to t_0+} \tilde{r}(t, t_1, r^*(t_1, t_0, 0)) \equiv \tilde{r}(t, t_0, 0).$$

Since $\lim_{t_1 \to t_0} r^*(t_1, t_0, 0) = 0$ and by assumption, $\tilde{r}(t, t_0, 0)$ is identically zero. This fact, in view of the relation (6.4.3), proves the continuity of solutions of (6.1.1) with respect to initial time. The proof is complete.

COROLLARY 6.4.1. *The function* $g(t, u) = Lu$, $L > 0$, *is admissible in Theorem 6.4.1.*

Using the arguments of Theorems 2.5.2 and 6.2.1, we can prove the following theorem on dependence on parameters. We merely state

THEOREM 6.4.2. *Let* $f \in C[J \times C_\rho \times R^m, R^n]$, *and, for* $\mu = \mu_0$, *let* $x_0(t) = x_0(t_0, \phi_0, \mu_0)(t)$ *be a solution of*

$$x' = f(t, x_t, \mu_0),$$

with an initial function ϕ_0 *at* $t = t_0$, *existing for* $t \geqslant t_0$. *Assume that*

$$\lim_{\mu \to \mu_0} f(t, \phi, \mu) = f(t, \phi, \mu_0)$$

uniformly in $(t, \phi) \in J \times C_\rho$, *and, for* $t \in J$, $\phi, \psi \in C_\rho$, $\mu \in R^m$,

$$\| f(t, \phi, \mu) - f(t, \psi, \mu)\| \leqslant g(t, \|\phi - \psi\|_0),$$

where $g \in C[J \times R_+, R_+]$. *Suppose that* $u(t) \equiv 0$ *is the maximal solution of* (6.1.2) *such that* $u(t_0) = 0$. *Then, given* $\epsilon > 0$, *there exists* a $\delta(\epsilon) > 0$ such that, for every μ, $\|\mu - \mu_0\| < \delta(\epsilon)$, the differential system

$$x' = f(t, x_t, \mu)$$

admits a unique solution $x(t) = x(t_0, \phi_0, \mu)(t)$ defined on some interval $[t_0, t_0 + a]$ such that

$$\|x(t) - x_0(t)\| < \epsilon, \qquad t_0 \leqslant t \leqslant t_0 + a.$$

6.5. Stability criteria

Let us consider the functional differential system (6.1.1). We shall assume that $f(t, 0) \equiv 0$, so that the system (6.1.1) possesses the trivial solution. Let us also suppose that the solutions $x(t_0, \phi_0)$ of (6.1.1) exist in the future.

DEFINITION 6.5.1. The trivial solution of (6.1.1) is said to be *stable*, if, for any $\epsilon > 0$ and $t_0 \in J$, there exists a $\delta > 0$ such that $\|\phi_0\|_0 \leqslant \delta$ implies

$$\|x_t(t_0, \phi_0)\|_0 < \epsilon, \qquad t \geqslant t_0.$$

DEFINITION 6.5.2. The trivial solution of (6.1.1) is said to be *asymptotically stable* if it is stable and, in addition, for any $\epsilon > 0$, $t_0 \in J$, there exist positive numbers δ_0, T such that $\|\phi_0\|_0 \leqslant \delta_0$ implies

$$\|x_t(t_0, \phi_0)\|_0 < \epsilon, \qquad t \geqslant t_0 + T.$$

Simple criteria for stability and asymptotic stability of the trivial solution of (6.1.1) are given in the following theorems.

THEOREM 6.5.1. Let $f \in C[J \times C_\rho, R^n]$, $g \in C[J \times [0, \rho), R_+]$, $g(t, 0) \equiv 0$, and, for $t \in J$, $\phi \in C_\rho$ such that

$$\|\phi_0\|_0 = \|\phi(0)\|, \qquad (6.5.1)$$

the inequality

$$\|f(t, \phi)\| \leqslant g(t, \|\phi(0)\|)$$

holds. If the trivial solution of the scalar differential equation (6.1.2) is stable, then the trivial solution of (6.1.1) is stable.

Proof. By Theorem 6.3.1, we have

$$\|x(t_0, \phi_0)(t)\| \leqslant r(t, t_0, \|\phi_0\|_0), \qquad t \geqslant t_0,$$

where $r(t, t_0, \|\phi_0\|_0)$ is the maximal solution of (6.1.2) through $(t_0, \|\phi_0\|_0)$. The fact $g \geqslant 0$ implies that $r(t, t_0, \|\phi_0\|_0)$ is nondecreasing in t, and therefore it follows that

$$\|x_t(t_0, \phi_0)\|_0 \leqslant r(t, t_0, \|\phi_0\|_0), \qquad t \geqslant t_0. \tag{6.5.2}$$

Assume that the trivial solution of (6.1.2) is stable. Then, given $0 < \epsilon < \rho$, $t_0 \in J$, there exists a $\delta > 0$ such that

$$r(t, t_0, \|\phi_0\|_0) < \epsilon, \qquad t \geqslant t_0,$$

provided $\|\phi_0\|_0 \leqslant \delta$. The conclusion is immediate from (6.5.2).

THEOREM 6.5.2. Let $f \in C[J \times C_\rho, R^n]$, $g \in C[J \times [0, \rho), R_+]$, and $g(t, 0) \equiv 0$. Assume that

$$A(t) \liminf_{h \to 0^-} h^{-1}[\|\phi(0) + hf(t, \phi)\| - \|\phi(0)\|] + \|\phi(0)\| D_- A(t)$$
$$\leqslant g(t, \|\phi(0)\| A(t))$$

for $t > t_0$ and $\phi \in C_\rho$ satisfying

$$\|\phi\|_0 \mid A_t \mid_0 = \|\phi(0)\| A(t), \tag{6.5.3}$$

where $A(t) \geqslant 1$ is continuous on $[t_0 - \tau, \infty)$ and $A(t) \to \infty$ as $t \to \infty$. Then, the stability of the trivial solution of (6.1.2) implies the asymptotic stability of the trivial solution of (6.1.1).

Proof. By Theorem 6.3.3, it follows that

$$\|x(t_0, \phi_0)(t)\| A(t) \leqslant r(t, t_0, \|\phi_0\|_0 \mid A_{t_0} \mid_0), \qquad t \geqslant t_0,$$

and, arguing as in Theorem 6.5.1, we have

$$\|x_t(t_0, \phi_0)\|_0 \mid A_t \mid_0 \leqslant r(t, t_0, \|\phi_0\|_0 \mid A_{t_0} \mid_0), \qquad t \geqslant t_0.$$

Since $A(t) \geqslant 1$, the stability of the trivial solution of (6.1.1) is a consequence of Theorem 6.5.1. Now let $\epsilon = \rho$ and designate by δ_0 the δ-obtained corresponding to ρ. Suppose that $\|\phi_0\|_0 \leqslant \delta_0$. Assume that, if possible, there exists a sequence $\{t_k\}$, $t_k \to \infty$ as $k \to \infty$ and a solution $x(t_0, \phi_0)$ of (6.1.1) with $\|\phi_0\|_0 \leqslant \delta_0$ such that

$$\|x_{t_k}(t_0, \phi_0)\|_0 \geqslant \epsilon.$$

Then, there results the inequality

$$\epsilon \mid A_{t_k} \mid_0 \leqslant r(t_k, t_0, \|\phi_0\|_0 \mid A_{t_0} \mid_0) < \rho.$$

Since $A(t) \to \infty$ as $t \to \infty$, the foregoing inequality leads to an absurdity for sufficiently large k. As a result, the asymptotic stability follows.

We now extend the preceding results to the perturbed systems. Corresponding to (6.1.1), let us consider

$$x' = f(t, x_t) + R(t, x_t), \tag{6.5.4}$$

where $R(t, \phi)$ is a perturbation such that the solutions $x(t_0, \phi_0)$ of (6.5.4) exist in the future.

THEOREM 6.5.3. Let

$$f, R \in C[J \times C_\rho, R^n], \quad R(t, 0) \equiv 0, \quad g \in C[J \times [0, \rho), R_+],$$

and $g(t, 0) \equiv 0$. Suppose that

$$\liminf_{h \to 0^-} h^{-1}[\|\phi(0) + hf(t, \phi)\| - \|\phi(0)\|] \leq 0 \tag{6.5.5}$$

for $t > t_0$ and $\phi \in C_\rho$ satisfying (6.5.1). Let the perturbation $R(t, \phi)$ verify the condition

$$\|R(t, \phi)\| \leq g(t, \|\phi(0)\|) \tag{6.5.6}$$

for $t \in J$, $\phi \in C_\rho$ such that (6.5.1) holds. Then, the stability of the trivial solution of (6.1.2) implies the stability of the trivial solution of the perturbed system (6.5.4).

Proof. Let $\phi \in C_\rho$ be such that (6.5.1) is satisfied. Then, we obtain

$$\liminf_{h \to 0^-} h^{-1}[\|\phi(0) + h\{f(t, \phi) + R(t, \phi)\}\| - \|\phi(0)\|] \leq g(t, \|\phi(0)\|),$$

using the conditions (6.5.5) and (6.5.6). If $x(t_0, \phi_0)$ is any solution of the perturbed system (6.5.4), we get, by Theorem 6.3.3, with $A(t) \equiv 1$, the inequality

$$\|x(t_0, \phi_0)(t)\| \leq r(t, t_0, \|\phi_0\|_0), \quad t \geq t_0,$$

where $r(t, t_0, \|\phi_0\|_0)$ is the maximal solution of (6.1.2). The stated result follows, arguing as in Theorem 6.5.1.

THEOREM 6.5.4. Let

$$f, R \in C[J \times C_\rho, R^n], \quad R(t, 0) \equiv 0, \quad g \in C[J \times [0, \rho), R_+],$$

and $g(t, 0) \equiv 0$. Assume that

$$A(t) \liminf_{h \to 0^-} h^{-1}[\|\phi(0) + hf(t, \phi)\| - \|\phi(0)\|] + \|\phi(0)\| D_-A(t) \leq 0 \tag{6.5.7}$$

for $t > t_0$ and $\phi \in C_\rho$ such that (6.5.3) holds, $A(t)$ being the same function as in Theorem 6.5.2. Suppose that $R(t, \phi)$ verifies the condition

$$A(t)\| R(t, \phi)\| \leq g(t, \|\phi(0)\| A(t)) \tag{6.5.8}$$

for $t \in J$ and $\phi \in C_\rho$ satisfying (6.5.3). Then, the stability of the trivial solution of (6.1.2) implies the asymptotic stability of the trivial solution of (6.1.1).

Proof. Let $\phi \in C_\rho$ be such that the relation (6.5.3) holds. Then, it follows that

$$A(t) \liminf_{h \to 0^-} h^{-1}[\|\phi(0) + h\{f(t, \phi) + R(t, \phi)\}\| - \|\phi(0)\|] + \|\phi(0)\| D_-A(t)$$
$$\leq g(t, \|\phi(0)\| A(t)),$$

because of the assumptions (6.5.7) and (6.5.8). It is now easy to establish the asymptotic stability, following the proof of Theorem 6.5.2.

6.6. Asymptotic behavior

Let us begin with a result that gives sufficient conditions for every solution of (6.1.1) to tend to a finite limit vector as $t \to \infty$.

THEOREM 6.6.1. Let $f \in C[J \times \mathscr{C}^n, R^n]$ and, for $(t, \phi) \in J \times \mathscr{C}^n$,

$$\|f(t, \phi)\| \leq g(t, \|\phi\|_0), \tag{6.6.1}$$

where $g \in C[J \times R_+, R_+]$ and $g(t, u)$ is monotonic nondecreasing in u for each $t \in J$. Assume that all the solutions $u(t)$ of (6.1.2) are bounded on $[t_0, \infty)$. Then, every solution of (6.1.1) tends to a finite limit ξ as $t \to \infty$.

Proof. By Theorem 6.1.2, every solution $x(t_0, \phi_0)$ of (6.1.1) exists on $[t_0, \infty)$. Let $x(t_0, \phi_0)$ be any solution of (6.1.1). Then, on account of Lemma 6.1.1, it follows that

$$\| x(t_0, \phi_0)(t)\| \leq r(t), \qquad t \geq t_0, \tag{6.6.2}$$

provided $\|\phi\|_0 = u_0$, where $r(t)$ is the maximal solution of (6.1.2).

Since, by assumption, every solution of (6.1.2) is bounded on $[t_0, \infty)$, we see from (6.6.2) that every solution of (6.1.1) is bounded on $[t_0, \infty)$. Furthermore, for any $t_1 > t_0$ and $t > t_1$, we have

$$\| x(t_0, \phi_0)(t) - x(t_0, \phi_0)(t_1) \| \leq \int_{t_1}^{t} \| f(s, x_s(t_0, \phi_0)) \| \, ds$$

$$\leq \int_{t_1}^{t} g(s, \| x_s \|_0) \, ds. \qquad (6.6.3)$$

Because $g(t, u) \geq 0$ and consequently $r(t)$ is monotonic nondecreasing in t, it results from the inequality (6.6.2) that

$$\| x_t(t_0, \phi_0) \|_0 \leq r(t), \qquad t \geq t_0.$$

This, together with (6.6.3) and the monotonicity of g in u, implies that

$$\| x(t_0, \phi_0)(t) - x(t_0, \phi_0)(t_1) \| \leq \int_{t_1}^{t} g(s, r(s)) \, ds$$

$$= r(t) - r(t_1). \qquad (6.6.4)$$

Moreover, by the assumption of boundedness of all solutions of (6.1.2), we deduce that $r(t)$ tends to a finite limit as $t \to \infty$. This means that, given an $\epsilon > 0$, it is possible to find a $t_1 > 0$ sufficiently large such that

$$0 \leq r(t) - r(t_1) < \epsilon, \qquad t > t_1.$$

Consequently, we obtain, as a result of (6.6.4),

$$\| x(t_0, \phi_0)(t) - x(t_0, \phi_0)(t_1) \| < \epsilon, \qquad t > t_1,$$

which proves that

$$\lim_{t \to \infty} x(t_0, \phi_0)(t) = \xi.$$

This completes the proof.

The next theorem deals with the asymptotic equivalence of two functional differential systems.

THEOREM 6.6.2. Let $u(t)$ be a positive solution of

$$u' > g(t, u)$$

for $t \geq t_0$ such that $\lim_{t\to\infty} u(t) = 0$, where $g \in C[J \times R_+, R]$. Suppose further that $f_1, f_2 \in C[J \times \mathscr{C}^n, R^n]$ and

$$\liminf_{h\to 0^-} h^{-1}[\|\phi(0) - \psi(0) + h[f_1(t, \phi) - f_2(t, \psi)]\| - \|\phi(0) - \psi(0)\|]$$
$$\leq g(t, \|\phi(0) - \psi(0)\|) \tag{6.6.5}$$

for $t > t_0$ and $\phi, \psi \in \Omega$, where

$$\Omega = [\phi, \psi \in \mathscr{C}^n : \|\phi(0) - \psi(0)\| = u(t), t \geq t_0].$$

Then, if the existence of solutions for all $t \geq t_0$ of the systems

$$x' = f_1(t, x_t), \tag{6.6.6}$$
$$y' = f_2(t, y_t) \tag{6.6.7}$$

is assured, the systems (6.6.6) and (6.6.7) are asymptotically equivalent.

Proof. Let us first suppose that $y(t_0, \psi_0)$ is a solution of (6.6.7) defined for $t \geq t_0$. Let $x(t_0, \phi_0)$ be some solution of (6.6.6) such that

$$\|\phi_0 - \psi_0\|_0 < u(t_0).$$

Define
$$m(t) = \|x(t_0, \phi_0)(t) - y(t_0, \psi_0)(t)\|.$$

Then,
$$m(t) < u(t), \quad t \geq t_0.$$

If this is false, let t_1 be the greatest lower bound of numbers $t \geq t_0$ for which $m(t) < u(t)$ is not satisfied. The continuity of the functions $m(t)$ and $u(t)$ guarantees that, at $t = t_1$,

$$m(t_1) = u(t_1) \tag{6.6.8}$$

and
$$m(t) < u(t), \quad t_0 \leq t < t_1. \tag{6.6.9}$$

This implies that
$$D_- m(t_1) \geq u'(t_1) > g(t_1, u(t_1)). \tag{6.6.10}$$

On the other hand, defining $\phi = x_{t_1}(t_0, \phi_0)$, $\psi = y_{t_1}(t_0, \psi_0)$, we see that

$$\phi(0) = x(t_0, \phi_0)(t_1), \quad \psi(0) = y(t_0, \psi_0)(t_1),$$

and therefore
$$m(t_1) = \|\phi(0) - \psi(0)\| = u(t_1),$$

because of (6.6.8). Thus, $\phi, \psi \in \Omega$ at $t = t_1$, and hence, using the condition (6.6.5), it is easy to deduce that

$$D_-m(t_1) \leqslant g(t_1, m(t_1)).$$

This being incompatible with (6.6.10), we conclude that $m(t) \leqslant u(t)$, $t \geqslant t_0$. The assumption that $\lim_{t \to \infty} u(t) = 0$ now implies that

$$\lim_{t \to \infty} x(t_0, \phi_0)(t) - y(t_0, \psi_0)(t) = 0. \tag{6.6.11}$$

If $x(t_0, \phi_0)$ is a given solution of (6.6.6), defined on $[t_0, \infty)$, arguing as before, we can assert that there is a solution $y(t_0, \psi_0)$ of (6.6.7) such that (6.6.11) is satisfied. It therefore follows that the systems (6.6.6) and (6.6.7) are asymptotically equivalent.

We shall now give an analog of Theorem 2.6.3 with respect to the following two systems:

$$x' = f_1(t, x), \tag{6.6.12}$$

$$y' = f_2(t, y_t). \tag{6.6.13}$$

THEOREM 6.6.3. Let $f_2 \in C[J \times \mathscr{C}^n, R^n]$ and $y(t_0, \phi_0)$ be any solution of (6.6.13) defined for $t \geqslant t_0$. Assume that $f_1 \in C[J \times R^n, R^n]$, $\partial f_1(t, x)/\partial x$ exists and is continuous on $J \times R^n$. If $x(t, t_0, \phi_0(0))$ is the solution of (6.6.12) such that $x(t_0, t_0, \phi_0(0)) = \phi_0(0)$ existing for $t \geqslant t_0$, then $y(t_0, \phi_0)$ satisfies the integral equation

$$y_{t_0} = \phi_0,$$

$$y(t_0, \phi_0)(t) = x(t, t_0, \phi_0(0))$$
$$+ \int_{t_0}^{t} \Phi(t, s, y(s))[f_2(s, y_s) - f_1(s, y(s))] \, ds, \qquad t \geqslant t_0,$$

where $\Phi(t, t_0, x_0) = \partial x(t, t_0, x_0)/\partial x_0$ and $y(t) = y(t_0, \phi_0)(t)$.

The proof of this theorem is very much the same as that of Theorem 2.6.3. It is important to note, however, that (6.6.12) is an ordinary differential system, whereas (6.6.13) is a functional differential system.

As an application of Theorem 6.6.3, we have

THEOREM 6.6.4. Assume that

(i) $f_1 \in C[J \times R^n, R^n]$, $\partial f_1(t, x)/\partial x$ exists and is continuous on $J \times R^n$, and $f_2 \in C[J \times \mathscr{C}^n, R^n]$;

(ii) $\Phi(t, t_0, x_0)$ is the fundamental matrix solution of the variational system

$$Z' = \frac{\partial f_1(t, x(t, t_0, x_0))}{\partial x} Z$$

such that $\Phi(t_0, t_0, x_0) =$ identity matrix I;

(iii) for a given solution $y(t) = y(t_0, \phi_0)(t)$ of (6.6.13), existing on $[t_0, \infty)$,

$$\int_t^\infty \Phi(t, s, y(s))[f_2(s, y_s) - f_1(s, y(s))]\, ds \to 0 \quad \text{as} \quad t \to \infty.$$

Then, there exists a solution $x(t)$ of (6.6.12), on $[t_0, \infty)$, satisfying the relation

$$\lim_{t\to\infty} [x(t) - y(t_0, \phi_0)(t)] = 0.$$

The proof can be constructed using the arguments employed in Theorem 2.10.3.

Finally, we may mention a result parallel to Theorem 2.14.10.

THEOREM 6.6.5. Assume that

(i) $f \in C[J \times R^n, R^n]$, $f(t, 0) \equiv 0$, and $f_x(t, x)$ exists and is continuous on $J \times R^n$;

(ii) $\mu[f_x(t, 0)] \leqslant -\sigma$, $\sigma > 0$, $t \in J$;

(iii) $R \in C[J \times \mathscr{C}^n, R^n]$, $R(t, 0) \equiv 0$, and there exists an $\alpha > 0$ such that, if $\|\phi\|_0 \leqslant \alpha$, $t \in J$,

$$\|R(t, \phi)\| \leqslant \gamma(t),$$

where $\gamma \in C[J, R_+]$ and

$$\int_t^{t+1} \gamma(s)\, ds \to 0 \quad \text{as} \quad t \to \infty.$$

Then, there exists a $T_0 \geqslant 0$ such that, if $t_0 \geqslant T_0$, the trivial solution of

$$x' = f(t, x) + R(t, x_t)$$

is asymptotically stable.

Proof. The proof is almost similar to that of Theorem 2.14.10. We only indicate the major changes. Proceeding as in Theorem 2.14.10, we arrive at the following step:

$$\|x(t_0, \phi_0)(t_1)\| = \delta(\epsilon)$$

and

$$\|x(t_0, \phi_0)(t)\| \leqslant \delta(\epsilon), \quad t \in [t_0, t_1].$$

This implies also that

$$\| x_t(t_0, \phi_0) \|_0 \leq \delta(\epsilon), \qquad t \in [t_0, t_1].$$

Hence, letting $\phi = x_t(t_0, \phi_0)$, we see that $\| \phi \|_0 \leq \alpha$, and therefore the assumption (iii) can be used. It only remains to follow the rest of the proof of Theorem 2.14.10, with necessary changes to complete the proof.

COROLLARY 6.6.1. *The function* $f(t, x) = Ax + F(t, x)$, *where*

(i) A *is a* $n \times n$ *constant matrix such that* $\mu(A) \leq -\sigma$;
(ii) $F \in C[J \times R^n, R^n]$ *and, given any* $\epsilon > 0$, *there exist* $\delta(\epsilon), T(\epsilon) > 0$ *such that* $\| F(t, x) \| \leq \epsilon \| x \|$ *provided* $\| x \| \leq \delta(\epsilon)$ *and* $t \geq T(\epsilon)$;

is admissible in Theorem 6.6.5.

6.7. A topological principle

In this section, the topological principle discussed in Sect. 2.9 is extended to functional differential system (6.1.1). We assume, in what follows, the uniqueness of solutions of (6.1.1). More specifically, it is assumed that $f \in C[J \times C_\rho, R^n]$ and $f(t, \phi)$ satisfies a Lipschitz condition in ϕ for a constant $L = L(\rho)$, so that we have the following estimate:

$$\| x(t_0, \phi_0)(t) - y(t_0, \psi_0)(t) \| \leq \| \phi_0 - \psi_0 \|_0 \exp[L(t - t_0)] \qquad (6.7.1)$$

on the common interval of existence of the solutions $x(t_0, \phi_0)$ and $y(t_0, \psi_0)$ of (6.1.1).

REMARK 6.7.1. In general, we cannot extend the solution $x(t_0, \phi_0)$ to the left of t_0, that is, we cannot guarantee the existence of a $\delta > 0$ such that the system (6.1.1) is satisfied with $x(t_0, \phi_0)(t)$ extended to $[t_0 - \delta - \tau, t^+)$. If $t^+ = \infty$, the solution $x(t_0, \phi_0)$ is defined in the future. Also, for every point $P_0 = (t_0, x_0) \in R^{n+1}$, there are, in general, infinitely many solutions of (6.1.1) satisfying $x(t_0, \phi_0)(t_0) = x_0$.

Let E be an open set in $\Omega = J \times S_\rho$, where $S_\rho = [x \in R^n: \| x \| < \rho]$, and ∂E be the boundary of E in Ω.

DEFINITION 6.7.1. *A point* $P_0 = (t_0, x_0) \in \partial E$ *is said to be a point of egress if there is at least one solution* $x(t) = x(t_0, \phi_0)(t)$ *of* (6.1.1) *with* $x(t_0) = x_0$ *defined on* $[t_0 - \tau - \delta, t_0]$, $\delta > 0$, *such that* $(t, x(t)) \in E$ *for* $t_0 - \delta \leq t < t_0$.

We define

$$S = [P \in \partial E \colon P \text{ is a point of egress}].$$

DEFINITION 6.7.2. A point $P_0 = (t_0, x_0) \in \partial E$ is said to be a point of *strict egress* if (i) for every solution $x(t)$ of (6.1.1) with $x(t_0) = x_0$, defined on $[t_0 - \tau, t^+)$, there is a $\delta > 0$ [t^+ and δ may depend on $x(t)$], such that $(t, x(t)) \in E^*$, where E^* is the exterior of E with respect to Ω, for $t_0 < t \leq t_0 + \delta$; and (ii) whenever $x(t)$ can be extended to some $[t_0 - \tau - \sigma, t^+)$ [$\sigma > 0$ depending on $x(t)$], there is an $\epsilon > 0, 0 < \epsilon < \sigma$, such that $(t, x(t)) \in E$ for $t_0 - \epsilon \leq t < t_0$.

We shall denote by S^* the set of points of strict egress.

We see from the preceding definitions that $P_0 = (t_0, x_0) \in S^*$ implies that $P_0 \in S$ if there is at least one $\phi_0 \in \mathscr{C}^n$, $\phi_0(0) = x_0$, such that the solution $x(t_0, \phi_0)(t)$ can be extended to the left. These definitions coincide with the definitions given in Sect. 2.9, if $\tau = 0$.

DEFINITION 6.7.3. A solution $x(t_0, \phi_0)$ of (6.1.1) is said to be *asymptotic* with respect to E if $(t, x(t)) \in E$ for $t_0 \leq t < t^+$.

Let $x(t) = x(t_0, \phi_0)(t)$, with $x(t_0) = \phi_0(0) \in E$. If $x(t)$ is not asymptotic with respect to E, then there exists a $t_1 > t_0$ such that $Q = (t_1, x(t_1)) \in S$ and $(t, x(t)) \in E$ for $t_0 \leq t < t_1$. We denote the point Q by $\pi(t_0, \phi_0)$. For every $t_0 > 0$, let $E(t_0)$ be the set of points $(t_0, x) \in E$. Again, for $t_0 > 0$, $\phi_0 \in \mathscr{C}^n$, define

$$G(t_0, \phi_0) = [(t_0, y_0) \in E(t_0): y_0 - \phi_0(0) + \phi_0 \in C_\rho$$

and $\pi(t_0, y_0 - \phi_0(0) + \phi_0)$ exists].

If $G(t_0, \phi_0)$ is not empty, we define a mapping taking $G(t_0, \phi_0)$ into S by $\pi(P_0) = \pi(t_0, y_0 - \phi_0(0) + \phi_0)$, where $P_0 = (t_0, y_0)$. Let $S \subset S^*$ and w be any set satisfying

$$S \subset w \subset S^*.$$

We then define a mapping $K: G(t_0, \phi_0) \cup w \to w$ as follows:

(i) $K(P_0) = \pi(P_0)$ if $P_0 \in G(t_0, \phi_0)$;
(ii) $K(P_0) = P_0$ if $P_0 \in w$.

LEMMA 6.7.1. *The mapping K is a retraction from $G(t_0, \phi_0) \cup w$ into w.*

Proof. Since $K(P_0) = P_0$ if $P_0 \in w$, it is enough to prove that K is continuous. This can be shown by using the estimate (6.7.1) and following the standard techniques.

We can now prove a theorem analogous to Theorem 2.9.1.

6.7. A TOPOLOGICAL PRINCIPLE

THEOREM 6.7.1. Let $\phi_0 \in \mathscr{C}^n$ and $t_0 > 0$. Assume that the following conditions hold:

(i) for every $(t_0, y_0) \in E(t_0)$, it follows that $y_0 - \phi_0(0) + \phi_0 \in C_\rho$;
(ii) $S \subset w \subset S^*$;
(iii) there exists a set Z, $Z \subset E(t_0) \cup w$ such that $Z \cap w$ is a retract of w and $Z \cap w$ is not a retract of Z.

Then, there exists at least one point $x_0 \in R^n$ such that

$$(t_0, x_0 + \phi_0(0)) \in Z - w, \quad x_0 + \phi_0 \in C_\rho,$$

and $x(t_0, x_0 + \phi_0)(t)$ is asymptotic with respect to E.

Note: If $\rho = \infty$, so that $S_\rho = R^n$, we can drop the condition (i). Furthermore, in the proof given below, it will be assumed that $\phi_0(0) = 0$, since there is no loss of generality in doing so.

Proof. By Lemma 6.7.1, the mapping K is a retraction from $G(t_0, \phi_0) \cup w$ into w. Since $Z \cap w$ is a retract of w, there is a retraction K^* from w into $Z \cap w$. Then, the mapping $T = K^* \cdot K$ is a retraction from $G(t_0, \phi_0) \cup w$ into $Z \cap w$. By the assumption (i), it follows that $x_0 + \phi_0 \in C_\rho$ for every $(t_0, x_0) \in Z - w$. Suppose that $x(t_0, x_0 + \phi_0)(t)$ is not asymptotic with respect to E for every $(t_0, x_0) \in Z - w$. Then, $\pi(t_0, x_0 + \phi_0)$ exists for every $(t_0, x_0) \in Z - w$, and hence

$$Z = (Z \cap w) \cup (Z - w) \subset w \cup G(t_0, \phi_0).$$

This implies that the restriction of the mapping T to Z is a retraction from Z into $w \cap Z$, which is a contradiction. The proof is therefore complete.

Corresponding to Theorem 2.9.2, we now state the following result, which gives sufficient conditions to ensure that $S \subset S^*$.

THEOREM 6.7.2. Let $f \in C[J \times C_\rho, R^n]$, $u \in C[\Omega, R^p]$, and $v \in C[\Omega, R^q]$. Let

(i) $E = [(t, x) \in \Omega : u_j(t, x) < 0$ and $v_k(t, x) < 0,$
$1 \leq j \leq p, \ 1 \leq k \leq q]$;

(ii) $L_\alpha = [(t, x) \in \Omega : u_\alpha(t, x) = 0$ and $u_j(t, x) \leq 0, \ v_k(t, x) \leq 0,$
$1 \leq j \leq p, \ 1 \leq k \leq q]$;

(iii) $M_\beta = [(t, x) \in \Omega : v_\beta(t, x) = 0$ and $u_j(t, x) \leq 0, \ v_k(t, x) \leq 0,$
$1 \leq j \leq p, \ 1 \leq k \leq q]$.

Assume that, for each $P_0 = (t_0, x_0) \in L_\alpha$, M_β and every solution $x(t)$ of (6.1.1) satisfying $x(t_0) = x_0$, it follows that

$$u_\alpha'(t, x(t))_{t=t_0} > 0,$$

$$v_\beta'(t, x(t))_{t=t_0} < 0.$$

Then, if $w = \bigcup_{i=1}^p L_i - \bigcup_{j=1}^q M_j$, we have $S \subset w \subset S^*$.

The proof can be constructed by making necessary changes in the proof of Theorem 2.9.2.

6.8. Systems with repulsive forces

Let $F \in C[J \times \mathscr{C}^n \times \mathscr{C}^n, R^n]$ and $F(t, \phi, \psi)$ be locally Lipschitzian in ϕ and ψ. We consider the second-order functional differential equation

$$x'' = F(t, x_t, y_t), \tag{6.8.1}$$

where $y(t) = x'(t)$. This can be written in an equivalent form:

$$\begin{aligned} x'(t) &= y(t), \\ y'(t) &= F(t, x_t, y_t). \end{aligned} \tag{6.8.2}$$

DEFINITION 6.8.1. A solution $x(t)$ of (6.8.1), defined for $t \geqslant t_0$, is said to be *asymptotic in the sense of Wintner* if $u^2(t) = x(t) \cdot x(t)$ is non-decreasing for $t \geqslant t_0$.

THEOREM 6.8.1. Let $F \in C[J \times \mathscr{C}^n \times \mathscr{C}^n, R^n]$ and $F(t, \phi, \psi)$ be locally Lipschitzian in ϕ, ψ, and, for all $\phi, \psi \in \mathscr{C}^n$, $\psi(0) \neq 0$, $t \geqslant T$,

$$\phi(0) \cdot F(t, \phi, \psi) + \psi(0) \cdot \psi(0) > 0.$$

Then, for every $t_0 > T$, $\phi_0, \psi_0 \in \mathscr{C}^n$, there exists a family of points $(x_0, y_0) \in R^{2n}$, depending at least on n parameters, such that the solutions $x(t_0, x_0 + \phi_0, y_0 + \psi_0)$ of (6.8.1) are asymptotic in the sense of Wintner.

Proof. To prove the theorem, we have to show the existence of a family of solutions $x(t)$ of (6.8.1) satisfying

$$[u^2(t)]' = 2x(t) \cdot x'(t) \leqslant 0 \quad \text{for} \quad t \geqslant t_0 > T.$$

Let $\Omega = [T, \infty) \times R^{2n}$. For some $b > 0$, $T < t_1 < t_0$, $\phi_0, \psi_0 \in \mathscr{C}^n$, let

$$E_b = [(t, x, y) \in R^{2n+1} : u(t, x, y) < 0, v(t, x, y) < 0],$$

where
$$u(t, x, y) = x \cdot y - b; \quad v(t, x, y) = t_1 - t.$$

By Theorem 6.7.2, we can conclude that the sets S_b, $S_b{}^*$ of egress and strict egress points obey
$$S_b \subset w_b \subset S_b{}^*,$$
where
$$w_b = [(t, x, y) \in R^{2n+1} : x \cdot y = b, t > t_1].$$

Let us take $Z_b \subset E_b(t_0) \cup w_b$, where Z_b is a segment connecting two points (t_0, ξ_1, η_1) and (t_0, ξ_2, η_2) located in two distinct components of w_b with $(t_0, 0, 0) \notin Z_b$. As a consequence of Theorem 6.7.1, it follows that there is at least one point (t_0, ξ, η) with
$$(t_0, \xi + \phi_0(0), \eta + \psi_0(0)) \in Z_b - w_b,$$
such that the solution
$$(x(t), y(t)) = [x(t_0, \gamma_b)(t), y(t_0, \gamma_b)(t)],$$
where $\gamma_b = (\xi + \phi_0, \eta + \psi_0)$, is asymptotic with respect to E_b, that is,
$$x(t) \cdot y(t) < b, \quad t \geqslant t_0.$$

Let a sequence $\{b_m\}$, $b_m > 0$, $b_m \to 0$ as $m \to \infty$, and the sets Z_{b_m} be chosen such that $Z_{b_{m+1}} \subset Z_{b_m}$. Then, for every m, there is (x_m, y_m), with $(t_0, x_m + \phi_0(0), y_m + \psi_0(0)) \in Z_{b_m} - w_{b_m}$, such that the solution $(x(t_0, \gamma_m)(t), y(t_0, \gamma_m)(t))$, where $\gamma_m = (x_m + \phi_0, y_m + \psi_0)$, is asymptotic with respect to E_{b_m}. By taking a convergent subsequence of $\{x_m, y_m\}$ with limit (x_0, y_0), it follows that $x(t) \cdot y(t) \leqslant 0$, $t \geqslant t_0$, where $x(t) = x(t_0, \gamma)(t)$, $y(t) = y(t_0, \gamma)(t)$, and $\gamma = (x_0 + \phi_0, y_0 + \psi_0)$.

Since $Z_{b_1} \cap Z_{\bar{b}_1}$ is empty for $b_1 \neq \bar{b}_1$, we see that it is possible to take (x_0, y_0) depending on n parameters. The proof is complete.

DEFINITION 6.8.2. Equation (6.8.1) is said to be a system with repulsive forces if
$$\phi(0) \cdot F(t, \phi, \psi) \geqslant 0, \quad t \geqslant T, \quad \phi, \psi \in \mathscr{C}^n.$$

COROLLARY 6.8.1. If Eq. (6.8.1) is a system with repulsive forces, then the conclusion of Theorem 6.8.1 is valid.

6.9. Functional differential inequalities

It is natural to expect that a comparison theorem for functional differential equations, analogous to Theorem 1.4.1, would be equally useful in some situations. To prove such a result, we need to show the existence of maximal solution for functional differential equations. We are thus led to the study of functional differential inequalities. With this motive, we shall consider a functional differential equation of the form

$$x' = f(t, x, x_t), \tag{6.9.1}$$

which is convenient for some later applications. The existence theorems 6.1.1 and 6.1.2 are valid for such an equation, with obvious changes.

Let us begin with the following basic theorem on fundamental differential inequalities, recalling that $\mathscr{C} = C[[-\tau, 0], R]$.

THEOREM 6.9.1. Let $f \in C[J \times R \times \mathscr{C}, R]$ and $f(t, x, \phi)$ be non-decreasing in ϕ for each (t, x). Let $x, y \in C[[t_0 - \tau, \infty), R]$ and

$$x_{t_0} < y_{t_0}. \tag{6.9.2}$$

Assume further that

$$\begin{aligned} D_-x(t) &\leqslant f(t, x(t), x_t), \\ D_-y(t) &> f(t, y(t), y_t), \end{aligned} \tag{6.9.3}$$

for $t \in (t_0, \infty)$. Then,

$$x(t) < y(t), \quad t \in [t_0, \infty). \tag{6.9.4}$$

Proof. If the assertion (6.9.4) is false, then the set

$$Z = [t \in [t_0, \infty): x(t) \geqslant y(t)]$$

is nonempty. Let $t_1 = \inf Z$. It is clear from (6.9.2) that $t_1 > t_0$. Moreover,

$$x(t_1) = y(t_1) \quad \text{and} \quad x(t) < y(t), \quad t \in [t_0, t_1). \tag{6.9.5}$$

Thus, we obtain for small $h < 0$

$$\frac{x(t_1 + h) - x(t_1)}{h} > \frac{y(t_1 + h) - y(t_1)}{h},$$

which, in turn, implies that

$$D_-x(t_1) \geqslant D_-y(t_1). \tag{6.9.6}$$

6.9. FUNCTIONAL DIFFERENTIAL INEQUALITIES

From the relations (6.9.2) and (6.9.5), we deduce that

$$x_{t_1} \leqslant y_{t_1},$$

which, in view of the monotonic character of $f(t, x, \phi)$ in ϕ, yields

$$f(t_1, x(t_1), x_{t_1}) \leqslant f(t_1, x(t_1), y_{t_1}). \tag{6.9.7}$$

On the other hand, the relations (6.9.3) and (6.9.6) lead to the inequality

$$f(t_1, x(t_1), x_{t_1}) > f(t_1, y(t_1), y_{t_1}),$$

which is incompatible with (6.9.7) because of (6.9.5). Thus the set Z is empty, and the result follows.

REMARK 6.9.1. The conclusion (6.9.4) remains valid even when the inequalities (6.9.3) are replaced by

$$D_{-}x(t) < f(t, x(t), x_t),$$
$$D_{-}y(t) \geqslant f(t, y(t), y_t).$$

DEFINITION 6.9.1. Let $x \in C[t_0 - \tau, \infty), R]$, and let $x_{+}'(t)$ exist for $t \in [t_0, \infty)$. If $x(u)$ satisfies the differential inequality

$$x_{+}'(t) < f(t, x(t), x_t), \qquad t \in [t_0, \infty),$$

it is said to be an *under function* with respect to (6.9.1).
On the other hand, if

$$x_{+}'(t) > f(t, x(t), x_t), \qquad t \in [t_0, \infty),$$

$x(u)$ is said to be an *over function* with respect to (6.9.1).

As in the case of ordinary differential equations, the next theorem shows that any solution of (6.9.1) can be bracketed between its under and over functions.

THEOREM 6.9.2. Let $y(u), z(u)$ be under and over functions with respect to (6.9.1). Let $f \in C[J \times R \times \mathscr{C}, R]$ and $f(t, x, \phi)$ be nondecreasing in ϕ, for each (t, x). Suppose that $x(t_0, \phi_0)$ is any solution of (6.9.1) defined on $[t_0, \infty)$, such that

$$y_{t_0} < \phi_0 < z_{t_0}.$$

Then,

$$y(t) < x(t_0, \phi_0)(t) < z(t), \qquad t \in [t_0, \infty).$$

The proof follows by a repeated application of Theorem 6.9.1.

We shall now consider the existence of maximal solution of (6.9.1).

DEFINITION 6.9.2. Let $r(t_0, \phi_0)$ be a solution of (6.9.1) defined on $[t_0, t_0 + a)$. For any other solution $x(t_0, \phi_0)$ of (6.9.1) defined on the same interval, if

$$x(t_0, \phi_0)(t) \leqslant r(t_0, \phi_0)(t), \qquad t \in [t_0, t_0 + a),$$

then $r(t_0, \phi_0)$ is said to be the maximal solution of (6.9.1).

A similar definition may be given for the minimal solution by reversing the preceding inequality.

THEOREM 6.9.3. *Let $f \in C[J \times R \times \mathscr{C}, R]$ and $f(t, x, \phi)$ be nondecreasing in ϕ for each (t, x). Then, given an initial function $\phi_0 \in \mathscr{C}$ at $t = t_0$, there exists an $\alpha_1 > 0$ such that Eq. (6.9.1) admits a unique maximal solution $r(t_0, \phi_0)$ defined on $[t_0, t_0 + \alpha_1)$.*

Proof. Following the proof of Theorem 6.1.1, we obtain $a, b > 0$. Suppose that $0 < \epsilon \leqslant b/2$. Consider the equation

$$x'(t) = f(t, x(t), x_t) + \epsilon. \tag{6.9.8}$$

Let $\phi_0 + \epsilon$ be the initial function at $t = t_0$. Observing that

$$|f(t, x, \psi)| \leqslant M + \frac{b}{2}$$

whenever $t \in [t_0, t_0 + a]$, $x \in S_\rho$, $\psi \in C_\rho$, $|x - y(t) - \epsilon| < b$ and $|\psi - y_t - \epsilon|_0 < b$, we deduce, on the basis of Theorem 6.1.1 that (6.9.8) has a solution $x(t_0, \phi_0, \epsilon)$ on the interval $[t_0 - \tau, t_0 + \alpha_1]$, where $\alpha_1 = \min[a, (b/(2M + b)]$.

For $0 < \epsilon_2 < \epsilon_1 \leqslant \epsilon$, we have

$$x_{t_0}(t_0, \phi_0, \epsilon_2) < x_{t_0}(t_0, \phi_0, \epsilon_1),$$

$$x'(t_0, \phi_0, \epsilon_2)(t) \leqslant f(t, x(t_0, \phi_0, \epsilon_2)(t), x_t(t_0, \phi_0, \epsilon_2)) + \epsilon_2,$$

$$x'(t_0, \phi_0, \epsilon_1)(t) > f(t, x(t_0, \phi_0, \epsilon_1)(t), x_t(t_0, \phi_0, \epsilon_1)) + \epsilon_2.$$

We can now apply Theorem 6.9.1 to get

$$x(t_0, \phi_0, \epsilon_2)(t) < x(t_0, \phi_0, \epsilon_1)(t), \qquad t \in [t_0, t_0 + \alpha_1).$$

Since the family of functions $x(t_0, \phi_0, \epsilon)(t)$ are equicontinuous and uniformly bounded on $[t_0, t_0 + \alpha_1)$, it follows by Ascoli–Arzela's theorem that there exists a decreasing sequence $\{\epsilon_n\}$, $\epsilon_n \to 0$ as $n \to \infty$, such that

$$\lim_{n\to\infty} x(t_0, \phi_0, \epsilon_n)(t) = r(t_0, \phi_0)(t)$$

uniformly on $[t_0, t_0 + \alpha_1)$. Clearly, $r_{t_0}(t_0, \phi_0) = \phi_0$. The uniform continuity of f implies that $f(t, x(t_0, \phi_0, \epsilon_n)(t), x_t(t_0, \phi_0, \epsilon_n))$ tends uniformly to $f(t, r(t_0, \phi_0)(t), r_t(t_0, \phi_0))$ as $n \to \infty$, and, thus, term-by-term integration is applicable to

$$x(t_0, \phi_0, \epsilon_n)(t) = \phi_0(0) + \epsilon_n$$
$$+ \int_{t_0}^{t} [f(s, x(t_0, \phi_0, \epsilon_n)(s), x_s(t_0, \phi_0, \epsilon_n)) + \epsilon_n] \, ds,$$

which, in turn, shows that the limit $r(t_0, \phi_0)(t)$ is a solution of (6.9.1) on $[t_0, t_0 + \alpha_1)$.

We shall now show that $r(t_0, \phi_0)$ is the desired maximal solution of (6.9.1) on $[t_0, t_0 + \alpha_1)$. Let $x(t_0, \phi_0)$ be any solution of (6.9.1) defined on $[t_0, t_0 + \alpha_1)$. We then have

$$x_{t_0}(t_0, \phi_0) < x_{t_0}(t_0, \phi_0, \epsilon),$$
$$x'(t_0, \phi_0, \epsilon)(t) \geq f(t, x(t_0, \phi_0, \epsilon)(t), x_t(t_0, \phi_0, \epsilon)) + \epsilon,$$
$$x'(t_0, \phi_0)(t) < f(t, x(t_0, \phi_0)(t), x_t(t_0, \phi_0)) + \epsilon,$$

for $0 < \epsilon < b/2$. By Remark 6.9.1, it follows that

$$x(t_0, \phi_0)(t) < x(t_0, \phi_0, \epsilon)(t), \qquad t \in [t_0, t_0 + \alpha_1).$$

Since

$$\lim_{\epsilon \to 0} x(t_0, \phi_0, \epsilon)(t) = r(t_0, \phi_0)(t)$$

uniformly on $[t_0, t_0 + \alpha_1)$, the theorem is proved.

REMARK 6.9.2. Under the assumption of Theorem 6.9.3, we can show the existence of the minimal solution also. The proof requires obvious changes.

We are now in a position to prove the following comparison theorem for functional differential inequalities. Let $\mathscr{C}_+$ denote the set of all nonnegative functions belonging to $\mathscr{C}$.

THEOREM 6.9.4. Let $m \in C[[t_0 - \tau, \infty), R_+]$, and satisfy the inequality

$$D_- m(t) \leq f(t, m(t), m_t), \qquad t > t_0,$$

where $f \in C[J \times R_+ \times \mathscr{C}_+, R]$. Assume that $f(t, x, \phi)$ is nondecreasing in ϕ for each (t, x) and that $r(t_0, \phi_0)$, $\phi_0 \in \mathscr{C}_+$, is the maximal solution of (6.9.1) existing for $t \geqslant t_0$. Then, $m_{t_0} \leqslant \phi_0$ implies

$$m(t) \leqslant r(t_0, \phi_0)(t), \qquad t \geqslant t_0. \tag{6.9.9}$$

Proof. Let $x(t_0, \phi_0, \epsilon)$ be a solution of (6.9.8) with an initial function $\phi_0 + \epsilon$ at $t = t_0$. Then, we have

$$m_{t_0} < \phi_0 + \epsilon,$$

$$D_- m(t) \leqslant f(t, m(t), m_t),$$

$$x'(t_0, \phi_0, \epsilon)(t) > f(t, x(t_0, \phi_0, \epsilon)(t), x_t(t_0, \phi_0, \epsilon)).$$

These inequalities imply, by Theorem 6.9.1, the relation

$$m(t) < x(t_0, \phi_0, \epsilon)(t), \qquad t \geqslant t_0.$$

The conclusion (6.9.9) results from the fact that

$$\lim_{\epsilon \to 0} x(t_0, \phi_0, \epsilon)(t) = r(t_0, \phi_0)(t) \qquad \text{uniformly for} \quad t \geqslant t_0.$$

The following theorem provides an estimate for the difference between a solution and an approximate solution of a functional differential system.

THEOREM 6.9.5. Let $g \in C[J \times R_+ \times \mathscr{C}_+, R_+]$ and $g(t, u, \sigma)$ be nondecreasing in σ for each (t, u). Assume that $f \in C[J \times R^n \times \mathscr{C}^n, R^n]$, and, for $t \in J$, $x, y \in R^n$, and $\phi, \psi \in \mathscr{C}^n$,

$$\| f(t, x, \phi) - f(t, y, \psi)\| \leqslant g(t, \| x - y \|, \| \phi - \psi \|). \tag{6.9.10}$$

Let $x(t_0, \phi_0)(t)$, $y(t) = y(t_0, \psi_0, \delta)(t)$ be a solution and a δ-approximate solution of the system

$$x' = f(t, x, x_t) \tag{6.9.11}$$

defined for $t \geqslant t_0$. Let $r(t_0, \sigma)$ be the maximal solution of

$$u' = g(t, u, u_t) + \delta,$$

with an initial function $\sigma \in \mathscr{C}_+$, at $t = t_0$, existing for $t \geqslant t_0$. Then, whenever $\| \phi_0 - \psi_0 \| \leqslant \sigma$, we have

$$\| x(t_0, \phi_0)(t) - y(t)\| \leqslant r(t_0, \sigma)(t), \qquad t \geqslant t_0.$$

Proof. Define $m(t) = \| x(t_0, \phi_0)(t) - y(t)\|$, so that $m_t = \| x_t(t_0, \phi_0) - y_t \|$, for $t \geq t_0$. Then, as in Theorem 6.2.1, it is easy to obtain the inequality

$$D_- m(t) \leq g(t, m(t), m_t) + \delta, \qquad t > t_0.$$

The desired result follows by Theorem 6.9.4.

A uniqueness theorem of Perron type is an immediate consequence of Theorem 6.9.4.

THEOREM 6.9.6. Let $g \in C[J \times R_+ \times \mathscr{C}_+, R_+]$ and $g(t, u, \sigma)$ be non-decreasing in σ for each (t, u). Assume that $f \in C[J \times R^n \times \mathscr{C}^n, R^n]$ and that (6.9.10) holds. If the maximal solution of

$$u' = g(t, u, u_t),$$

with the initial function $\sigma = 0$, is identically zero, then there is atmost one solution of the system (6.9.11) on $[t_0, t_0 + a)$.

To give yet another interesting application of Theorem 6.9.4, let us consider the differential-difference system

$$y'(t) = f(t, y(t), y(t - \tau)), \qquad \tau > 0. \tag{6.9.12}$$

If τ is small, it is natural to expect that it can be neglected, and we can consider the ordinary differential system

$$x' = f(t, x, x). \tag{6.9.13}$$

THEOREM 6.9.7. Assume that

  (i) $f \in C[J \times R^n \times R^n, R^n]$, $f_x(t, x, x)$ exists and is continuous for $t \in J$, $x \in R^n$, and

$$\mu[f_x(t, x, x)] \leq -\alpha, \qquad t \geq 0, \qquad x \in R^n;$$

  (ii) for $t \in J$, $x, y, z \in R^n$,

$$\| f(t, x, y) - f(t, x, z) \| \leq L \| y - z \|;$$

  (iii) for $t \in J$ and $y \in C[[t_0 - \tau, \infty), R^n]$,

$$\| f(t, y(t), y(t - \tau)) \| \leq N \sup_{t-\tau \leq s \leq t} \| y(s) \|,$$

and $\Phi(t, t_0, x_0)$ is the solution of the variational system

$$z' = f_x(t, x(t), x(t)) z, \tag{6.9.14}$$

such that $\Phi(t_0, t_0, x_0)$ is the identity matrix I, where $x(t) = x(t, t_0, x_0)$ is any solution of (6.9.13).

Then, if $0 < \tau < \alpha/LN$, every solution of (6.9.12), defined on $[t_0, \infty)$, tends to zero exponentially, as $t \to \infty$.

Proof. Let $y(t) = y(t_0, \phi_0)(t)$ be any solution of (6.9.12), defined on $[t_0, \infty)$. Then, by Theorem 6.6.3, it follows that

$$y_{t_0}(t_0, \phi_0) = \phi_0,$$

$$y(t_0, \phi_0)(t) = x(t, t_0, \phi_0(0))$$
$$+ \int_{t_0}^{t} \Phi(t, s, y(s))[f(s, y(s), y(s-\tau)) - f(s, y(s), y(s))]\, ds,$$
(6.9.15)

for $t \geqslant t_0$. Moreover, on the basis of Theorem 2.6.4, we deduce that

$$\|x(t, t_0, \phi_0(0))\| \leqslant \left\| \int_0^1 \Phi(t, t_0, \phi_0(0)s)\, ds \right\| \cdot \|\phi_0(0)\|$$
$$\leqslant \max_{0 \leqslant s \leqslant 1} \|\Phi(t, t_0, \phi_0(0)s)\| \, \|\phi_0(0)\|.$$
(6.9.16)

Also, using assumption (i), we have

$$\|\Phi(t, t_0, \phi_0(0)s)\| \leqslant e^{-\alpha(t-t_0)}, \qquad t \geqslant t_0, \qquad 0 \leqslant s \leqslant 1. \quad (6.9.17)$$

Thus, we obtain, on account of (6.9.15), (6.9.16), and (6.9.17),

$$\|y_{t_0}\|_0 \leqslant \|\phi_0\|_0,$$

$$\|y(t_0, \phi_0)(t)\| \leqslant \|\phi_0\|_0 \, e^{-\alpha(t-t_0)} + L \int_{t_0}^{t} e^{-\alpha(t-s)} \|y(s-\tau) - y(s)\|\, ds,$$

because of condition (ii). Observe that

$$\|y(s-\tau) - y(s)\| = \left\| \int_{s-\tau}^{s} \frac{dy(\theta)}{d\theta}\, d\theta \right\|$$
$$\leqslant \int_{s-\tau}^{s} \|f(\theta, y(\theta), y(\theta-\tau))\|\, d\theta$$
$$\leqslant N \int_{s-\tau}^{s} \sup_{s-\tau \leqslant \theta \leqslant s} \|y(\theta)\|\, d\theta$$
$$\leqslant N\tau \sup_{s-2\tau \leqslant \theta \leqslant s} \|y(\theta)\|.$$

6.9. FUNCTIONAL DIFFERENTIAL INEQUALITIES

Hence,
$$\|y(t_0, \phi_0)(t)\| \leq e^{-\alpha t}\left[\|\phi_0\|_0 e^{\alpha t_0} + M \int_{t_0}^{t} e^{\alpha s} \sup_{s-2\tau \leq \theta \leq s} \|y(\theta)\| \, ds\right], \quad (6.9.18)$$

where $M = LN\tau$. We now define
$$u(t) = e^{-\alpha t}\left[\|\phi_0\|_0 e^{\alpha t_0} + M \int_{t_0}^{t} e^{\alpha s} \sup_{s-2\tau \leq \theta \leq s} \|y(\theta)\| \, ds\right],$$

so that
$$u'(t) = -\alpha u(t) + M \sup_{t-2\tau \leq \theta \leq t} \|y(\theta)\|.$$

This, together with the fact that
$$\sup_{t-2\tau \leq \theta \leq t} \|y(\theta)\| \leq \sup_{t-2\tau \leq \theta \leq t} u(\theta),$$

yields the functional differential inequality
$$u'(t) \leq -\alpha u(t) + M \sup_{-2\tau \leq \theta \leq 0} u_t.$$

By Theorem 6.9.4, we therefore get
$$u(t) \leq r(t_0, \sigma)(t), \qquad t \geq t_0, \quad (6.9.19)$$

provided $u_{t_0} = \|\phi_0\| \leq \sigma$, where σ is the initial function at $t = t_0$, defined on $-2\tau \leq s \leq 0$, and $r(t) = r(t_0, \sigma)(t)$ is the maximal solution of
$$u' = -\alpha u + M \sup_{-2\tau \leq \theta \leq 0} u_t. \quad (6.9.20)$$

All that is required is to find the explicit form of $r(t)$. Let
$$r(t) = \sigma(0) e^{-\gamma(t-t_0)}.$$

Then, we get, from (6.9.20), the equation
$$-\gamma = -\alpha + M e^{\gamma \tau}.$$

It is now easy to see from this that $r(t)$ is the solution of (6.9.20) with $\gamma > 0$, provided $\alpha > M$, which implies that
$$0 < \tau < \alpha/LN.$$

The conclusion of the theorem is now immediate from the inequality (6.9.19), noting that
$$\|y(t_0, \phi_0)(t)\| \leq u(t).$$

The proof is complete.

6.10. Notes

Theorem 6.1.1 is adapted from the work of Driver [2, 3]. See also Lakshmikantham and Shendge [1] and Yoshizawa [3]. The remaining results of Sect. 6.1 are due to Lakshmikantham and Shendge [1]. See also Driver [2, 3] and Sugiyama [1, 5]. Theorems 6.2.1, 6.2.2, and 6.2.3 are taken from Lakshmikantham [7]. Theorem 6.2.4 is new. Section 6.3 contains results adapted from Lakshmikantham [7].

The results of Sect. 6.4 are new. See also Driver [4], Sugiyama [4] for particular cases. Concerning linear functional differential equations, see Hale [4] and also Sugiyama [6].

The stability criteria discussed in Sect. 6.5 are taken from Lakshmikantham [7]. All the results of Sect. 6.6 are new. The extension of Wazewski's topological principle and applications given in Sects. 6.7 and 6.8 are due to Onuchic [1]. Section 6.9 consists of the results of Lakshmikantham and Shendge [1], except Theorem 6.9.7, which is new. For the notion of extremal solutions, see also Sugiyama [8].

Delay-differential equations have been discussed in a recent monograph of Oğuztöreli [1], which also treats time-lag control systems. See also Driver [1, 2, 3] and Lakshmikantham [7] concerning this area. A systematic presentation of fundamental theory in neutral functional differential equations may be found in the work of Driver [4]. The books of Bellman and Cooke [4] and Halanay [22] are famous contributions in this field. The survey papers by Bellman and Danskin [2], Bellman et al. [1], El'sgol'ts and Myshkis [1], El'sgol'ts, Zverkin, Kamenskii, and Norkin [1, 2], Hale [7], and Kamenskii [5] give important information about the development of the theory. The books of El'sgol'ts [7], Krasovskii [5], Myshkis [6], Pinney [1], and Hahn [1] also discuss equations of this type.

For related work, see also Bullock [1], Cooke [1–8], El'sgol'ts [1–6], El'sgol'ts, Myshkis, and Shimanov [1], Fodcuk [1], Franklin [1], Hale [2, 3, 9, 10], Hale and Perello [1], Hastings [1, 2], Jones [2–5], Kakutani and Markus [1], Karasik [2], Myshkis [1–7], Perello [1], Shimanov [1–10], Stokes [1], Wright [1–4], and Zverkin [1–3].

Chapter 7

7.0. Introduction

In extending the second method of Lyapunov to the differential equations with time lag, one has a choice of treating the solutions as the elements of a function space or as elements of Euclidean space, for all future time. Each approach has certain advantages, although it appears natural that the proper setting for the study of functional differential equations is a function space. If we choose to study from the point of view that the solutions define curves in a function space, the concept of Lyapunov functional may be used to discuss many problems including stability and boundedness of functional differential systems. Moreover, if Lyapunov functionals are employed for direct theorems on stability, the converse problem of showing the existence of Lyapunov functionals can be solved in a manner analogous to that in ordinary differential equations. The main advantage in using Lyapunov functionals consists, however, in applying the converse theorems for the study of perturbed systems. In this chapter, we study a variety of problems of functional differential systems by means of Lyapunov functionals and the theory of differential inequalities.

7.1. Stability criteria

We shall consider the functional differential system

$$x' = f(t, x_t), \qquad (7.1.1)$$

where $f \in C[J \times C_\rho, R^n]$, C_ρ being, as before, the set

$$C_\rho = [\phi \in \mathscr{C}^n : \|\phi\|_0 < \rho].$$

We shall assume that $f(t, 0) \equiv 0$, so that the system (7.1.1) possesses a trivial solution. Suppose also that $f(t, \phi)$ is smooth enough to guarantee the existence of solutions of (7.1.1) in the future. Let $x(t_0, \phi_0)$ be any solution of (7.1.1) with an initial function $\phi_0 \in C_\rho$ at $t = t_0$.

DEFINITION 7.1.1. The trivial solution of (7.1.1) is said to be (S_1) *equistable* if, for each $\epsilon > 0$, $t_0 \in J$, there exists a positive function $\delta = \delta(t_0, \epsilon)$, which is continuous in t_0 for each ϵ, such that the inequality

$$\|\phi_0\|_0 \leqslant \delta$$

implies

$$\|x_t(t_0, \phi_0)\|_0 < \epsilon, \qquad t \geqslant t_0.$$

On the strength of this definition and that of Definition 3.2.1, we can formulate the other notions of stability (S_2) to (S_{10}) of the trivial solution of the functional differential system (7.1.1). The same is true concerning other various types of stability and boundedness definitions.

DEFINITION 7.1.2. For any $V \in C[J \times C_\rho, R_+]$, define

$$D^+V(t, x_t(t_0, \phi_0)) = \limsup_{h \to 0^+} h^{-1}[V(t + h, x_{t+h}(t_0, \phi_0))$$
$$- V(t, x_t(t_0, \phi_0))]. \qquad (7.1.2)$$

Sometimes, we also define

$$D^+V(t, \phi) = \limsup_{h \to 0^+} h^{-1}[V(t + h, x_{t+h}(t, \phi)) - V(t, \phi)]. \qquad (7.1.3)$$

where it is understood that $x(t, \phi)$ is any solution of (7.1.1) with an initial function ϕ at time t.

REMARK 7.1.1. If the uniqueness of solutions of (7.1.1) is assured, both the definitions (7.1.2) and (7.1.3) are identical, since letting $\phi = x_t(t_0, \phi_0)$ and noting that $x_{t+h}(t_0, \phi_0) = x_{t+h}(t, \phi)$, $h \geqslant 0$, because of uniqueness, (7.1.2) reduces to (7.1.3), and vice versa.

We now state certain fundamental propositions regarding Lyapunov stability of the trivial solution of (7.1.1).

THEOREM 7.1.1. Assume that there exist a functional $V(t, \phi)$ and a function $g(t, u)$ fulfilling the following properties:

(i) $V \in C[J \times C_\rho, R_+]$, and, for $t \geqslant t_0$,

$$D^+V(t, x_t(t_0, \phi_0)) \leqslant g(t, V(t, x_t(t_0, \phi_0)));$$

7.1. STABILITY CRITERIA

(ii) $g \in C[J \times R_+, R]$, and $g(t, 0) = 0$;

(iii) there exist functions $b \in \mathcal{K}$ and $a \in C[J \times [0, \rho), R_+]$, $a(t, u) \in \mathcal{K}$ for each $t \in J$, such that

$$b(\|\phi\|_0) \leqslant V(t, \phi) \leqslant a(t, \|\phi\|_0), \qquad (t, \phi) \in J \times C_\rho.$$

Then, the trivial solution of the functional differential system (7.1.1) is

(1°) equistable if the trivial solution of (6.1.2) is equistable;

(2°) equi-asymptotically stable if the trivial solution of (6.1.2) is equi-asymptotically stable.

Proof. Suppose that the trivial solution of (6.1.2) is equistable. Let $0 < \epsilon < \rho$, $t_0 \in J$ be given. Then, given $b(\epsilon) > 0$, $t_0 \in J$, there exists a positive function $\delta = \delta(t_0, \epsilon)$ that is continuous in t_0 for each ϵ such that $u_0 \leqslant \delta$ implies

$$u(t, t_0, u_0) < b(\epsilon), \qquad t \geqslant t_0. \tag{7.1.4}$$

Choose $u_0 = a(t_0, \|\phi_0\|_0)$ so that $V(t_0, \phi_0) \leqslant u_0$. By the assumption on $a(t, u)$, it follows that there exists a $\delta_1 = \delta_1(t_0, \epsilon)$, which is continuous in t_0 for each ϵ, such that

$$\|\phi_0\|_0 \leqslant \delta_1 \quad \text{and} \quad a(t_0, \|\phi_0\|_0) \leqslant \delta$$

hold simultaneously. The choice of u_0, the condition (i), and Theorem 1.4.1 imply the inequality

$$V(t, x_t(t_0, \phi_0)) \leqslant r(t, t_0, u_0), \qquad t \geqslant t_0, \tag{7.1.5}$$

where $r(t, t_0, u_0)$ is the maximal solution of (6.1.2). This, together with (7.1.4) and assumption (iii), yields, for $t \geqslant t_0$,

$$b(\|x_t(t_0, \phi_0)\|_0) \leqslant V(t, x_t(t_0, \phi_0)) \leqslant r(t, t_0, u_0) < b(\epsilon),$$

which implies that

$$\|x_t(t_0, \phi_0)\|_0 < \epsilon, \qquad t \geqslant t_0,$$

provided $\|\phi_0\|_0 \leqslant \delta_1$. This proves (1°).

To prove (2°), it is enough to show that (S_3) holds. For this purpose, let $0 < \epsilon < \rho$ and $t_0 \in J$ be given. Then, since (S_3^*) is satisfied, given $b(\epsilon) > 0$, $t_0 \in J$, there exist positive integers $\delta_0 = \delta_0(t_0)$ and $T = T(t_0, \epsilon)$ such that

$$u(t, t_0, u_0) < b(\epsilon), \qquad t \geqslant t_0 + T, \tag{7.1.6}$$

provided $u_0 \leq \delta_0$. Choosing $u_0 = a(t_0, \|\phi_0\|_0)$, we obtain, as previously, a $\hat{\delta}_0 = \hat{\delta}_0(t_0)$ such that

$$\|\phi_0\|_0 \leq \hat{\delta}_0 \quad \text{and} \quad a(t_0, \|\phi_0\|_0) \leq \delta_0$$

are satisfied at the same time. Moreover, it follows that $V(t_0, \phi_0) \leq u_0$. Let $\bar{\delta}_0(t_0, \rho)$ be the number obtained by taking $\epsilon = \rho$ and let $\delta_0^* = \min[\hat{\delta}_0, \bar{\delta}_0]$. Assume that $\|\phi_0\|_0 \leq \delta_0^*$. Then, using (7.1.5), (7.1.6), and the fact that $b(\|\phi\|_0) \leq V(t, \phi)$, we get

$$b(\|x_t(t_0, \phi_0)\|_0) \leq V(t, x_t(t_0, \phi_0)) \leq r(t, t_0, u_0)$$
$$< b(\epsilon), \quad t \geq t_0 + T,$$

which, in turn, yields that

$$\|x_t(t_0, \phi_0)\|_0 < \epsilon, \quad t \geq t_0 + T,$$

whenever $\|\phi_0\|_0 \leq \delta_0^*$. This establishes (2°), and the proof of the theorem is complete.

COROLLARY 7.1.1. *The function $g(t, u) \equiv 0$ is admissible in Theorem 7.1.1, to give equistability of the trivial solution of (7.1.1).*

COROLLARY 7.1.2. *Let $g(t, u) = \lambda(t) w(u)$, where $\lambda \in C[J, R]$, $w(u) \geq 0$ is continuous for $u \geq 0$, $w(u) > 0$ for $u > 0$. If*

$$-\int_0^{u_0} \frac{ds}{w(s)} \leq \int_{t_0}^t \lambda(s) \, ds < \int_{u_0}^\infty \frac{ds}{w(s)}$$

for some $u_0 > 0$, every $t_0 \geq 0$ and $t_0 \leq t < \infty$, then $g(t, u)$ is admissible in Theorem 7.1.1 to assure equistability of the trivial solution of (7.1.1). If, on the other hand, there exists a T, $t_0 \leq T < \infty$, verifying the property

$$\int_{t_0}^T \lambda(s) \, ds = -\int_0^{u_0} \frac{ds}{w(s)},$$

then $g(t, u)$ is admissible in Theorem 7.1.1 to guarantee equi-asymptotic stability of the trivial solution of (7.1.1).

THEOREM 7.1.2. *Assume that there exist a functional $V(t, \phi)$ and a function $g(t, u)$ satisfying conditions (i) and (ii) of Theorem 7.1.1. Suppose further that there exist functions $a, b \in \mathcal{K}$ such that, for $(t, \phi) \in J \times C_\rho$,*

$$b(\|\phi\|_0) \leq V(t, \phi) \leq a(\|\phi\|_0).$$

Then, if one of the notions (S_1^*) through (S_8^*) holds, the corresponding one of the notions (S_1) through (S_8) holds for the system (7.1.1).

It is not difficult to construct the proof corresponding to each of the statements of this theorem, on the basis of the arguments used in Theorem 7.1.1 and respective theorems in Sect. 3.3. and 3.4. We leave the details.

COROLLARY 7.1.3. *The function* $g(t, u) = -C(u)$, $C \in \mathcal{K}$, *is admissible in Theorem* 7.1.2 *to assure that the trivial solution of* (7.1.1) *is uniformly asymptotically stable.*

Proof. By Corollary 3.4.2, it follows that the trivial solution of the scalar differential equation (6.1.2) is uniformly asymptotically stable, and hence Theorem 7.1.2 guarantees the stated result.

REMARK 7.1.2. Notice that, in Theorems 7.1.1 and 7.1.2, we have asked that $V(t, \phi)$ satisfies the condition

$$b(\| \phi \|_0) \leqslant V(t, \phi), \quad (t, \phi) \in J \times C_\rho, \quad b \in \mathcal{K}.$$

The same results can be proved, with minor modifications, even when we assume a weaker condition, namely,

$$b(\| \phi(0) \|) \leqslant V(t, \phi), \quad (t, \phi) \in J \times C_\rho, \quad b \in \mathcal{K}.$$

This latter assumption is more convenient in applications. Observe, however, that in the next section, when we consider the converse problem of showing the existence of Lyapunov functions, we do obtain functionals verifying the former condition.

The following theorem, which is parallel to Theorem 3.4.9, is of interest in itself. Its proof can be constructed on the basis of the proofs of Theorem 3.4.9 and the foregoing theorems in this section.

THEOREM 7.1.3. *Assume that there exists a functional* $V(t, \phi)$ *satisfying the following properties:*

(i) $V \in C[J \times C_\rho, R_+]$, *and, for* $(t, \phi) \in J \times C_\rho$,

$$b(\| \phi(0) \|) \leqslant V(t, \phi) \leqslant a(\| \phi \|_0), \quad a, b \in \mathcal{K}. \quad (7.1.7)$$

(ii) $D^+V(t, x_t(t_0, \phi_0)) \leqslant -C(\| x(t_0, \phi_0)(t) \|)$, $t \geqslant t_0$, $C \in \mathcal{K}$.

Then, the trivial solution of (7.1.1) *is uniformly asymptotically stable.*

CHAPTER 7

Let us consider the scalar differential-difference equation

$$x'(t) = -ax(t) - bx(t-\tau), \qquad (7.1.8)$$

where τ, $a > 0$ and b are constants. For each $t > t_0 + \tau$, we have

$$x(t-\tau) = x(t) - \int_{-\tau}^{0} x'(t+s)\, ds.$$

Equation (7.1.8) can therefore be written as

$$x'(t) = -(a+b)\, x(t) - ab \int_{-\tau}^{0} x(t+s)\, ds$$
$$\qquad - b^2 \int_{-\tau}^{0} x(t+s-\tau)\, ds. \qquad (7.1.9)$$

Let us take as a Lyapunov functional

$$V(t, \phi) = \phi^2(0) + a \int_{-\tau}^{0} \phi^2(s)\, ds.$$

Observe that $V(t, \phi)$ fulfills the condition (7.1.7), where $b(u) = u^2$ and $a(u) = (1 + a\tau)\, u^2$. If $x(t_0, \phi_0)(t)$ is the solution of (7.1.8) corresponding to a given initial function ϕ_0, then $V(t, x_t(t_0, \phi_0))$ is a differentiable function of t, and, therefore, a simple calculation yields the inequality

$$V'(t, x_t(t_0, \phi_0)) \leqslant -a(a - |b|)[x^2(t_0, \phi_0)(t) + x^2(t_0, \phi_0)(t-\tau)].$$

It is now easy to see from Theorem 7.1.3 that $|b| \leqslant a$, $|b| < a$ imply uniform stability and uniform asymptotic stability of the trivial solution of (7.1.8), respectively. It is observed that the particular Lyapunov functional just used has given a region of stability which is independent of the lag τ and the sign of b.

If, on the other hand, we take the functional

$$V(t, \phi) = \phi^2(0) + \frac{a}{\tau} \int_{-\tau}^{0} \int_{-\tau}^{0} \phi^2(s)\, ds\, du + \frac{b}{\tau} \int_{-\tau}^{0} \int_{-\tau}^{0} \phi^2(s)\, ds\, du$$

and assume that $b \geqslant 0$, we see that this functional verifies condition (i) of Theorem 7.1.3. To verify condition (ii), we compute $V'(t, x_t(t_0, \phi_0))$ using (7.1.9). After some calculations, we obtain the following relation:

$$V'(t, x_t(t_0, \phi_0)) \leqslant -\tau^{-1} \int_{-\tau}^{0} [a(1 - b\tau)\{x^2(t_0, \phi_0)(t) + x^2(t_0, \phi_0)(t+s)\}$$
$$+ b(1 - b\tau)\{x^2(t_0, \phi_0)(t) + x^2(t_0, \phi_0)(t+s-\tau)\}]\, ds.$$

It follows from Theorem 7.1.3 that $0 \leq b\tau \leq 1$, $0 \leq b\tau < 1$ yield uniform stability and uniform asymptotic stability of the trivial solution of (7.1.8), respectively.

The foregoing discussion shows that, by choosing a suitable Lyapunov functional, it is possible to get the information about the qualitative structure of the region of stability in different ways.

Finally, we shall state a result whose proof is analogous to that of Theorem 3.4.10.

THEOREM 7.1.4. *Assume that there exist a functional $V(t, \phi)$ and a function $g(t, u)$ enjoying the following properties:*

(i) $V \in C[J \times C_\rho, R_+]$, *and, for $t \geq t_0$,*

$$D^+V(t, x_t(t_0, \phi_0)) \leq g(t, V(t, x_t(t_0, \phi_0)));$$

(ii) *there exist functions $a, b \in \mathscr{K}$ such that, for $(t, \phi) \in J \times C_\rho$,*

$$b(\|\phi\|_0) \leq V(t, \phi) \leq a(\|\phi\|_0);$$

(iii) $g \in C[J \times R_+, R]$, *and, for every pair of numbers α, β such that $0 < \alpha \leq \beta < \rho$, there exist $\theta = \theta(\alpha, \beta) \geq 0$, $k = k(\alpha, \beta) > 0$ satisfying*

$$g(t, u) \leq -k, \qquad t \geq \theta, \qquad \alpha \leq u \leq \beta;$$

(iv) *for a $\lambda \in C[J, R_+]$, $f(t, \phi)$ satisfies the condition*

$$\|f(t, \phi)\| \leq \lambda(t) \|\phi\|_0, \qquad (t, \phi) \in J \times C_\rho.$$

Then, the trivial solution of (7.1.1) is uniformly asymptotically stable.

7.2. Converse theorems

As in the case of ordinary differential equations, we may define the notion of generalized exponential asymptotic stability.

DEFINITION 7.2.1. *The trivial solution of (7.1.1) is said to be (S_{11}) generalized exponentially asymptotically stable if*

$$\|x_t(t_0, \phi_0)\|_0 \leq K(t_0) \|\phi_0\|_0 \exp[p(t_0) - p(t)], \qquad t \geq t_0, \qquad (7.2.1)$$

where $K(t) > 0$ is continuous for $t \in J$, $p \in \mathscr{K}$ for $t \in J$, and $p(t) \to \infty$ as $t \to \infty$.

The particular case when $K(t) \equiv K > 0$, $p(t) = \alpha t$, $\alpha > 0$ is referred to as the *exponential asymptotic stability* of the trivial solution of (7.1.1).

THEOREM 7.2.1. Assume that $f(t, \phi)$ is linear in ϕ and that the trivial solution of the system (7.1.1) is generalized exponentially asymptotically stable. Suppose further that $p(t)$ is continuously differentiable on J. Then, there exists a functional $V(t, \phi)$ satisfying the following properties:

(1°) $V \in C[J \times C_\rho, R_+]$, and V is Lipschitzian in ϕ with the function $K(t)$;
(2°) $\|\phi\|_0 \leqslant V(t, \phi) \leqslant K(t) \|\phi\|_0$, $t \in J$, $\phi \in C_\rho$;
(3°) $D^+V(t, \phi) \leqslant -p'(t) V(t, \phi)$, $t \in J$, $\phi \in C_\rho$.

Proof. Consider the functional

$$V(t, \phi) = \sup_{\sigma \geqslant 0} \| x_{t+\sigma}(t, \phi)\|_0 \exp[p(t + \sigma) - p(t)]. \qquad (7.2.2)$$

Clearly, from (7.2.1), it follows that $V(t, \phi)$ verifies the property (2°). Moreover,

$$D^+V(t, \phi) = \limsup_{h \to 0^+} h^{-1}[V(t + h, x_{t+h}(t, \phi)) - V(t, \phi)]$$

$$= \limsup_{h \to 0^+} h^{-1}[\sup_{\sigma \geqslant 0} \| x_{t+h+\sigma}(t + h, x_{t+h}(t, \phi))\|_0$$

$$\times \exp\{p(t + h + \sigma) - p(t + h)\}$$

$$- \sup_{\sigma \geqslant 0} \| x_{t+\sigma}(t, \phi)\|_0 \exp\{p(t + \sigma) - p(t)\}];$$

$$D^+V(t, \phi) = \limsup_{h \to 0^+} h^{-1}[\sup_{\sigma \geqslant h} \| x_{t+\sigma}(t, \phi)\|_0 \exp\{p(t + \sigma) - p(t)\}$$

$$- \sup_{\sigma \geqslant 0} \| x_{t+\sigma}(t, \phi)\|_0 \exp\{p(t + \sigma) - p(t)\}]$$

$$\leqslant \limsup_{h \to 0^+} h^{-1}[\sup_{\sigma \geqslant 0} \| x_{t+\sigma}(t, \phi)\|_0 \exp\{p(t + \sigma) - p(t)\}$$

$$\times \{\exp(p(t) - p(t + h)) - 1\}]$$

$$= -p'(t) V(t, \phi).$$

This proves (3°). Let us note that, in proving the foregoing inequality, we have used the uniqueness of solutions of (7.1.1).

Now let $\phi_1, \phi_2 \in C_\rho$. Then, using the fact that $f(t, \phi)$ is linear in ϕ and the inequality (7.2.1), we get

$$|V(t, \phi_1) - V(t, \phi_2)| = |\sup_{\sigma \geq 0} \|x_{t+\sigma}(t, \phi_1)\|_0 \exp\{p(t+\sigma) - p(t)\}$$
$$- \sup_{\sigma \geq 0} \|x_{t+\sigma}(t, \phi_2)\|_0 \exp\{p(t+\sigma) - p(t)\}|$$
$$\leq \sup_{\sigma \geq 0} \|x_{t+\sigma}(t, \phi_1) - x_{t+\sigma}(t, \phi_2)\|_0 \exp\{p(t+\sigma) - p(t)\}$$
$$= \sup_{\sigma \geq 0} \|x_{t+\sigma}(t, \phi_1 - \phi_2)\|_0 \exp\{p(t+\sigma) - p(t)\}$$
$$\leq K(t) \|\phi_1 - \phi_2\|_0.$$

The continuity of $V(t, \phi)$ may be proved as in Theorem 3.6.1, with minor changes. This completes the proof of the theorem.

A similar result is true, even when $f(t, \phi)$ is nonlinear in ϕ, provided $f(t, \phi)$ satisfies a Lipschitz condition.

THEOREM 7.2.2. Assume that the trivial solution of the system (7.1.1) is generalized exponentially asymptotically stable. Let the function $p(t)$ occurring in (7.2.1) be continuously differentiable for $t \in J$. Suppose that $f(t, \phi)$ verifies a Lipschitz condition in ϕ with a constant $L > 0$. Let the function $K(t)$ be bounded, and, for some q, $0 < q < 1$, let there exist a $T > 0$ such that

$$K(t) \exp\{-q\, p(t+T) - p(t)\} \leq 1, \quad t \in J. \tag{7.2.3}$$

Then, there exists a functional $V(t, \phi)$ possessing the following properties:

(1°) $V \in C[J \times C_{\rho_0}, R_+]$, and, for $t \in J$, $\phi_1, \phi_2 \in C_{\rho_0}$,

$$|V(t, \phi_1) - V(t, \phi_2)| \leq e^{LT} \sup_{0 \leq \sigma \leq T} \exp\{p(t+\sigma) - p(t)\} \|\phi_1 - \phi_2\|_0;$$

(2°) $\|\phi\|_0 \leq V(t, \phi) \leq K(t) \|\phi\|_0$, $t \in J$, $\phi \in C_{\rho_0}$;

(3°) $D^+ V(t, \phi) \leq -(1-q) p'(t) V(t, \phi)$, $t \in J$, $\phi \in C_{\rho_0}$.

Proof. Let q, T be given satisfying (7.2.3). Define

$$V(t, \phi) = \sup_{\sigma \geq 0} \|x_{t+\sigma}(t, \phi)\|_0 \exp[(1-q)\{p(t+\sigma) - p(t)\}].$$

Since $K(t)$ is assumed to be bounded, let $\rho_0 = \rho/M$, where $M = \sup_{t \in J} K(t)$. Then, it is clear that $V \in C[J \times C_{\rho_0}, R_+]$. The relations (2°) and (3°) can be proved, following the proof of Theorem 7.2.1.

To show that $V(t, \phi)$ satisfies the stated Lipschitz condition, notice that
$$\| x_{t+\sigma}(t, \phi) \|_0 \exp[(1 - q)\{p(t + \sigma) - p(t)\}]$$
$$\leqslant K(t) \exp[-q\{p(t + \sigma) - p(t)\}] \| \phi \|_0 ,$$
and, because of (7.2.3),
$$V(t, \phi) = \sup_{0 \leqslant \sigma \leqslant T} \| x_{t+\sigma}(t, \phi) \|_0 \exp[(1 - q)\{p(t + \sigma) - p(t)\}].$$

Consequently, for $\phi_1, \phi_2 \in C_{\rho_0}$ and $t \in J$,
$$| V(t, \phi_1) - V(t, \phi_2) |$$
$$\leqslant \sup_{0 \leqslant \sigma \leqslant T} \| x_{t+\sigma}(t, \phi_1) - x_{t+\sigma}(t, \phi_2) \|_0 \exp[(1 - q)\{p(t + \sigma) - p(t)\}]$$
$$\leqslant e^{LT} \sup_{0 \leqslant \sigma \leqslant T} \exp[(1 - q)\{p(t + \sigma) - p(t)\}] \| \phi_1 - \phi_2 \|_0 ,$$
on the basis of the estimate (6.2.5). This proves the stated result.

The next theorem is a result similar to Theorem 7.2.2, whose conditions and the arguments of proof are slightly different.

THEOREM 7.2.3. Assume that

(i) for any two solutions $x(0, \phi_0)$, $x(0, \psi_0)$ of (7.1.1), the lower estimate
$$\| \phi_0 - \psi_0 \|_0 \exp\left[-\int_0^t p_1(s)\, ds\right] \leqslant \| x_t(0, \phi_0) - x_t(0, \psi_0) \|_0 , \quad t \geqslant 0,$$
holds, where $p_1 \in C[J, R]$;

(ii) there exists a $p \in \mathscr{K}$ for $t \in J$, $p(t) \to \infty$ as $t \to \infty$, $p'(t)$ exists, and
$$\| x_t(0, \phi_0) \|_0 \leqslant K \| \phi_0 \|_0 e^{-p(t)}, \quad t \geqslant 0, \quad K > 0,$$
where $x(0, \phi_0)$ is a solution of (7.1.1) with an initial function ϕ_0 at $t = 0$;

(iii) the system (7.1.1) is smooth enough to ensure uniqueness and continuous dependence of solutions.

Then, there exists a functional $V(t, \phi)$ satisfying the following properties:

(1°) $V \in C[J \times C_\rho, R_+]$, $V(t, \phi)$ is Lipschitzian in ϕ for a continuous function $K(t) \geqslant 0$;

(2°) $\| \phi \|_0 \leqslant V(t, \phi) \leqslant K(t) \| \phi \|_0$;

(3°) $D^+ V(t, \phi) = -p'(t) V(t, \phi)$.

7.2. CONVERSE THEOREMS

Proof. Let us denote $\phi = x_t(0, \phi_0)$, so that, in view of assumption (iii), we have $\phi_0 = x_0(t, \phi)$. We now define the functional $V(t, \phi)$ by

$$V(t, \phi) = Ke^{-p(t)} \| x_0(t, \phi) \|_0.$$

Then, it is clear that $V \in C[J \times C_\rho, R_+]$, and, because of condition (ii), there results

$$\| \phi \|_0 \leqslant V(t, \phi).$$

In view of uniqueness, we have, for small $h > 0$,

$$V(t+h, x_{t+h}(t, \phi)) = Ke^{-p(t+h)} \| x_0(t+h, x_{t+h}(t, \phi)) \|_0$$
$$= Ke^{-p(t+h)} \| x_0(t, \phi) \|_0,$$

and, consequently, it follows that

$$D^+V(t, \phi) = -p'(t)V(t, \phi).$$

Letting $\phi = x_t(0, \phi_0)$, $\psi = x_t(0, \psi_0)$, we obtain the inequality

$$\| x_0(t, \phi) - x_0(t, \psi) \|_0 \leqslant \| \phi - \psi \|_0 \exp\left[\int_0^t p_1(s)\, ds\right], \quad t \geqslant 0,$$

because of uniqueness and assumption (i).

Thus, we have

$$| V(t, \phi) - V(t, \psi) | = Ke^{-p(t)} \| x_0(t, \phi) - x_0(t, \psi) \|_0$$
$$\leqslant K \exp\left[-p(t) + \int_0^t p_1(s)\, ds\right] \| \phi - \psi \|_0$$
$$\equiv K(t) \| \phi - \psi \|_0, \qquad (7.2.4)$$

provided we define $K(t) = K \exp[-p(t) + \int_0^t p_1(s)\, ds]$.

Finally, the upper estimate in (2°) results, by setting $\psi = 0$. The proof of the theorem is therefore complete.

COROLLARY 7.2.1. *If $\int_0^t p_1(s)\, ds \leqslant p(t)$, in addition to the assumptions of Theorem 7.2.3, then $K(t)$ is to be replaced by $K > 0$ in the conclusion of Theorem 7.2.3.*

THEOREM 7.2.4. *Assume that*

(i) *condition (i) of Theorem 7.2.3 holds;*

(ii) *the system (7.1.1) is smooth enough to ensure the uniqueness and* continuous dependence or solutions, and there exist functions λ_1, $\lambda_2 \in \mathcal{K}$ such that

$$\lambda_2(\|\phi_0\|_0) \leq \|x_t(0,\phi_0)\|_0 \leq \lambda_1(\|\phi_0\|_0), \qquad t \geq 0; \tag{7.2.5}$$

(iii) $g \in C[J \times R_+, R]$, and, for $t \in J$, $u_1 \geq u_2$,

$$g(t, u_1) - g(t, u_2) \leq p_2(t)(u_1 - u_2), \tag{7.2.6}$$

where $p_2 \in C[J, R]$, and the solutions $u(t, 0, u_0)$ of (6.1.2) verify the estimate

$$\gamma_2(u_0) \leq u(t, 0, u_0) \leq \gamma_1(u_0), \qquad t \geq 0, \tag{7.2.7}$$

for $\gamma_1, \gamma_2 \in \mathcal{K}$.

Then, there exists a functional $V(t, \phi)$ satisfying the following conditions:

(1°) $V \in C[J \times C_\rho, R_+]$, and $V(t, \phi)$ is Lipschitzian in ϕ for a continuous function $K(t) \geq 0$;

(2°) $b(\|\phi\|_0) \leq V(t, \phi) \leq a(\|\phi\|_0)$, $a, b \in \mathcal{K}$, $(t, \phi) \in J \times C_\rho$;

(3°) $D^+V(t, \phi) = g(t, V(t, \phi))$, $(t, \phi) \in J \times C_\rho$.

Proof. By condition (7.2.6), the uniqueness of solutions of (6.1.2) is assured. Let $u(t, 0, u_0)$ and $x(0, \phi_0)$ be the solutions of (6.1.2) and (7.1.1), respectively, satisfying the assumptions of the theorem. Denote $\phi = x_t(0, \phi_0)$, so that, by uniqueness, we have $\phi_0 = x_0(t, \phi)$. Let us now define

$$V(t, \phi) = u(t, 0, \|x_0(t, \phi)\|_0).$$

It is clear that $V \in C[J \times C_\rho, R_+]$. Furthermore, for small $h > 0$,

$$V(t+h, x_{t+h}(t, \phi)) = u(t+h, 0, \|x_0(t+h, x_{t+h}(t, \phi))\|_0)$$
$$= u(t+h, 0, \|x_0(t, \phi)\|_0).$$

Thus, for $(t, \phi) \in J \times C_\rho$,

$$D^+V(t, \phi) = \limsup_{h \to 0^+} h^{-1}[V(t+h, x_{t+h}(t, \phi)) - V(t, \phi)]$$
$$= \limsup_{h \to 0^+} h^{-1}[u(t+h, 0, \|x_0(t, \phi)\|_0) - u(t, 0, \|x_0(t, \phi)\|_0)]$$
$$= u'(t, 0, \|x_0(t, \phi)\|_0)$$
$$= g(t, u(t, 0, \|x_0(t, \phi)\|_0))$$
$$= g(t, V(t, \phi)),$$

proving (3°).

7.2. CONVERSE THEOREMS

Since $\phi = x_t(0, \phi_0)$ and $\phi_0 = x_0(t, \phi)$, the relation (7.2.5) yields

$$\lambda_1^{-1}(\|\phi\|_0) \leqslant \|x_0(t, \phi)\|_0 \leqslant \lambda_2^{-1}(\|\phi\|_0),$$

$\lambda_1^{-1}, \lambda_2^{-1}$ being the inverse functions of λ_1, λ_2, respectively. Hence, using this inequality and (7.2.7) successively, we obtain

$$V(t, \phi) \geqslant \gamma_2(\|x_0(t, \phi)\|_0)$$
$$\geqslant \gamma_2(\lambda_1^{-1}(\|\phi\|_0))$$
$$\equiv b(\|\phi\|_0),$$

and

$$V(t, \phi) \leqslant \gamma_1(\|x_0(t, \phi)\|_0)$$
$$\leqslant \gamma_1(\lambda_2^{-1}(\|\phi\|_0))$$
$$\equiv a(\|\phi\|_0).$$

Evidently $a, b \in \mathcal{K}$, and hence (2°) is verified.

Finally, for $t \in J, \phi_1, \phi_2 \in C_\rho$,

$$|V(t, \phi_1) - V(t, \phi_2)| = |u(t, 0, \|x_0(t, \phi_1)\|_0) - u(t, 0, \|x_0(t, \phi_2)\|_0)|$$
$$\leqslant \|x_0(t, \phi_1) - x_0(t, \phi_2)\|_0 \exp\left[\int_0^t p_2(s) \, ds\right],$$

using the condition (7.2.6). Furthermore, as observed in the proof of Theorem 7.2.3, we have, as a consequence of assumption (i),

$$\|x_0(t, \phi_1) - x_0(t, \phi_2)\|_0 \leqslant \|\phi_1 - \phi_2\|_0 \exp\left[\int_0^t p_1(s) \, ds\right].$$

These considerations imply that

$$|V(t, \phi_1) - V(t, \phi_2)| \leqslant K(t) \|\phi_1 - \phi_2\|_0,$$

where $K(t) = \exp[\int_0^t [p_1(s) + p_2(s)] \, ds]$. The proof is complete.

REMARK 7.2.1. We note that, since $p_2(t)$ need not be nonnegative, there is a possibility that $K(t)$ may be bounded by a constant.

On the basis of Theorem 7.2.4, it is possible to state and prove other converse theorems involving differential inequalities, parallel to certain theorems in Sect. 3.6. We shall only state two converse theorems with respect to uniform asymptotic stability.

THEOREM 7.2.5. Assume that

(i) the system (7.1.1) is smooth enough to ensure the uniqueness and continuous dependence of solutions;

(ii) $\beta_1 \|\phi_0\|_0 \sigma_1(t - t_0) \leq \| x_t(t_0, \phi_0)\|_0 \leq \beta_2 \|\phi_0\|_0 \sigma_2(t - t_0), t \geq t_0$, where $\beta_1, \beta_2 > 0$ are constants.

Then, there exists a functional $V(t, \phi)$ verifying the following properties:

(1°) $V \in C[J \times C_\rho, R_+]$, and there exist two functions $a, b \in \mathscr{K}$ such that
$$b(\|\phi\|_0) \leq V(t, \phi) \leq a(\|\phi\|_0), \qquad (t, \phi) \in J \times C_\rho;$$

(2°) $d/dt\, [V(t, x_t(t_0, \phi_0))] \leq -\alpha V(t, x_t(t_0, \phi_0)), t \geq t_0$.

Proof. For some fixed $T > 0$, which we shall choose later, define
$$V(t, \phi) = \int_t^{t+T} \| x_s(t, \phi)\|_0^2 \, ds.$$

Since, by assumption (ii), we have
$$\| x_s(t, \phi)\|_0 \leq \beta_2 \|\phi\|_0 \sigma_2(0), \qquad s \geq t,$$

it follows that
$$V(t, \phi) \leq \int_t^{t+T} \beta_2^2 \|\phi\|_0^2 \sigma_2^2(0) \, ds$$
$$= \beta_2^2 \sigma_2^2(0) T \|\phi\|_0^2$$
$$\equiv a(\|\phi\|_0). \tag{7.2.8}$$

Moreover, similar arguments yield
$$V(t, \phi) \geq \int_t^{t+T} [\beta_1 \|\phi\|_0 \sigma_1(s - t)]^2 \, ds$$
$$= \beta_1^2 \|\phi\|_0^2 \int_0^T \sigma_1^2(u) \, du$$
$$= b(\|\phi\|_0).$$

We have thus proved (1°).

To prove the validity of (2°), notice that
$$V(t, x_t(t_0, \phi_0)) = \int_t^{t+T} \| x_s(t, x_t(t_0, \phi_0))\|_0^2 \, ds$$
$$= \int_t^{t+T} \| x_s(t_0, \phi_0)\|_0^2 \, ds.$$

Hence,

$$d/dt\,[V(t, x_t(t_0, \phi_0))] = \|x_{t+T}(t_0, \phi_0)\|_0^2 - \|x_t(t_0, \phi_0)\|_0^2$$
$$= [\|x_{t+T}(t, x_t(t_0, \phi_0))\|_0 - \|x_t(t_0, \phi_0)\|_0]$$
$$\times [\|x_{t+T}(t_0, \phi_0)\|_0 + \|x_t(t_0, \phi_0)\|_0]$$
$$\leqslant [\beta_2 \sigma_2(T)\|x_t(t_0, \phi_0)\|_0 - \|x_t(t_0, \phi_0)\|_0]$$
$$\times [\|x_{t+T}(t_0, \phi_0)\|_0 + \|x_t(t_0, \phi_0)\|_0].$$

Let us now fix T so that $\sigma_2(T) < (2\beta_2)^{-1}$. This choice is possible, since $\sigma_2 \in \mathscr{L}$. It then results that

$$d/dt\,[V(t, x_t(t_0, \phi_0))] \leqslant -\tfrac{1}{2} \|x_t(t_0, \phi_0)\|_0^2.$$

This, together with (7.2.8), yields, setting $(2\alpha)^{-1} = \beta_2^2 \sigma_2^2(0)\,T$,

$$d/dt\,[V(t, x_t(t_0, \phi_0))] \leqslant -\alpha V(t, x_t(t_0, \phi_0)),$$

and proves the theorem.

THEOREM 7.2.6. *Let the trivial solution of* (7.1.1) *be uniformly asymptotically stable. Suppose that*

$$\|f(t, \phi) - f(t, \psi)\| \leqslant L(t)\|\phi - \psi\|_0,$$

for $(t, \phi), (t, \psi) \in J \times C_\rho$, *where* $L(t) \geqslant 0$ *is continuous on* J *and*

$$\int_t^{t+u} L(s)\,ds \leqslant Ku, \qquad u \geqslant 0.$$

Then, there exists a functional $V(t, \phi)$ *with the following properties:*

(1°) $V \in C[J \times C_\rho, R_+]$, *and* $V(t, \phi)$ *satisfies*

$$|V(t, \phi) - V(t, \psi)| \leqslant M\|\phi - \psi\|_0,$$

for $t \in J, \phi, \psi \in C_{\delta(\delta_0)}$;

(2°) $b(\|\phi\|_0) \leqslant V(t, \phi) \leqslant a(\|\phi\|_0)$, $a, b \in \mathscr{K}$;
(3°) $D^+V(t, \phi) \leqslant -C[V(t, \phi)]$, $C \in \mathscr{K}$.

The proof of this theorem can be constructed parallel to Theorem 3.6.9 with essential changes. We leave the details.

7.3. Autonomous systems

In this section, we consider some stability and instability results for autonomous systems of the form

$$x'(t) = f(x_t), \qquad t \in J, \tag{7.3.1}$$

where $f \in C[C_\rho, R^n]$ and $f(\phi)$ is locally Lipschitzian in ϕ. It is quite natural to consider the system (7.3.1) as defining motions or paths in $\mathscr{C}^n$. In fact, we can define a *motion* through ϕ as the set of functions in $\mathscr{C}^n$ given by $\bigcup_{t \in J} x_t(0, \phi)$, assuming that the solutions $x_t(0, \phi)$ exist on J. We shall, in what follows, abbreviate $x_t(0, \phi)$ by $x_t(\phi)$.

DEFINITION 7.3.1. An element $\psi \in \mathscr{C}^n$ is said to be in the ω-limit set of ϕ, $\Omega(\phi)$ if $x_t(\phi)$ is defined for $[-\tau, \infty)$ and there is a sequence of non-negative real numbers $\{t_n\}$, $t_n \to \infty$ as $n \to \infty$, such that

$$\| x_{t_n}(\phi) - \psi \|_0 \to 0 \qquad \text{as} \qquad n \to \infty.$$

DEFINITION 7.3.2. A set $M \subset \mathscr{C}^n$ is said to be an *invariant set* if, for any ϕ in M, there exists a function $x(\phi)$ depending on ϕ, defined on $(-\infty, \infty)$, $x_t(\phi) \in M$ for $t \in (-\infty, \infty)$, $x_0(\phi) = \phi$, such that, if $x^*(\sigma, x_\sigma)$ is the solution of (7.3.1) with the initial function x_σ at σ, then $x^*(\sigma, x_\sigma) = x_t(\phi)$ for all $t \geqslant \sigma$.

We notice that to any element of an invariant set there corresponds a solution that must be defined on $(-\infty, \infty)$.

LEMMA 7.3.1. Let $x(\phi)$ be a solution of the system (7.3.1) with an initial function ϕ at $t = 0$, defined on $[-\tau, \infty)$, and let

$$\| x_t(\phi) \|_0 \leqslant \rho_1 < \rho \qquad \text{for} \qquad t \in J.$$

Then, the family of functions $\{x_t(\phi)\}$, $t \in J$, belongs to a compact subset of $\mathscr{C}^n$, that is, the motion through ϕ belongs to a compact subset of $\mathscr{C}^n$.

The proof of this lemma follows from the fact that, for any $\rho_1 < \rho$, there exists a constant $L > 0$ such that $\| f(\phi) \| \leqslant L$ for all ϕ satisfying $\| \phi \|_0 \leqslant \rho_1$.

LEMMA 7.3.2. Let $\phi \in C_\rho$ be such that the solution $x(\phi)$ of the system (7.3.1) is defined on $[-\tau, \infty)$ and $\| x_t(\phi) \|_0 \leqslant \rho_1 < \rho$, $t \in J$. Then, the ω-limit set $\Omega(\phi)$ is nonempty, compact, connected invariant set, and

$$d[x_t(\phi), \Omega(\phi)] \to 0 \qquad \text{as} \qquad t \to \infty. \tag{7.3.2}$$

Proof. By Lemma 7.3.1, the family of functions $x_t(\phi)$, $t \in J$, belongs to a compact subset $S \subset \mathscr{C}^n$, and, furthermore, S could be chosen to be the set of $\psi \in \mathscr{C}^n$ such that $\|\psi\|_0 \leqslant \rho_1$, $\|\psi'\|_0 \leqslant k$, for some constant k. This proves that $\Omega(\phi)$ is nonempty and bounded.

If $\psi \in \Omega(\phi)$, then there exists a sequence $\{t_n\}$, $t_n \to \infty$ as $n \to \infty$, such that $\|x_{t_n}(\phi) - \psi\|_0 \to 0$ as $n \to \infty$. For any integer N, there exists a subsequence of $\{t_n\}$, which we keep the same designation, and a function $g_\alpha(\phi)$ defined for $-N \leqslant \alpha \leqslant N$, such that $\|x_{t_n+\alpha}(\phi) - g_\alpha(\phi)\|_0 \to 0$ as $n \to \infty$ uniformly for $\alpha \in [-N, N]$. By the diagonalization process, we can choose the t_n so that $\|x_{t_n+\alpha}(\phi) - g_\alpha(\phi)\|_0 \to 0$ as $n \to \infty$ uniformly on all compact subsets of $(-\infty, \infty)$. In particular, the sequence $\{x_{t_n+\alpha}(\phi)\}$ defines a function $g_\alpha(\phi)$, for $-\infty < \alpha < \infty$. It is easy to see that $g_\alpha(\phi)$ satisfies (7.3.1). Since $g_0(\phi) = \psi$, it follows that the solution $x_t(\psi)$ of (7.3.1) with initial value ψ at $t = 0$ is defined for $t \in (-\infty, \infty)$ and, furthermore, is in $\Omega(\phi)$, since

$$\|x_{t_n+t}(\phi) - x_t(\psi)\|_0 \to 0 \quad \text{as} \quad n \to \infty$$

for any fixed t. This shows that $\Omega(\phi)$ is invariant. It is clear that $\Omega(\phi)$ is connected.

To show that $\Omega(\phi)$ is closed, suppose ψ_n in $\Omega(\phi)$ approaches ψ as $n \to \infty$. There exists an increasing sequence $\{t_n\} = \{t_n(\psi_n)\}$ such that $t_n \to \infty$ as $n \to \infty$, and $\|x_{t_n}(\phi) - \psi_n\|_0 \to 0$ as $n \to \infty$. Given any $\epsilon > 0$, choose n so large that

$$\|\psi_n - \psi\|_0 < \epsilon/2 \quad \text{and} \quad \|x_{t_n}(\phi) - \psi_n\|_0 < \epsilon/2.$$

Then,

$$\|x_{t_n}(\phi) - \psi\|_0 < \epsilon,$$

for sufficiently large n, which shows that $\psi \in \Omega(\phi)$, and hence $\Omega(\phi)$ is closed. But, clearly, $\Omega(\phi) \subset S$, and, since S is compact, it follows that $\Omega(\phi)$ is compact.

To prove (7.3.2), suppose that there is an increasing sequence $\{t_n\}$, $t_n \to \infty$ as $n \to \infty$, and an $\alpha > 0$ such that

$$\|x_{t_n}(\phi) - \psi\|_0 \geqslant \alpha, \quad \psi \in \Omega(\phi).$$

Since $x_{t_n}(\phi)$ belongs to a compact subset of $\mathscr{C}^n$, there exists a subsequence that converges to an element ψ in $\mathscr{C}^n$, and thus ψ is in $\Omega(\phi)$. This is a contradiction to the foregoing inequality and completes the proof of the lemma.

REMARK 7.3.1. In the proof of Lemma 7.3.2, we have only used the fact that $x_t(\phi)$ is continuous in t, ϕ and that $x_t(\phi)$ belongs to a compact subset of $\mathscr{C}^n$. Therefore, the Lipschitz condition on f could have been replaced by $\|\phi\|_0 \leqslant \rho_1 < \rho$ implies $\|f(\phi)\| \leqslant L$ for some L.

THEOREM 7.3.1. Let

(i) $V \in C[C_\rho, R]$ and $\Omega_\alpha = [\phi \in C_\rho : V(\phi) < \alpha]$;
(ii) there exist a constant K such that $\|\phi(0)\| \leqslant K$, $V(\phi) \geqslant 0$, and, for $\phi \in \Omega_\alpha$, $D^+V(\phi) \leqslant 0$;
(iii) E be the set of all points in Ω_α, where $D^+V(\phi) = 0$, and M be the largest invariant set in E.

Then, every solution of (7.3.1) with initial value in Ω_α approaches M as $t \to \infty$.

Proof. The conditions on V imply that $V(x_t(\phi))$ is a nonincreasing function of t and bounded below within Ω_α. Hence, $\phi \in \Omega_\alpha$ implies $x_t(\phi) \in \Omega_\alpha$ and $\|x(\phi)(t)\| \leqslant K$ for all $t \geqslant 0$, which shows that $\|x_t(\phi)\|_0 \leqslant K$ for all $t \geqslant 0$; that is, $x_t(\phi)$ is bounded, and Lemma 7.3.2 yields that $\Omega(\phi)$ is an invariant set. But $V(x_t(\phi))$ has a limit $\alpha_0 < \alpha$, as $t \to \infty$ and $V = \alpha_0$ on $\Omega(\phi)$. Hence, $\Omega(\phi)$ is in Ω_α and $D^+V(\phi) = 0$ on $\Omega(\phi)$. Consequently, the fact that $\Omega(\phi)$ is invariant implies that $\Omega(\phi)$ is in M, and, by Lemma 7.3.2, $x_t(\phi)$ tends to M as $t \to \infty$. This completes the proof.

The conditions $\|\phi(0)\| \leqslant K$ and $V \geqslant 0$ of the foregoing theorem may be replaced by the assumption that the region where $V(\phi) < \alpha$ is compact.

COROLLARY 7.3.1. If the conditions of Theorem 7.3.1 are satisfied and $D^+V(\phi) < 0$, for all $\phi \neq 0$ in Ω_α, then every solution of (7.3.1) with initial value in Ω_α approaches zero as $t \to \infty$.

THEOREM 7.3.2. Assume that

(i) $V \in C[\mathscr{C}^n, R_+]$ and $D^+V(\phi) \leqslant 0$, $\phi \in \mathscr{C}^n$;
(ii) E is the set of all points ϕ in $\mathscr{C}^n$ for which $D^+V(\phi) = 0$, and M is the largest invariant set in E.

Then, all solutions of (7.3.1), which are bounded for $t \geqslant 0$, approach M as $t \to \infty$.

If, in addition, there exists a continuous, nonnegative function $b(u)$ on the interval $[0, \infty)$ such that $b(u) \to \infty$ as $u \to \infty$ and

$$b(\|\phi(0)\|) \leqslant V(\phi), \quad \phi \in \mathscr{C}^n,$$

then all solutions of (7.3.1) are bounded for $t \geqslant 0$.

Proof. The first part of the theorem proceeds essentially as in Theorem 7.3.1. For the second part, let $\phi_0 \in \mathscr{C}^n$. Then, there is a constant N such that $V(\phi) > V(\phi_0)$, for $\|\phi(0)\| \geq N$. Since $V(x_t(\phi_0))$ is a nonincreasing function of t, it follows that $\|x(\phi_0)(t)\| < N$, for $t \geq 0$, which implies $\|x_t(\phi_0)\|_0 < N$, $t \geq 0$.

COROLLARY 7.3.2. *If $f(0) = 0$, all the conditions of Theorem 7.3.2 are satisfied, and $V(0) = 0$, $D^+V(\phi) < 0$ for $\phi \neq 0$, then all solutions of (7.3.1) approach zero as $t \to \infty$, and the origin is globally asymptotically stable.*

We next give a theorem on instability of the trivial solution of (7.3.1).

THEOREM 7.3.3. *Suppose that $V(\phi)$ is a continuous, bounded scalar function on C_ρ and that there exists a γ and an open set E in $\mathscr{C}^n$ such that the following conditions are satisfied:*

(i) $V(\phi) > 0$ on E, $V(\phi) = 0$ on that part of the boundary of E in C_γ;

(ii) 0 belongs to the closure of $E \cap C_\gamma$;

(iii) $V(\phi) \leq a(\|\phi(0)\|)$, on $E \cap C_\gamma$, where, $a \in \mathscr{K}$;

(iv) $D_+V(\phi) = \liminf_{h \to 0^+} h^{-1}[V(x_h(\phi)) - V(\phi)] \geq 0$, on the closure of $E \cap C_\gamma$ and the set U of ϕ in the closure of $E \cap C_\gamma$ such that $D_+V(\phi) = 0$, contains no invariant set of (7.3.1) except $\phi = 0$.

Under these conditions, the trivial solution of (7.3.1) is unstable, and the trajectory of each solution of (7.3.1), with initial value in $E \cap C_\gamma$, intersects $\bar{C}_\gamma$ at some finite time.

Proof. Suppose $\phi_0 \in E \cap C_\gamma$. By hypothesis (iii),

$$\|\phi_0(0)\| \geq a^{-1}(V(\phi_0)),$$

and (iii) and (iv) imply that $x_t(\phi_0)$ satisfies

$$\|x(\phi_0)(t)\| \geq a^{-1}(V(x_t(\phi_0)))$$
$$\geq a^{-1}(V(\phi_0)),$$

as long as $x_t(\phi_0) \in E \cap C_\gamma$. If $x_t(\phi_0)$ leaves $E \cap C_\gamma$, then it must cross the boundary ∂C_γ of C_γ. In fact, it must cross either ∂E or ∂C_γ, but it cannot cross ∂E inside C_γ since $V = 0$ on that part of ∂E inside C_γ and $V(x_t(\phi_0)) \geq V(\phi_0) > 0$, $t \geq 0$. Now, suppose that $x_t(\phi_0)$ never reaches ∂C_γ. Then, $x_t(\phi_0)$ belongs to a compact subset of the closure of $E \cap C_\gamma$, for $t \geq 0$. Consequently, $x_t(\phi_0)$ approaches $\Omega(\phi_0)$, the ω-limit set of ϕ_0

and $\Omega(\phi_0) \subset$ closure of $E \cap C_\gamma$. Since $V(x_t(\phi_0))$ is nondecreasing and bounded above, it follows that $V(x_t(\phi_0)) \to B$, a constant, as $t \to \infty$, and, thus, $D^+V(x_t(\psi)) = 0$ for $\psi \in \Omega(\phi_0)$. Since $\psi \in \Omega(\phi_0)$ implies that

$$\|\psi(0)\| \leqslant a^{-1}(V(\phi_0)) > 0,$$

we have a contradiction to hypothesis (iv). Consequently, there is a $t_1 > 0$ such that $\|x(\phi_0)(t_1)\| = \gamma$. Hypothesis (ii) implies instability, since ϕ_0 can be chosen arbitrarily close to zero. This completes the proof of the theorem.

7.4. Perturbed systems

We shall be interested, in this section, in the perturbed functional differential system

$$x'(t) = f(t, x_t) + R(t, x_t), \tag{7.4.1}$$

where $f, R \in C[J \times C_\rho, R^n]$ and $f(t, \phi)$, $R(t, \phi)$ satisfy a Lipschitz condition in ϕ for each $t \in J$.

THEOREM 7.4.1. Suppose that the trivial solution of (7.1.1) is exponentially asymptotically stable and $f(t, \phi)$ is linear in ϕ. Assume further that

$$\|R(t, \phi)\| \leqslant \eta \|\phi\|_0, \quad t \in J, \quad \phi \in C_\rho, \tag{7.4.2}$$

η being a sufficiently small positive number. Then, the trivial solution of the perturbed system also enjoys the exponential asymptotic stability.

Proof. By Theorem 7.2.1, there exists a $V(t, \phi)$ such that, for $(t, \phi) \in J \times C_\rho$,

(i) $V \in C[J \times C_\rho, R_+]$, and V is Lipschitzian in ϕ with a constant $K > 0$;
(ii) $\|\phi\|_0 \leqslant V(t, \phi) \leqslant K\|\phi\|_0$;
(iii) $D^+V(t, \phi) \leqslant -\alpha V(t, \phi), \alpha > 0$.

Let $y(t_0, \phi_0)$ be any solution of (7.4.1) such that $\|\phi_0\|_0 < \rho/2K$. Define

$$m(t) = V(t, y_t(t_0, \phi_0)).$$

Whenever $\|\phi_0\|_0 < \rho/2K$, we have, because of (ii), $m(t_0) < \rho/2$. We assert that $m(t) < \rho, t \geqslant t_0$. Let this be false. Then, there exist two numbers t_1 and t_2 such that

$$m(t_2) = \rho/2, \quad m(t_1) = \rho, \quad \text{and} \quad m(t) \geqslant \rho/2, \quad t \in [t_2, t_1].$$

7.4. PERTURBED SYSTEMS

We see from these relations that

$$D^+m(t_2) \geqslant 0. \tag{7.4.3}$$

On the other hand, it follows, from (ii), that

$$\|y_t(t_0, \phi_0)\|_0 \leqslant \rho \quad \text{for} \quad t_0 \leqslant t \leqslant t_1.$$

Let $\phi = y_{t_2}(t_0, \phi_0)$, and, because of uniqueness,

$$y_{t_2+h}(t_0, \phi_0) = y_{t_2+h}(t_2, \phi), \quad h \geqslant 0.$$

Thus,

$$D^+m(t_2) = \limsup_{h \to 0^+} h^{-1}[V(t_2 + h, y_{t_2+h}(t_0, \phi_0)) - V(t_2, y_{t_2}(t_0, \phi_0))]$$

$$= \limsup_{h \to 0^+} h^{-1}[V(t_2 + h, y_{t_2+h}(t_2, \phi)) - V(t_2 + h, x_{t_2+h}(t_2, \phi))$$

$$+ V(t_2 + h, x_{t_2+h}(t_2, \phi)) - V(t_2, \phi)],$$

where $x(t_2, \phi)$ is the solution of (7.1.1) with an initial function ϕ at $t = t_2$. Using the Lipschitzian character of V, the assumption (7.4.2), and the preceding relation, we obtain

$$D^+m(t_2) \leqslant K\eta\|\phi\|_0 - \alpha m(t_2). \tag{7.4.4}$$

Since $\eta > 0$ is sufficiently small, there exists a $\gamma > 0$ such that $K\eta < \alpha - \gamma$, and, hence, the fact that $\|\phi_0\|_0 \leqslant V(t, \phi)$, together with the inequality (7.4.4), implies that

$$D^+m(t_2) \leqslant m(t_2)[-\alpha + K\eta]$$
$$\leqslant -\gamma m(t_2)$$
$$< 0,$$

since $m(t_2) = \rho/2 > 0$. This contradicts (7.4.3) and proves that $m(t) < \rho$, $t \geqslant t_0$. Consequently, by ii, it follows that

$$\|y_t(t_0, \phi_0)\|_0 < \rho, \quad t \geqslant t_0,$$

whenever $\|\phi_0\|_0 < \rho/2K$. Thus, setting $\phi = y_t(t_0, \phi_0)$ and arguing as before, we arrive at the inequality

$$D^+V(t, y_t(t_0, \phi_0)) \leqslant -\gamma V(t, y_t(t_0, \phi_0)),$$

and therefore, by Theorem 1.4.1,

$$V(t, y_t(t_0, \phi_0)) \leqslant V(t_0, \phi_0) \exp[-\gamma(t - t_0)], \quad t \geqslant t_0.$$

Because of ii, from this inequality results a further inequality

$$\|y_t(t_0, \phi_0)\|_0 \leqslant \rho/2 \exp[-\gamma(t - t_0)], \qquad t \geqslant t_0,$$

proving the stated result.

THEOREM 7.4.2. Assume that the trivial solution of (7.1.1) is generalized exponentially asymptotically stable and that $f(t, \phi)$ is linear in ϕ. Suppose that $g \in C[J \times R_+, R_+]$ and $g(t, 0) \equiv 0$, $g(t, u)$ is nondecreasing in u for each $t \in J$, and

$$\|R(t, \phi)\| \leqslant g(t, \|\phi\|_0), \qquad t \in J, \qquad \phi \in C_\rho. \qquad (7.4.5)$$

If $y(t_0, \phi_0)$ be the solution of (7.4.1) such that $u_0 = K(t_0)\|\phi_0\|_0$, then

$$\|y_t(t_0, \phi_0)\|_0 \leqslant r(t, t_0, u_0), \qquad t \geqslant t_0, \qquad (7.4.6)$$

where $r(t, t_0, u_0)$ is the maximal solution of

$$u' = -p'(t)u + K(t)g(t, u), \qquad u(t_0) = u_0, \qquad (7.4.7)$$

existing for $t \geqslant t_0$ and satisfying $r(t, t_0, u_0) < \rho, t \geqslant t_0$, whenever $u_0 < \rho$.

Proof. Let $y(t_0, \phi_0)$ be any solution of (7.4.1) such that $\|\phi_0\|_0 < \rho/K(t_0)$. Setting $\phi = y_t(t_0, \phi_0)$, we have

$$y_{t+h}(t_0, \phi_0) = y_{t+h}(t, \phi), \qquad h \geqslant 0,$$

because of uniqueness of solutions. Suppose now that $x_{t+h}(t, \phi), h \geqslant 0$, is the solution of (7.1.1) through (t, ϕ). If $\|y_t(t_0, \phi_0)\|_0 < \rho, t \geqslant t_0$, we should have

$$D^+V(t, \phi)_{(7.4.1)} = \limsup_{h \to 0^+} h^{-1}[V(t + h, y_{t+h}(t, \phi)) - V(t + h, x_{t+h}(t, \phi))$$
$$+ V(t + h, x_{t+h}(t, \phi)) - V(t, \phi)]$$
$$\leqslant D^+V(t, \phi)_{(7.1.1)} + \limsup_{h \to 0^+} [K(t + h)/h]\|y_{t+h}(t, \phi) - x_{t+h}(t, \phi)\|_0$$
$$\leqslant D^+V(t, \phi)_{(7.1.1)} + K(t)\|y'(t, \phi)(t) - x'(t, \phi)(t)\|$$
$$\leqslant -p'(t)V(t, \phi) + K(t)\|R(t, \phi)\|.$$

Thus, since $\phi = y_t(t_0, \phi_0)$, it follows that

$$D^+V(t, y_t(t_0, \phi_0)) \leqslant -p'(t)V(t, y_t(t_0, \phi)) + K(t)g(t, V(t, y_t(t_0, \phi_0))).$$

By Theorem 1.4.1, it now results that
$$V(t, y_t(t_0, \phi_0)) \leq r(t, t_0, u_0), \qquad t \geq t_0,$$
choosing $u_0 = K(t_0)\|\phi_0\|_0$, $r(t, t_0, u_0)$ being the maximal solution of (7.4.7). Furthermore, from (ii) in the proof of Theorem 7.4.1,
$$\|y_t(t_0, \phi_0)\|_0 \leq V(t, y_t(t_0, \phi_0)).$$
Also, by assumption, $r(t, t_0, u_0) < \rho$, whenever $u_0 < \rho$. Our choices $\|\phi_0\|_0 < \rho/K(t_0)$ and $u_0 = K(t_0) \|\phi_0\|_0$ imply that $u_0 < \rho$. Hence,
$$\|y_t(t_0, \phi_0)\|_0 < \rho, \qquad t \geq t_0.$$
Thus, the estimate (7.4.6) holds.

THEOREM 7.4.3. Under the assumptions of Theorem 7.4.2, the stability properties of the trivial solution of the scalar differential equation (7.4.7) imply the corresponding stability properties of the trivial solution of the perturbed system (7.4.1).

The proof of this theorem is immediate from the relation (7.4.6). However, the following special cases are of importance.

COROLLARY 7.4.1. The function $g(t, u) = \lambda(t) u$, where $\lambda \in C[J, R_+]$, is admissible in Theorem 7.4.3, provided there exists a continuous function $q(t) > 0$, $t \in J$, such that

$$\exp\left[p(t_0) - p(t) + \int_{t_0}^{t} K(s)\lambda(s)\, ds\right] \leq q(t_0), \qquad t \geq t_0. \qquad (7.4.8)$$

Proof. It is enough to show that, under the assumptions of the corollary, the trivial solution $u = 0$ of (7.4.7) is equistable. For, the general solution $u(t, t_0, u_0)$ of
$$u' = -p'(t)u + K(t)\lambda(t)u, \qquad u(t_0) = u_0,$$
is given by
$$u(t, t_0, u_0) = u_0 \exp\left[p(t_0) - p(t) + \int_{t_0}^{t} K(s)\lambda(s)\, ds\right], \qquad t \geq t_0,$$
and, hence, equistability follows from (7.4.8).

COROLLARY 7.4.2. The functions $g(t, u) = \lambda(t)u$, $\lambda \in C[J, R_+]$, $p(t) = \alpha t$, $\alpha > 0$, and $K(t) = K > 0$ are admissible in Theorem 7.4.3, provided

$$\limsup_{t \to \infty} \left[(t - t_0)^{-1} \int_{t_0}^{t} \lambda(s)\, ds\right] < \alpha/K. \qquad (7.4.9)$$

Proof. In this case, the general solution $u(t, t_0, u_0)$ of (7.4.7) is given by

$$u(t, t_0, u_0) = u_0 \exp\left[-\alpha(t - t_0) + K \int_{t_0}^{t} \lambda(s)\, ds\right], \qquad t \geq t_0,$$

and, therefore, the condition (7.4.9) shows that the trivial solution of (7.4.7) is uniformly asymptotically stable. Hence, by Theorem 7.4.3, the trivial solution of (7.4.1) is also uniformly asymptotically stable.

The foregoing results can be extended to the case when $f(t, \phi)$ is nonlinear, on the basis of Theorem 7.2.2. Furthermore, as in the case of ordinary differential equations, we can show that, if the trivial solution of the unperturbed system is uniformly asymptotically stable, then it has certain stability properties under different classes of perturbations. For example, the concepts of total stability may be formulated parallel to the definitions of Sect. 3.8, and corresponding results may be proved. Likewise, boundedness, Lagrange stability, integral stability, and partial stability can be discussed. We shall omit such results as exercises to the reader. In the following sections, we shall only concentrate on extreme and perfect stability criteria and existence of almost periodic solutions.

7.5. Extreme stability

Associated with the system (7.1.1), let us consider the product system

$$x' = f(t, x_t), \qquad y' = f(t, y_t). \tag{7.5.1}$$

DEFINITION 7.5.1. The system (7.1.1) is said to be *extremely uniformly stable* if, for every $\epsilon > 0$, $t_0 \in J$, there exists a $\delta(\epsilon) > 0$ such that

$$\|\phi_0 - \psi_0\|_0 \leq \delta(\epsilon)$$

implies

$$\|x_t(t_0, \phi_0) - y_t(t_0, \psi_0)\|_0 < \epsilon, \qquad t \geq t_0;$$

extremely quasi-uniform asymptotically stable if, for every $\epsilon > 0$, $\eta > 0$, and $t_0 \in J$, there exists a $T(\epsilon, \eta) > 0$ such that

$$\|\phi_0 - \psi_0\|_0 \leq \eta$$

implies

$$\|x_t(t_0, \phi_0) - y_t(t_0, \psi_0)\|_0 < \epsilon, \qquad t \geq t_0 + T(\epsilon, \eta).$$

If the preceding two concepts hold simultaneously for $\phi_0, \psi_0 \in \mathscr{C}^n$, we shall say that the system (7.1.1) is *extremely uniformly completely stable*.

7.5. EXTREME STABILITY

The following theorem provides necessary conditions for the system (7.1.1) to be extremely uniformly completely stable.

THEOREM 7.5.1. Assume that

(i) $f \in C[J \times \mathscr{C}^n, R^n]$, and, for every $\alpha > 0$, if $\|\phi\|_0 \leqslant \alpha$, $\|\psi\|_0 \leqslant \alpha$,
$$\|f(t,\phi) - f(t,\psi)\| \leqslant L(\alpha)\|\phi - \psi\|_0 ;$$

(ii) the solutions of (7.1.1) are uniformly bounded;
(iii) the system (7.1.1) is extremely uniformly completely stable.

Then, there exists a functional $V(t, \phi, \psi)$ satisfying the following conditions:

(1°) $V \in C[J \times \mathscr{C}^n \times \mathscr{C}^n, R_+]$, and $V(t, \phi, \psi)$ satisfies
$$|V(t, \phi_1, \psi_1) - V(t, \phi_2, \psi_2)| \leqslant M(\eta)[\|\phi_1 - \phi_2\|_0 + \|\psi_1 - \psi_2\|_0],$$
for $\phi_i, \psi_i \in C_n$, $(i = 1, 2)$, $t \geqslant 0$, where $M(\eta)$ is a positive continuous function;

(2°) there exist functions $a, b \in \mathscr{K}$ such that
$$b(\|\phi - \psi\|_0) \leqslant V(t, \phi, \psi) \leqslant a(\|\phi - \psi\|_0);$$

(3°)
$$D^+V(t, \phi, \psi)_{(7.5.1)} = \limsup_{h \to 0^+} h^{-1}[V(t+h, x_{t+h}(t, \phi), y_{t+h}(t, \psi)) - V(t, \phi, \psi)]$$
$$\leqslant -V(t, \phi, \psi).$$

Proof. Let η be an arbitrary nonnegative number. Consider the case $\phi_0, \psi_0 \in C_n$. Then, corresponding to each $\epsilon > 0$, $t_0 \in J$, there exists a $T(\epsilon, \eta) > 0$ such that, if $t \geqslant t_0 + T(\epsilon, \eta)$, then
$$\|x_t(t_0, \phi_0) - y_t(t_0, \psi_0)\|_0 < \epsilon.$$

We assume that, if $\epsilon > 1$, $T(\epsilon, \eta) = T(1, \eta)$. By the uniform boundedness of solutions, it follows that
$$\|x_t(t_0, \phi_0)\|_0 \leqslant \gamma(\eta), \qquad \|y_t(t_0, \psi_0)\|_0 \leqslant \gamma(\eta)$$
for all $t \geqslant t_0$. Furthermore, if $\|\phi\|_0 \leqslant \gamma(\eta)$, $\|\psi\|_0 \leqslant \gamma(\eta)$, we have
$$\|f(t, \phi) - f(t, \psi)\| \leqslant L(\gamma(\eta))\|\phi - \psi\|_0.$$

We can assume that $T(\epsilon, \eta)$, $\gamma(\eta)$, and $L(\gamma(\eta))$ are continuous. Moreover, if we consider a positive function $A(\epsilon, \eta)$ for $\epsilon > 0$ and $\eta \geqslant 0$, such that
$$A(\epsilon, \eta) = \exp[\{L(\gamma(\eta)) + 1\} T(\epsilon, \eta)] + 2\gamma(\eta) \exp[T(\epsilon, \eta)], \qquad (7.5.2)$$

there exist two continuous functions $g(\epsilon)$ and $M(\eta)$ such that $g(\epsilon) > 0$ for $\epsilon > 0$, $g(0) = 0$, $M(\eta) > 0$, and

$$g(\epsilon) A(\epsilon, \eta) \leqslant M(\eta). \tag{7.5.3}$$

We now define $V_k(t, \phi, \psi)$ as follows:

$$V_k(t, \phi, \psi) = g(k^{-1}) \sup_{\sigma \geqslant 0} G_k(\| x_{t+\sigma}(t, \phi) - y_{t+\sigma}(t, \psi)\|_0) e^\sigma, \tag{7.5.4}$$

for $k = 1, 2, 3, \ldots$, where

$$G_k(u) = \begin{cases} u - k^{-1}, & u \geqslant k^{-1}, \\ 0, & 0 \leqslant u < k^{-1}. \end{cases} \tag{7.5.5}$$

It is easy to see that

$$V_k(t, \phi, \psi) = 0 \quad \text{if} \quad \phi = \psi. \tag{7.5.6}$$

$$g(k^{-1}) G_k(\|\phi - \psi\|_0) \leqslant V_k(t, \phi, \psi), \tag{7.5.7}$$

and

$$V_k(t, \phi, \psi) \leqslant g(k^{-1}) 2\gamma(\eta) \exp[T(k^{-1}, \eta)] \\ \leqslant M(\eta), \tag{7.5.8}$$

by (7.5.3). Moreover, for $\phi_0, \psi_0 \in \mathscr{C}^n$, there exists a $\beta \in \mathscr{K}$ such that

$$\| x_t(t_0, \phi_0) - y_t(t_0, \phi_0)\|_0 \leqslant \beta(\| \phi_0 - \psi_0\|_0), \quad t \geqslant t_0,$$

because the system (7.1.1) is extremely uniformly stable. We thus have

$$V_k(t, \phi, \psi) \leqslant g(k^{-1}) G_k(\beta(\|\phi - \psi\|_0)) \exp[T(k^{-1}, \|\phi - \psi\|_0)] \\ \leqslant \beta(\|\phi - \psi\|_0) M(\|\phi - \psi\|_0). \tag{7.5.9}$$

We shall next show that $V_k(t, \phi, \psi)$ satisfies a Lipschitz condition with respect to ϕ and ψ. For any $\phi, \phi_1, \psi, \psi_1 \in C_n$,

$$| V_k(t, \phi, \psi) - V_k(t, \phi, \psi_1)| \\ \leqslant g(k^{-1}) \sup_{\sigma \geqslant 0} e^\sigma | G_k(\| x_{t+\sigma}(t, \phi) - y_{t+\sigma}(t, \psi)\|_0 \\ \qquad - G_k(\| x_{t+\sigma}(t, \phi_1) - y_{t+\sigma}(t, \psi_1)\|_0)| \\ \leqslant g(k^{-1}) \sup_{T(k^{-1}, \eta) \geqslant \sigma \geqslant 0} e^\sigma [\| x_{t+\sigma}(t, \phi) - x_{t+\sigma}(t, \phi_1)\|_0 \\ \qquad + \| y_{t+\sigma}(t, \psi) - y_{t+\sigma}(t, \psi_1)\|_0] \\ \leqslant g(k^{-1}) \exp[T(k^{-1}, \eta)][\exp\{L(\gamma(\eta)) T(k^{-1}, \eta)\}](\|\phi - \phi_1\|_0 + \|\psi - \psi_1\|_0) \\ \leqslant M(\eta)[\|\phi - \phi_1\|_0 + \|\psi - \psi_1\|_0]. \tag{7.5.10}$$

7.5. EXTREME STABILITY

By (7.5.10) and the continuity of $x_t(t_0, \phi_0)$ in t, it follows that $V_k(t, \phi, \psi)$ is continuous in (t, ϕ, ψ).

Let us now consider $D^+V_k(t, \phi, \psi)$ with respect to the product system (7.5.1). Since

$$V_k(t+h, x_{t+h}(t,\phi), y_{t+h}(t,\psi)) = g(k^{-1}) \sup_{\sigma \geq 0} G_k(\| x_{t+h+\sigma}(t,\phi) - y_{t+h+\sigma}(t,\psi)\|_0)e^{\sigma}$$

$$= g(k^{-1}) \sup_{\sigma \geq h} G_k(\| x_{t+\sigma}(t,\phi) - y_{t+\sigma}(t,\psi)\|_0)e^{\sigma-h}$$

$$\leq V_k(t, x_t(t,\phi), y_t(t,\psi))e^{-h},$$

it follows that

$$D^+V_k(t,\phi,\psi)_{(7.5.1)} \leq \limsup_{h \to 0^+} V_k(t,\phi,\psi) \frac{e^{-h}-1}{h}$$

$$\leq -V_k(t,\phi,\psi). \tag{7.5.11}$$

The desired Lyapunov functional may now be defined by

$$V(t,\phi,\psi) = \sum_{k=1}^{\infty} (2^k)^{-1} V_k(t,\phi,\psi). \tag{7.5.12}$$

In view of (7.5.8), this $V(t, \phi, \psi)$ can be defined for all $t \in J, \phi, \psi \in \mathscr{C}^n$. From (7.5.7) and (7.5.9), it is easy to see that there exist functions $a, b \in \mathscr{K}$ satisfying (2°). Furthermore, because of the inequality

$$| V(t,\phi,\psi) - V(t,\phi_1,\psi_1)| \leq \sum_{k=1}^{\infty} (2^k)^{-1} | V_k(t,\phi,\psi) - V_k(t,\phi_1,\psi_1)|,$$

we obtain, for $\phi, \phi_1, \psi, \psi_1 \in C_n$,

$$| V(t,\phi,\psi) - V(t,\phi_1,\psi_1)| \leq M(\eta)[\|\phi - \phi_1\|_0 + \|\psi - \psi_1\|_0],$$

which proves that $V(t, \phi, \psi)$ satisfies the Lipschitz condition as described in (1°).

Finally, we shall show that (3°) is satisfied. By the definition,

$$D^+V(t,\phi,\psi)_{(7.5.1)} = \limsup_{h \to 0^+} \sum_{k=1}^{\infty} (2^k)^{-1} h^{-1}[V_k(t+h, x_{t+h}(t,\phi), y_{t+h}(t,\psi))$$

$$- V_k(t, x_t(t,\phi), y_t(t,\psi))].$$

For any integer $N > 0$,

$$D^+V(t,\phi,\psi)_{(7.5.1)} \leq \sum_{k=1}^{N} (2^k)^{-1} \limsup_{h \to 0^+} h^{-1}[V_k(t+h, x_{t+h}(t,\phi), y_{t+h}(t,\psi)) - V_k(t,\phi,\psi)]$$

$$+ \limsup_{h \to 0^+} \sum_{k=N+1}^{\infty} (2^k)^{-1}h^{-1}[V_k(t+h, x_{t+h}(t,\phi), y_{t+h}(t,\psi)) - V_k(t,\phi,\psi)].$$

Since

$$h^{-1}[V_k(t+h, x_{t+h}(t,\phi), y_{t+h}(t,\psi)) - V_k(t,\phi,\psi)] \leq V_k(t,\phi,\psi)\frac{e^{-h}-1}{h},$$

if h is small enough,

$$D^+V(t,\phi,\psi)_{(7.5.1)} \leq \sum_{k=1}^{N} (2^k)^{-1} \limsup_{h \to 0^+} h^{-1}[V_k(t+h, x_{t+h}(t,\phi), y_{t+h}(t,\psi)) - V_k(t,\phi,\psi)],$$

and therefore, by (7.5.11),

$$D^+V(t,\phi,\psi)_{(7.5.1)} \leq \sum_{k=1}^{N} (2^k)^{-1}(-V_k(t,\phi,\psi)).$$

As N is arbitrary, this implies

$$D^+V(t,\phi,\psi)_{(7.5.1)} \leq \sum_{k=1}^{\infty} (2^k)^{-1}(-V_k(t,\phi,\psi))$$

$$\leq -V(t,\phi,\psi).$$

The proof is complete.

Sufficient conditions for the extreme complete stability of the system (7.1.1) are given by the following result.

THEOREM 7.5.2. Assume that

(i) $f \in C[J \times \mathscr{C}^n, R^n]$, and, for every $\phi_0 \in \mathscr{C}^n$, the solutions $x(t_0, \phi_0)$ exist in the future;

(ii) $V \in C[J \times \mathscr{C}^n \times \mathscr{C}^n, R_+]$, and, for $t \geq t_0$,

$$D^+V(t, x_t(t_0, \phi_0), y_t(t_0, \psi_0)) \leq g(t, V(t, x_t(t_0, \phi_0), y_t(t_0, \psi_0)));$$

(iii) $g \in C[J \times R_+, R]$, and $g(t, 0) \equiv 0$;

(iv) there exist functions $a, b \in \mathcal{K}$ on the interval $[0, \infty)$ such that, for $t \geq 0$, $\phi, \psi \in \mathscr{C}^n$,

$$b(\|\phi - \psi\|_0) \leq V(t, \phi, \psi) \leq a(\|\phi - \psi\|_0).$$

Then, if the trivial solution of (6.1.2) is uniformly completely stable, the system (7.1.1) is extremely uniformly completely stable.

Proof. Assume that the trivial solution of (6.1.2) is uniformly completely stable. This means that (S_2^*) and (S_8^*) hold at the same time. If $\epsilon > 0$, $t_0 \in J$ are given, there exists a $\delta = \delta(\epsilon) > 0$ such that $u_0 \leq \delta$ implies

$$u(t, t_0, u_0) < b(\epsilon), \qquad t \geq t_0.$$

Choose $\delta_1 = a^{-1}(\delta)$ and $u_0 = V(t_0, \phi_0, \psi_0)$. Then, Theorem 1.4.1 yields, because of conditions (i) and (ii),

$$V(t, x_t(t_0, \phi_0), y_t(t_0, \psi_0)) \leq r(t, t_0, u_0), \qquad t \geq t_0,$$

where $r(t, t_0, u_0)$ is the maximal solution of (6.1.2). We therefore obtain

$$b(\|x_t(t_0, \phi_0) - y_t(t_0, \psi_0)\|_0) < b(\epsilon), \qquad t \geq t_0,$$

which implies that

$$\|x_t(t_0, \phi_0) - y_t(t_0, \psi_0)\|_0 < \epsilon, \qquad t \geq t_0,$$

provided $\|\phi_0 - \psi_0\|_0 \leq \delta_1$. This proves extreme uniform stability of the system (7.1.1).

Let now $\alpha > 0$, $\epsilon > 0$, and $t_0 \in J$ be given. Suppose $\|\phi_0 - \psi_0\|_0 \leq \alpha$. Let $\alpha_1 = a(\alpha)$. Since (S_8^*) holds, given $\alpha_1 > 0$, $b(\epsilon) > 0$, and $t_0 \in J$, there exists a positive number $T = T(t_0, \alpha, \epsilon)$ such that, if $u_0 \leq \alpha_1$,

$$u(t, t_0, u_0) < b(\epsilon), \qquad t \geq t_0 + T.$$

As previously, it results that

$$b(\|x_t(t_0, \phi_0) - y_t(t_0, \psi_0)\|_0) < b(\epsilon), \qquad t \geq t_0 + T,$$

and this shows that, whenever $\|\phi_0 - \psi_0\|_0 \leq \alpha$, we have

$$\|x_t(t_0, \phi_0) - y_t(t_0, \psi_0)\|_0 < \epsilon, \qquad t \geq t_0 + T.$$

It therefore follows that the system (7.1.1) is extremely uniformly completely stable.

COROLLARY 7.5.1. *The function $g(t, u) = -\alpha u$, $\alpha > 0$, is admissible in Theorem 7.5.2.*

7.6. Almost periodic systems

We shall continue to consider the functional differential system (7.1.1). For the purpose of this section, however, we take

$$f \in C[(-\infty, \infty) \times C_\rho, R^n].$$

All the results that follow are extensions of the results of Sect. 3.18 to functional differential systems.

DEFINITION 7.6.1. A functional $f \in C[(-\infty, \infty) \times C_\rho, R^n]$ is said to be *almost periodic in t uniformly with respect to* $\phi \in S$ for any compact set $S \subset C_\rho$ if, given any $\eta > 0$, it is possible to find an $l(\eta)$ such that, in any interval of length $l(\eta)$, there is a τ such that the inequality

$$\|f(t+\tau, \phi) - f(t, \phi)\| \leq \eta$$

is satisfied for $t \in (-\infty, \infty)$, $\phi \in S$.

We shall first prove a uniqueness result.

THEOREM 7.6.1. Assume that

  (i) $f \in C[(-\infty, \infty) \times C_\rho, R^n]$ and $f(t, \phi)$ is almost periodic in t uniformly with respect to $\phi \in S$, S being any compact set in C_ρ;

  (ii) $V \in C[J \times C_\rho \times C_\rho, R_+]$, $V(t, \phi, \phi) \equiv 0$, $V(t, \phi, \psi)$ satisfies a Lipschitz condition in ϕ, ψ for a constant $M = M(\rho) > 0$, and

$$b(\|\phi - \psi\|_0) \leq V(t, \phi, \psi), \qquad b \in \mathcal{K};$$

  (iii) $g \in C[J \times R_+, R]$, $g(t, 0) \equiv 0$, and, for $t \geq 0$,

$$D^+ V(t, \phi, \psi) \leq g(t, V(t, \phi, \psi)),$$

where $D^+ V(t, \phi, \psi)$ is defined with respect to the product system $x' = f(t, x_t)$, $y' = f(t, y_t)$;

  (iv) the maximal solution of (6.1.2), through the point $(\tau_0, 0)$, $\tau_0 \geq 0$, is identically zero.

Then, there exists a unique solution of the almost periodic system (7.1.1), to the right of $t_0 \in (-\infty, \infty)$.

Proof. Since $f(t, \phi)$ is continuous, there exists at least one solution for a given $t_0 \in (-\infty, \infty)$ and a $\phi_0 \in C_\rho$. Suppose that, for some $t_0 \in (-\infty, \infty)$ and $\phi_0 \in C_\rho$, there exist two solutions $x(t_0, \phi_0)$ and $y(t_0, \phi_0)$ of (7.1.1). Then, at some $t_1 > t_0$, we should have

$$\| x_{t_1}(t_0, \phi_0) - y_{t_1}(t_0, \phi_0)\|_0 = \epsilon, \tag{7.6.1}$$

7.6. ALMOST PERIODIC SYSTEMS

where we may assume $\epsilon < \rho$. For $t_0 \leq t \leq t_1$, there exists a positive constant $\rho_1 < \rho$ such that

$$\|x_t(t_0, \phi_0)\|_0 \leq \rho_1, \qquad \|y_t(t_0, \phi_0)\|_0 \leq \rho_1.$$

These solutions are uniformly continuous functions and bounded by ρ_1 on the interval $t_0 - \tau \leq t \leq t_1$, and, hence, there exists a compact set $S \subset C_\rho$ such that

$$x_t(t_0, \phi_0) \in S, \qquad y_t(t_0, \phi_0) \in S \quad \text{for} \quad t \in [t_0, t_1].$$

By Lemma 1.3.1, given $b(\epsilon)/2$ and a compact set $[\tau_0, T]$, there is a $\delta = \delta(\epsilon) > 0$ such that

$$r(t, \tau_0, 0, \delta) \leq b(\epsilon)/2, \qquad t \in [\tau_0, T], \tag{7.6.2}$$

where $r(t, \tau_0, 0, \delta)$ is the maximal solution of

$$u' = g(t, u) + \delta, \qquad u(\tau_0) = 0.$$

Let θ be a $\delta/2M$ translation number for $f(t, \phi)$ such that $t_0 + \theta \geq 0$, that is,

$$\|f(t + \theta, \phi) - f(t, \phi)\| < \delta/2M, \qquad t \in (-\infty, \infty), \tag{7.6.3}$$

provided ϕ belongs to a compact set $S \subset C_\rho$. We consider the function

$$m(t) = V(t + \theta, x_t(t_0, \phi_0), y_t(t_0, \phi_0)), \qquad t \in [t_0, t_1].$$

Then,

$$D^+ m(t) = \limsup_{h \to 0^+} h^{-1}[V(t + h + \theta, x_{t+h}(t_0, \phi_0), y_{t+h}(t_0, \phi_0))$$
$$- V(t + \theta, x_t(t_0, \phi_0), y_t(t_0, \phi_0))]$$
$$\leq \limsup_{h \to 0^+} h^{-1}[V(t + h + \theta, x^*_{t+h+\theta}(t + \theta, x_t), y^*_{t+h+\theta}(t + \theta, y_t))$$
$$- V(t + \theta, x_t, y_t)] + \limsup_{h \to 0^+} h^{-1} M(\rho)[\|x_{t+h} - x^*_{t+h+\theta}(t + \theta, x_t)\|_0$$
$$+ \|y_{t+\theta} - y^*_{t+h+\theta}(t + \theta, y_t)\|_0],$$

where $x^*(t + \theta, x_t)$, $y^*(t + \theta, y_t)$ are the solutions of (7.1.1) such that $x^*_{t+\theta}(t + \theta, x_t) = x_t = x_t(t_0, \phi_0)$ and $y^*_{t+\theta}(t + \theta, y_t) = y_t = y_t(t_0, \phi_0)$, respectively. We thus have

$$D^+ m(t) \leq g(t + \theta, m(t)) + M(\rho)[\|f(t, x_t) - f(t + \theta, x_t)\|$$
$$+ \|f(t, y_t) - f(t + \theta, y_t)\|]$$
$$\leq g(t + \theta, m(t)) + \delta, \qquad t \in [t_0, t_1],$$

because of (7.6.3). Defining $\tau_0 = t_0 + \theta$, we get, by Theorem 1.4.1, the inequality

$$m(t) \leqslant r(t + \theta, \tau_0, 0, \delta), \qquad t \in [t_0, t_1],$$

and, at $t = t_1$,

$$m(t_1) \leqslant r(t_1 + \theta, \tau_0, 0, \delta) \leqslant b(\epsilon)/2. \tag{7.6.4}$$

Assumption (ii) and (7.6.1), on the other hand, lead to

$$m(t_1) \geqslant b(\epsilon),$$

contradicting (7.6.4). It therefore follows that there is a unique solution for the system (7.1.1) to the right of t_0.

COROLLARY 7.6.1. *The function $g(t, u) \equiv 0$ is admissible in Theorem 7.6.1.*

DEFINITION 7.6.2. If, for any $\rho > 0$, there exists an $M(\rho) > 0$ such that $\|f(t, \phi)\| \leqslant M(\rho)$, whenever $\phi \in C_\rho$, we shall say that $f(t, \phi)$ is bounded.

The next theorem gives the sufficient conditions for perfect stability criteria of the trivial solution of (7.1.1).

THEOREM 7.6.2. *Suppose that*

(i) $V \in C[J \times C_\rho, R_+]$, $V(t, \phi)$ *is Lipschitzian in ϕ for a constant $L = L(\rho) > 0$, and, for $(t, \phi) \in J \times C_\rho$,*

$$b(\|\phi\|_0) \leqslant V(t, \phi), \qquad b \in \mathcal{K}; \tag{7.6.5}$$

(ii) $g \in C[J \times R_+, R]$, $g(t, 0) \equiv 0$, *and, for $(t, \phi) \in J \times C_\rho$,*

$$D^+V(t, \phi)_{(7.1.1)} \leqslant g(t, V(t, \phi));$$

(iii) $f \in C[(-\infty, \infty) \times C_\rho, R^n]$, $f(t, 0) \equiv 0$, $f(t, \phi)$ *is bounded, and $f(t, \phi)$ is almost periodic in t uniformly with respect to $\phi \in S$, S being any compact set in C_ρ.*

Then, the null solution of (7.1.1) is

(1°) *perfectly equistable if the trivial solution of (6.1.2) is strongly equistable;*

(2°) *perfectly uniform stable if the trivial solution of (6.1.2) is strongly uniform stable;*

(3°) perfectly equi-asymptotically stable if the trivial solution of (6.1.2) is strongly equi-asymptotically stable;

(4°) perfectly uniformly asymptotically stable if the trivial solution of (6.1.2) is strongly uniformly asymptotically stable.

Proof. We shall prove only the statement corresponding to (4°). Let us suppose that the trivial solution of (6.1.2) is strongly uniformly asymptotically stable. Let $0 < \epsilon < \rho$ and $t_0 \in (-\infty, \infty)$ be given. Then, given $b(\epsilon) > 0$, $\tau_0 \in J$, and any compact interval $K = [\tau_0, t^*]$, there exist an $\eta = \eta(\epsilon) > 0$ and a $\delta = \delta(\tau_0, \epsilon) > 0$ such that

$$u(t, \tau_0, u_0, \eta) < b(\epsilon), \qquad t \in [\tau_0, t^*], \qquad (7.6.6)$$

whenever $u_0 < \delta$, where $u(t, \tau_0, u_0, \eta)$ is any solution of (3.18.6). Choose $L\delta_1 = \delta$ and $u_0 = L\|\phi_0\|_0$, L being the Lipschitz constant for $V(t, \phi)$. Consider a solution $x(t_0, \phi_0)$ of (7.1.1) such that $t_0 \in (-\infty, \infty)$, $\|\phi_0\|_0 < \delta_1$. Suppose that, at some t, we have

$$\|x_t(t_0, \phi_0)\|_0 = \epsilon.$$

Then, there exist t_1 and t_2 such that $t_0 < t_1 < t_2$, $\|x_{t_1}(t_0, \phi_0)\|_0 = \delta_1$, $\|x_{t_2}(t_0, \phi_0)\|_0 = \epsilon$, and that

$$\delta_1 < \|x_t(t_0, \phi_0)\|_0 < \epsilon, \qquad t_1 < t < t_2.$$

Clearly, there exists a compact set $S \subset C_\rho$ such that $x_t(t_0, \phi_0) \in S$ for $t_0 \leq t \leq t_2$. Let θ be an η/L translation number of $f(t, \phi)$ for $\phi \in S$ such that $t_0 + \theta \geq 0$, that is,

$$\|f(t, \phi) - f(t + \theta, \phi)\| < \eta/L, \qquad t \in (-\infty, \infty), \qquad \phi \in S. \qquad (7.6.7)$$

Consider the function

$$m(t) = V(t + \theta, x_t(t_0, \phi_0)), \qquad t \geq t_0.$$

Then we have

$$D^+ m(t) = \limsup_{h \to 0^+} h^{-1}[V(t + \theta + h, x_{t+h}(t_0, \phi_0)) - V(t + \theta, x_t(t_0, \phi_0))]$$

$$\leq \limsup_{h \to 0^+} h^{-1}[V(t + \theta + h, x^*_{t+\theta+h}(t + \theta, x_t(t_0, \phi_0)))$$

$$\quad - V(t + \theta, x_t(t_0, \phi_0))]$$

$$\quad + \limsup_{h \to 0^+} h^{-1}[V(t + \theta + h, x_{t+h}(t_0, \phi_0))$$

$$\quad - V(t + \theta + h, x^*_{t+\theta+h}(t + \theta, x_t(t_0, \phi_0)))],$$

where $x^*(t+\theta, x_t(t_0, \phi_0))$ is a solution of (7.1.1) such that $x_{t+\theta}^*(t+\theta, x_t(t_0, \phi_0)) = x_t(t_0, \phi_0)$. It, therefore, follows that

$$D^+m(t) \leq g(t+\theta, m(t)) + \limsup_{h\to 0^+} h^{-1}L \parallel x_{t+h}(t_0, \phi_0)$$
$$- x_{t+\theta+h}^*(t+\theta, x_t(t_0, \phi_0))\parallel_0$$
$$\leq g(t+\theta, m(t)) + L \parallel f(t, x_t(t_0, \phi_0)) - f(t+\theta, x_t(t_0, \phi_0))\parallel.$$

Since $x_t(t_0, \phi_0) \in S$ for $t_0 \leq t \leq t_2$, we obtain, using (7.6.7),

$$D^+m(t) \leq g(t+\theta, m(t)) + \eta, \qquad t \in [t_0, t_2],$$

and hence, by Theorem 1.4.1,

$$m(t) \leq r(t+\theta, \tau_0, u_0, \eta), \qquad t \in [t_0, t_2], \tag{7.6.8}$$

letting $\tau_0 = t_0 + \theta$, where $r(t+\theta, \tau_0, u_0, \eta)$ is the maximal solution of (3.18.6). At $t = t_2$, we are led to an absurdity

$$b(\epsilon) \leq V(t_2+\theta, x_{t_2}(t_0, \phi_0)) \leq r(t_2+\theta, \tau_0, u_0, \eta) < b(\epsilon),$$

in view of relations (7.6.5), (7.6.6), and (7.6.8). Thus, the perfect uniform stability of the trivial solution of (7.1.1) is proved.

By assumption, given $b(\epsilon) > 0$, $\tau_0 \geq 0$, there exist positive numbers δ_0, $\eta = \eta(\epsilon)$ and $T = T(\epsilon)$ such that

$$u(t, \tau_0, u_0, \eta) < b(\epsilon), \qquad t \geq \tau_0 + T, \tag{7.6.9}$$

whenever $u_0 \leq \delta_0$. Choose $u_0 = L \parallel \phi_0 \parallel_0$ and $L\hat{\delta}_0 = \delta_0$, and let $\delta_0^* = \min[\hat{\delta}_0, \tilde{\delta}_0]$, where $\tilde{\delta}_0 = \delta(\rho)$. Suppose now that $x(t_0, \phi_0)$ is any solution of (7.1.1) such that $\parallel \phi_0 \parallel_0 \leq \delta_0^*$, $t_0 \in (-\infty, \infty)$. Since $\parallel x_t(t_0, \phi_0)\parallel_0 < \rho$ for all $t \geq t_0$ and $f(t, \phi)$ is bounded and consequently $x'(t_0, \phi_0)(t)$ is bounded by some constant, there exists a compact set S such that $x_t(t_0, \phi_0) \in S$ for all $t \geq t_0$. As before, let θ be an η/L-translation number of $f(t, \phi)$ so that (7.6.7) is satisfied. Considering the function

$$m(t) = V(t+\theta, x_t(t_0, \phi_0)) \qquad \text{for} \quad t \geq t_0,$$

it is easy to obtain, as previously,

$$m(t) \leq r(t+\theta, \tau_0, u_0, \eta), \qquad t \geq t_0.$$

This, together with (7.6.5) and (7.6.9), yields

$$b(\parallel x_t(t_0, \phi_0)\parallel_0) \leq m(t) \leq r(t+\theta, \tau_0, u_0, \eta) < b(\epsilon),$$

7.6. ALMOST PERIODIC SYSTEMS

for $t + \theta \geqslant \tau_0 + T$, and, consequently,

$$\| x_t(t_0, \phi_0)\|_0 < \epsilon, \qquad t \geqslant t_0 + T,$$

whenever $\|\phi_0\|_0 \leqslant \delta_0{}^*$.
This completely establishes (4°).

COROLLARY 7.6.2. The function $g(t, u) = -\alpha u$, $\alpha > 0$, is admissible in Theorem 7.6.2 to yield perfect uniform asymptotic stability of the trivial solution of (7.1.1).

As remarked in Sect. 3.18, if the functional $f(t, \phi)$ is not almost periodic and $f \in C[J \times C_\rho, R^n]$, then, from the strong stability properties of the trivial solution of (6.1.2), we may deduce strong stability properties of the trivial solution of (7.1.1), on the basis of Theorem 7.6.2.

Finally, the following theorem assures the existence of an almost periodic solution.

THEOREM 7.6.3. Suppose that

(i) $V \in C[J \times C_\rho \times C_\rho, R_+]$, $V(t, \phi, \psi)$ is Lipschitzian in ϕ and ψ for a constant $L = L(\rho) > 0$, and, for $t \in J$, $\phi, \psi \in C_\rho$,

$$b(\|\phi - \psi\|_0) \leqslant V(t, \phi, \psi) \leqslant a(\|\phi - \psi\|_0), \qquad a, b \in \mathcal{K}; \qquad (7.6.10)$$

(ii) $g \in C[J \times R_+, R]$, and, for $t \in J$, $\phi, \psi \in C_\rho$,

$$D^+ V(t, \phi, \psi) \leqslant g(t, V(t, \phi, \psi));$$

(iii) $f \in C[(-\infty, \infty) \times C_\rho, R^n]$, $f(t, \phi)$ is bounded, almost periodic in t uniformly with respect to $\phi \in S$, S being any compact subset in C_ρ, and $f(t, \phi)$ is smooth enough to ensure the existence and uniqueness of solutions of (7.1.1);

(iv) for any $b(\epsilon) > 0$, $\alpha > 0$, and $\xi_0 \in J$, there exist positive numbers $\eta = \eta(\epsilon)$, $T = T(\epsilon, \alpha)$ such that, if $u_0 \leqslant \alpha$ and $\xi \geqslant \xi_0 + T$,

$$u(\xi, \xi_0, u_0, \eta) < b(\epsilon), \qquad (7.6.11)$$

where $u(\xi, \xi_0, u_0, \eta)$ is any solution of

$$u' = g(\xi, u) + \eta, \qquad u(\xi_0) = u_0, \qquad \xi_0 \geqslant 0; \qquad (7.6.12)$$

(v) there exists a solution $x(t_0, \phi_0)$ of (7.1.1) such that

$$\| x_t(t_0, \phi_0)\|_0 \leqslant B < \rho, \qquad t \geqslant t_0, \qquad t_0 \in (-\infty, \infty).$$

Then, (7.1.1) admits a bounded almost periodic solution, with a bound B.

Proof. The proof runs, naturally, parallel to the proof of Theorem 3.18.5. Hence, we shall only indicate necessary changes. Let $x(t_0, \phi_0)$ be the solution of (7.1.1) such that $\| x_t(t_0, \phi_0)\|_0 \leq B$. Since $f(t, \phi)$ is assumed to be bounded, we have, consequently, that $\| x'(t_0, \phi_0)(t)\|$ is bounded for a constant B_1, for $t \geq t_0$. Let S be the compact subset of C_ρ consisting of functions that are bounded by B and are Lipschitzian for a constant B_1. Let $\{\tau_k\}$ be any sequence such that $\tau_k \to \infty$ as $k \to \infty$ and

$$f(t + \tau_k, \phi) - f(t, \phi) \to 0 \quad \text{as} \quad k \to \infty,$$

uniformly for $t \in (-\infty, \infty)$, $\phi \in S$. Let β be any number, and let U be any compact subset of $[\beta, \infty)$. Let $0 < \epsilon < \rho$, and choose $\alpha = a(2B)$. Then, let η and T be the numbers defined in assumption (iv), for this choice. Let $k_0 = k_0(\beta)$ be the smallest value of k such that

$$\beta + \tau_{k_0} \geq t_0 + \tau + T.$$

Choose an integer $n_0 = n_0(\epsilon, \beta) \geq k_0$ so large that, for $k_2 \geq k_1 \geq n_0$,

$$\|f(t + \tau_{k_1}, \phi) - f(t + \tau_{k_2}, \phi)\| \leq \eta/3L, \qquad (7.6.13)$$

for all $t \in (-\infty, \infty)$, $\phi \in S$. Let θ be an $\eta/3L$-translation number for $f(t, \phi)$ such that $t_0 + \theta \geq 0$, that is,

$$\|f(t + \theta, \phi) - f(t, \phi)\| \leq \eta/3L, \qquad (7.6.14)$$

for $t \in (-\infty, \infty)$, $\phi \in S$.

Consider the function, for $t \geq t_0$,

$$m(t) = V(t + \theta, x_t, x_{t_1}),$$

where $t_1 = t + \tau_{k_2} - \tau_{k_1}$ and $x_t = x_t(t_0, \phi_0)$. Then,

$$D^+ m(t) \leq \limsup_{h \to 0^+} h^{-1}[V(t + \theta + h, x_{t+h}, x_{t_1+h})$$

$$- V(t + \theta + h, x^*_{t+\theta+h}(t + \theta, x_t), y^*_{t+\theta+h}(t + \theta, x_{t_1})]$$

$$+ \limsup_{h \to 0^+} h^{-1}[V(t + \theta + h, x^*_{t+\theta+h}(t + \theta, x_t), y^*_{t+\theta+h}(t + \theta, x_{t_1}))$$

$$- V(t + \theta, x_t, x_{t_1})],$$

where $x^*(t + \theta, x_t)$, $y^*(t + \theta, x_{t_1})$ are the solutions of (7.1.1) such that

$x^*_{t+\theta}(t+\theta, x_t) = x_t$, $y^*_{t+\theta}(t+\theta, x_{t_1}) = x_{t_1}$, respectively. Thus, in view of the Lipschitzian character of $V(t, \phi, \psi)$ and assumption (ii), we get

$$D^+m(t) \leqslant g(t+\theta, m(t)) + L \limsup_{h \to 0^+} h^{-1}[\| x_{t+h} - x^*_{t+\theta+h}(t+\theta, x_t)\|_0$$
$$+ \| x_{t_1+h} - y^*_{t+\theta+h}(t+\theta, x_{t_1})\|_0]$$
$$\leqslant g(t+\theta, m(t)) + L[\| x'(t_0, \phi_0)(t) - x^{*\prime}(t+\theta, x_t)(t+\theta)\|$$
$$+ \| x'(t_0, \phi_0)(t_1) - y^{*\prime}(t+\theta, x_{t_1})(t+\theta)\|]$$
$$\leqslant g(t+\theta, m(t) + L[\| f(t, x_t) - f(t+\theta, x_t)\|$$
$$+ \| f(t + \tau_{k_2} - \tau_{k_1}, x_{t_1}) - f(t, x_{t_1})\| + \| f(t, x_{t_1})$$
$$- f(t+\theta, x_{t_1})\|].$$

Since $t + \tau_{k_1} \geqslant t_0 + \tau$, for $t \in U$, we obtain, using the relations (7.6.13) and (7.6.14),

$$D^+m(t + \tau_{k_1}) \leqslant g(t + \tau_{k_1} + \theta, m(t + \tau_{k_1})) + \eta,$$

which implies, by Theorem 1.4.1, if $u_0 = m(t_0)$,

$$m(t + \tau_{k_1}) \leqslant r(t + \tau_{k_1} + \theta, t_0 + \theta, u_0, \eta),$$

where $r(\xi, \xi_0, u_0, \eta)$ is the maximal solution of (7.6.12). By assumption (iv), it follows that

$$r(\xi, \xi_0, u_0, \eta) < b(\epsilon) \quad \text{if} \quad \xi \geqslant \xi_0 + T.$$

But, for all $t \in U$, $t + \tau_{k_1} \geqslant t_0 + \tau + T$. Hence, identifying $\xi = t + \tau_{k_1} + \theta$, $\xi_0 = t_0 + \theta$, we get

$$m(t + \tau_{k_1}) < b(\epsilon), \quad t \in U.$$

Consequently, for all $t \in U$, $k_2 \geqslant k_1 \geqslant n_0$, we have, in view of (7.6.10),

$$\| x_{t+\tau_{k_1}} - x_{t+\tau_{k_2}}\|_0 < \epsilon,$$

which, in turn, leads to the inequality

$$\| x(t_0, \phi_0)(t + \tau_{k_1}) - x(t_0, \phi_0)(t + \tau_{k_2})\| < \epsilon, \quad t \in U.$$

This proves the existence of a function $w(t)$ defined on $[\beta, \infty)$ and bounded by B. Since β is arbitrary, $w(t)$ is defined for $t \in (-\infty, \infty)$, and we have

$$x(t_0, \phi_0)(t + \tau_{k_1}) - w(t) \to 0 \quad \text{as} \quad k_1 \to \infty,$$

uniformly on all compact subsets of $(-\infty, \infty)$.

Following closely the rest of the proof of Theorem 3.18.5, we can show that $w(t)$ satisfies (7.1.1) and is almost periodic.

This completes the proof.

COROLLARY 7.6.3. If, in addition to the hypothesis of Theorem 7.6.3, the trivial solution of (6.1.2) is strongly uniformly asymptotically stable, then the system (7.1.1) admits an almost periodic solution that is perfectly uniformly asymptotically stable. In particular, $g(t, u) = -\alpha u$, $\alpha > 0$, is admissible.

7.7. Notes

The results of Sect. 7.1 are adapted from the work of Driver [3]. See also Halanay [22] and Krasovskii [5]. Theorem 7.1.4 is new. Theorems 7.2.1 and 7.2.2 are taken from Hale [1]. See also Yoshizawa [3]. Theorems 7.2.3 and 7.2.4 are new. Theorems 7.2.5 and 7.2.6 are based on Halanay [22].

The results on autonomous systems in Sect. 7.3 are taken from the work of Hale [8], which may also be referred to for a number of illustrative examples. For the results on perturbed systems of Sect. 7.4, see Corduneanu [2], Halanay [22], and Hale [1].

Theorem 7.5.1 is due to Yoshizawa [1], whereas Theorem 7.5.2 is new. Section 7.6 contains the work of Lakshmikantham and Leela [3]. See also Hale [6] and Yoshizawa [2, 3].

For closely related results, see Driver [3], Halanay [22], Hale [5], J. Kato [1], Krasovskii [5], Lakshmikantham and Leela [2], Liberman [1], Miller [1], Razumikhin [2, 6], Reklishkii [1–5], Seifert [1], Sugiyama [8], and Yoshizawa [3].

Chapter 8

8.0. Introduction

In what follows, we wish to treat the solutions of the functional differential system (7.1.1) as elements of euclidean space for all future time except at the initial moment. Our main tool, in this chapter, is therefore a Lyapunov function instead of a functional. The derivative of a Lyapunov function with respect to the functional differential system will be a functional, which may be estimated either by means of a function or a functional. While estimating the derivative of the Lyapunov function in terms of a function, a basic question is to select a minimal class of functions for which this can be done. Thus, by using the theory of ordinary differential inequalities and choosing the minimal sets of functions suitably, several results are obtained. If, on the other hand, the estimation of the derivative of the Lyapunov function by means of a functional is considered, the selection of a minimal set of functions is unnecessary. Nevertheless, this technique crucially depends on the notion of maximal solution for functional differential equations and the theory of functional differential inequalities. This method also offers a unified approach, analogous to the use of general comparison principle in ordinary differential equations. Moreover, it is important to note that the knowledge of solutions is not demanded in either case.

8.1. Basic comparison theorems

Let $V \in C[[-\tau, \infty) \times S_\rho, R_+]$, and let $\phi \in C_\rho$. We define $D^+V(t, \phi(0), \phi)$, $D_-V(t, \phi(0), \phi)$ with respect to the functional differential system (7.1.1) as follows:

$$D^+V(t, \phi(0), \phi) = \limsup_{h \to 0^+} h^{-1}[V(t+h, \phi(0) + hf(t, \phi)) - V(t, \phi(0))],$$

$$D_-V(t, \phi(0), \phi) = \liminf_{h \to 0^-} h^{-1}[V(t+h, \phi(0) + hf(t, \phi)) - V(t, \phi(0))]. \qquad (8.1.1)$$

We need, subsequently, the following subsets of $\mathscr{C}^n$, defined by

$$\Omega_1 = [\phi \in C_\rho : |V_t|_0 = V(t, \phi(0)), t \in J], \tag{8.1.2}$$

$$\Omega_0 = [\phi \in C_\rho : V(t+s, \phi(s)) < L(V(t, \phi(0))), t \in J], \tag{8.1.3}$$

and

$$\Omega_A = [\phi \in C_\rho : |V_t A_t|_0 = V(t, \phi(0))A(t)), t \in J], \tag{8.1.4}$$

where $A(t) > 0$ is continuous on $[-\tau, \infty)$,

(i) $|V_t|_0 = \sup_{\tau - \leqslant s \leqslant 0} V(t+s, \phi(s))$;

(ii) $L(u)$ is continuous on R_+, nondecreasing in u, and $L(u) > u$, for $u > 0$; and

(iii)
$$|V_t A_t|_0 = \sup_{-\tau \leqslant s \leqslant 0} V(t+s, \phi(s))A(t+s). \tag{8.1.5}$$

We now state a few fundamental comparison results.

THEOREM 8.1.1. Let $V \in C[[-\tau, \infty) \times S_\rho, R_+]$ and $V(t, x)$ be locally Lipschitzian in x. Assume that the functional $D_- V(t, \phi(0), \phi)$, defined by (8.1.1), verifies the inequality

$$D_- V(t, \phi(0), \phi) \leqslant g(t, V(t, \phi(0))), \qquad t > t_0, \qquad \phi \in \Omega_1, \tag{8.1.6}$$

where $g \in C[J \times R_+, R_+]$, and $r(t, t_0, u_0)$ is the maximal solution of the scalar differential equation (6.1.2), existing to the right of $t_0 \geqslant 0$. Let $x(t_0, \phi_0)$ be any solution of (7.1.1) defined in the future, satisfying

$$\sup_{-\tau \leqslant s \leqslant 0} V(t_0, \phi_0(s)) \leqslant u_0. \tag{8.1.7}$$

Then,

$$V(t, x(t_0, \phi_0)(t)) \leqslant r(t, t_0, u_0), \qquad t \geqslant t_0. \tag{8.1.8}$$

Proof. Let $x(t_0, \phi_0)$ be any solution of (7.1.1) with an initial function $\phi_0 \in C_\rho$ at $t = t_0$. Define the function

$$m(t) = V(t, x(t_0, \phi_0)(t)).$$

For $\epsilon > 0$ sufficiently small, consider the differential equation

$$u' = g(t, u) + \epsilon, \qquad u(t_0) = u_0 \geqslant 0, \tag{8.1.9}$$

whose solutions $u(t, \epsilon) = u(t, t_0, u_0, \epsilon)$ exist as far as $r(t, t_0, u_0)$ exists,

8.1. BASIC COMPARISON THEOREMS

to the right of t_0. Since
$$\lim_{\epsilon \to 0} u(t, \epsilon) = r(t, t_0, u_0),$$
the truth of the desired inequality (8.1.8) is immediate, if we can establish that
$$m(t) < u(t, \epsilon), \quad t \geqslant t_0.$$
Supposing that this is not true and proceeding as in the proof of Theorem 6.3.3, we can see that there exists a $t_1 > t_0$ such that

(i) $m(t) \leqslant u(t, \epsilon), t_0 \leqslant t \leqslant t_1$;
(ii) $m(t_1) = u(t_1, \epsilon), t = t_1$.

From (i) and (ii), we get the inequality
$$D_- m(t_1) \geqslant u'(t_1, \epsilon) = g(t_1, u(t_1 \epsilon)) + \epsilon. \tag{8.1.10}$$
Since $g(t, u) + \epsilon$ is positive, the solutions $u(t, \epsilon)$ are monotonic increasing in t, and therefore, by (i) and (ii),
$$\mid m_{t_1} \mid_0 = m(t_1) = u(t_1, \epsilon).$$
Setting $\phi = x_{t_1}(t_0, \phi_0)$ and noting that $\phi(0) = x(t_0, \phi_0)(t_1)$, it follows that
$$\mid V_{t_1} \mid_0 = V(t_1, \phi(0)).$$
This means that $\phi \in \Omega_1$, and, consequently, using the Lipschitzian character of $V(t, x)$ in x and the relation (8.1.6), we obtain, after simple computation, the inequality
$$D_- m(t_1) \leqslant g(t_1, m(t_1)).$$
This is incompatible with (8.1.10), on account of (ii). It therefore follows that (8.1.8) is true, and the proof is complete.

The following corollary is a useful tool in itself in certain situations.

COROLLARY 8.1.1. Let $V \in C[[-\tau, \infty) \times S_\rho, R_+]$ and $V(t, x)$ be locally Lipschitzian in x. Assume that, for $t > t_0, \phi \in \Omega_0$,
$$D_- V(t, \phi(0), \phi) \leqslant 0.$$
Let $x(t_0, \phi_0)$ be any solution of (7.1.1) such that $x(t_0, \phi_0)(t) \in S_\rho$ for $t \in [t_0, t_1] \subset J$. Then,
$$V(t, x(t_0, \phi_0)(t) \leqslant \sup_{-\tau \leqslant s \leqslant 0} V(t_0, \phi_0(s)), \quad t \in [t_0, t_1].$$

Proof. Proceeding as in Theorem 8.1.1 with $g \equiv 0$, we arrive at the inequality

$$V(t, x(t_0, \phi_0)(t)) \leq V(t_2, x(t_0, \phi_0)(t_2)),$$

where $t_2 \in (t_0, t_1)$. Since $V(t_2, x(t_0, \phi_0)(t_2)) > 0$, the assumptions on $L(u)$ imply that

$$V(t_2 + s, x(t_0, \phi_0)(t_2 + s)) < L(V(t_2, x(t_0, \phi_0)(t_2))),$$

which shows that $x_t(t_0, \phi_0) \in \Omega_0$, $t_0 \leq t \leq t_2$. The rest of the proof is similar to the proof of Theorem 8.1.1.

The next comparison theorem gives a better estimate.

THEOREM 8.1.2. *Let the assumptions of Theorem 8.1.1 hold except that the inequality (8.1.6) is replaced by*

$$D^+V(t, \phi(0), \phi) + C(\|\phi(0)\|) \leq g(t, V(t, \phi(0))), \qquad (8.1.11)$$

for $t \geq t_0$, $\phi \in C_\rho$, where the function $C \in \mathcal{K}$. Assume further that $g(t, u)$ is monotone nondecreasing in u for each t. Then (8.1.7) implies

$$V(t, x(t_0, \phi_0)(t)) + \int_{t_0}^{t} C(\|x(t_0, \phi_0)(s)\|) \, ds \leq r(t, t_0, u_0), \qquad t \geq t_0. \qquad (8.1.12)$$

Proof. We let, for $t \geq t_0$,

$$m(t) = V(t, x(t_0, \phi_0)(t)) + \int_{t_0}^{t} C(\|x(t_0, \phi_0)(s)\|) \, ds.$$

Set $\phi = x_t(t_0, \phi_0)$ so that $\phi(0) = x(t_0, \phi_0)(t)$.

We then obtain, using the condition (8.1.11), the inequality

$$D^+m(t_1) \leq g(t_1, m(t_1)).$$

Here, we have used the monotonicity of $g(t, u)$ in u and the fact that

$$V(t, x(t_0, \phi_0)(t)) \leq m(t),$$

while applying the assumption (8.1.11). With these changes, it is easy to prove the stated result, following the arguments in the proof of Theorem 8.1.1.

The following variant of Theorem 8.1.1 is more useful in certain situations.

8.1. BASIC COMPARISON THEOREMS

THEOREM 8.1.3. Assume that the hypotheses of Theorem 8.1.1 hold except that the inequality (8.1.6) is replaced by

$$A(t) D_- V(t, \phi(0), \phi) + V(t, \phi(0)) D_- A(t) \leqslant g(t, V(t, \phi(0)) A(t)), \quad (8.1.13)$$

for $t > t_0$, $\phi \in \Omega_A$, where $A(t) > 0$ is continuous on $[-\tau, \infty)$.
Then,

$$\sup_{-\tau \leqslant s \leqslant 0} V(t_0, \phi_0(s)) A(t_0) \leqslant u_0$$

implies the estimate

$$A(t) V(t, x(t_0, \phi_0)(t)) \leqslant r(t, t_0, u_0), \quad t \geqslant t_0. \quad (8.1.14)$$

Proof. We set

$$L(t, x) = V(t, x) A(t).$$

Let $\phi \in \Omega_A$. For small $h > 0$, we have

$$L(t + h, \phi(0) + hf(t, \phi)) - L(t, \phi(0))$$
$$= V(t + h, \phi(0) + hf(t, \phi))[A(t + h) - A(t)]$$
$$+ A(t)[V(t + h, \phi(0) + hf(t, \phi)) - V(t, \phi(0))],$$

and therefore, in view of the assumption (8.1.13), it follows that

$$D_- L(t, \phi(0), \phi) \leqslant g(t, L(t, \phi(0))),$$

for $t > t_0$, $\phi \in \Omega_1$, where Ω_1, in this case, is to be defined with $L(t, x)$ replacing $V(t, x)$ in (8.1.2). It is clear that $L(t, x)$ is locally Lipschitzian in x, and, thus, all the assumptions of Theorem 8.1.1 are satisfied, with $L(t, x)$ in place of $V(t, x)$. The conclusion is now immediate from Theorem 8.1.1.

On the basis of the comparison theorem for functional differential inequalities developed in Sect. 6.10, we are now in a position to prove the following result, which plays an equally vital role in studying the behavior of solutions of functional differential systems.

THEOREM 8.1.4. $V \in C[-\tau, \infty) \times S_\rho, R_+]$, and $V(t, x)$ is locally Lipschitzian in x. Assume that, for $t \in J$, $\phi \in C_\rho$,

$$D^+ V(t, \phi(0), \phi) \leqslant g(t, V(t, \phi(0)), V_t), \quad (8.1.15)$$

where $V_t = V(t + s, \phi(s))$, $-\tau \leqslant s \leqslant 0$, $g \in C[J \times R_+ \times \mathscr{C}_+, R]$,

$g(t, u, \sigma)$ is nondecreasing in σ for each (t, u), and $r(t_0, \sigma_0)$ is the maximal solution of the functional differential equation

$$u' = g(t, u, u_t) \tag{8.1.16}$$

with an initial function $\sigma_0 \in \mathscr{C}_+$, at $t = t_0$, existing for $t \geqslant t_0$. If $x(t_0, \phi_0)$ is any solution of (7.1.1) defined in the future such that

$$V_{t_0} = V(t_0 + s, \phi_0(s)) \leqslant \sigma_0, \tag{8.1.17}$$

then we have

$$V(t, x(t_0, \phi_0)(t)) \leqslant r(t_0, \sigma_0)(t), \qquad t \geqslant t_0. \tag{8.1.18}$$

Proof. Let $x(t_0, \phi_0)$ be any solution of (7.1.1) such that

$$\| x_t(t_0, \phi_0)\|_0 < \rho.$$

Set $\phi = x_t(t_0, \phi_0)$, which implies that $\phi(0) = x(t_0, \phi_0)(t)$. Define

$$m(t) = V(t, x(t_0, \phi_0)(t)),$$

so that

$$m_t = V(t + s, \phi(s)).$$

Since (8.1.17) holds, we have $m_{t_0} \leqslant \sigma_0$. Moreover, for small $h > 0$,

$$m(t + h) - m(t) \leqslant K \| x(t_0, \phi_0)(t + h) - x(t_0, \phi_0)(t) + hf(t, x_t(t_0, \phi_0))\|$$
$$+ V(t + h, \phi(0) + hf(t, \phi)) - V(t, \phi(0)),$$

because of the fact that $V(t, x)$ satisfies a Lipschitz condition in x. This, together with (8.1.15), yields the inequality

$$D^+m(t) \leqslant g(t, m(t), m_t). \tag{8.1.19}$$

Define

$$v(t) = m(t) - \int_{t_0}^{t} g(\xi, m(\xi), m_\xi)\, d\xi, \qquad t \geqslant t_0.$$

Then, it follows that $D^+v(t) \leqslant 0$, in view of (8.1.19). By Lemma 1.2.1, $v(t)$ is nondecreasing, and therefore $D_-v(t) \leqslant 0$, which implies that

$$D_-m(t) \leqslant g(t, m(t), m_t), \qquad t > t_0.$$

The desired result (8.1.18) now follows from Theorem 6.10.4.

8.2. Stability criteria

We shall, in what follows, give sufficient conditions for various stability notions in terms of Lyapunov functions. This will be accomplished in two different ways. In one approach, the theory of ordinary differential inequalities will be used, as before, whereas in the other, the theory of functional differential inequalities play a major role. Since, for the purposes of this chapter, it is convenient to interpret the solutions of (7.1.1) as elements of euclidean space, the definitions of stability and boundedness have to be modified accordingly. For example, the definition 7.1.1 would appear in the following form.

DEFINITION 8.2.1. The trivial solution of (7.1.1) is said to be (S_1) *equistable* if, for each $\epsilon > 0$, $t_0 \in J$, there exists a positive function $\delta = \delta(t_0, \epsilon)$ that is continuous in t_0 for each ϵ, such that, whenever

$$\|\phi_0\|_0 \leq \delta,$$

we have

$$\|x(t_0, \phi_0)(t)\| < \epsilon, \quad t \geq t_0.$$

With this understanding, we can prove the following results.

THEOREM 8.2.1. Let there exist functions $V(t, x)$ and $g(t, u)$ enjoying the following properties:

(i) $V \in C[-\tau, \infty) \times S_\rho, R_+]$, $V(t, x)$ is positive definite, locally Lipschitzian in x, and

$$V(t, x) \leq a(t, \|x\|), \quad (t, x) \in J \times S_\rho, \tag{8.2.1}$$

where $a \in C[J \times [0, \rho), R_+]$, and $a \in \mathscr{K}$ for each $t \in J$;

(ii) $g \in C[J \times R_+, R_+]$, $g(t, 0) \equiv 0$, and, for $t > t_0$, $\phi \in \Omega_1$,

$$D_- V(t, \phi(0), \phi) \leq g(t, V(t, \phi(0))).$$

Then the trivial solution of (7.1.1) is

(1°) equistable if the trivial solution of (6.1.2) is equistable;

(2°) uniform stable if the trivial solution of (6.1.2) is uniform stable and, in addition, $V(t, x)$ is decrescent.

Proof. Let $x(t_0, \phi_0)$ be any solution of (7.1.1). Choose

$$u_0 = a(t_0, \|\phi_0\|_0)$$

so that $V(t_0, \phi_0) \leq u_0$, by (8.2.1). An application of Theorem 8.1.1 yields the estimate

$$V(t, x(t_0, \phi_0)(t)) \leq r(t, t_0, u_0), \qquad t \geq t_0, \qquad (8.2.2)$$

where $r(t, t_0, u_0)$ is the maximal solution of (6.1.2). Also, because of the positive definiteness of $V(t, x)$, we have

$$b(\| x \|) \leq V(t, x), \qquad (t, x) \in J \times S_\rho, \qquad b \in \mathcal{K}. \qquad (8.2.3)$$

Let $0 < \epsilon < \rho$ and $t_0 \in J$ be given. Assume that the null solution of (6.1.2) is equistable. Then, given $b(\epsilon) > 0$, $t_0 \in J$, there exists a $\delta = \delta(t_0, \epsilon) > 0$ satisfying

$$u(t, t_0, u_0) < b(\epsilon), \qquad t \geq t_0, \qquad (8.2.4)$$

provided $u_0 \leq \delta$. Moreover, there exists a $\delta_1 = \delta_1(t_0, \epsilon)$ such that

$$\| \phi_0 \|_0 \leq \delta_1 \quad \text{and} \quad a(t_0, \| \phi_0 \|_0) \leq \delta \qquad (8.2.5)$$

hold together, because of the assumption on $a(t, u)$. It now follows, from $\| \phi_0 \| \leq \delta_1$, that

$$b(\| x(t_0, \phi_0)(t) \|) \leq V(t, x(t_0, \phi_0)(t))$$
$$\leq r(t, t_0, u_0)$$
$$< b(\epsilon), \qquad t \geq t_0,$$

and, consequently,

$$\| x(t_0, \phi_0)(t) \| < \epsilon, \qquad t \geq t_0,$$

whenever $\| \phi_0 \|_0 \leq \delta_1$. Thus, (1°) is proved.

If $V(t, x)$ is decrescent, there exists a function $a \in \mathcal{K}$ satisfying

$$V(t, x) \leq a(\| x \|), \qquad (t, x) \in J \times S_\rho.$$

Hence, if we assume that the trivial solution of (6.1.2) is uniform stable, it is easy to see, from the foregoing proof, that δ_1 does not depend on t_0, proving (2°). The proof is complete.

COROLLARY 8.2.1. *The function $g(t, u) \equiv 0$ is admissible in Theorem 8.2.1.*

THEOREM 8.2.2. *Assume that there exist functions $V(t, x)$, $g(t, u)$, and $A(t)$ satisfying the following conditions:*

(i) $A(t) > 0$ *is continuous on* $[-\tau, \infty)$, *and* $A(t) \to \infty$ *as* $t \to \infty$;

(ii) $V \in C[[-\tau, \infty) \times S_\rho, R_+]$, $V(t, x)$ is positive definite, locally Lipschitzian in x, and verifies (8.2.1);

(iii) $g \in C[J \times R_+, R_+], g(t, 0) \equiv 0$, and, for $t > t_0, \phi \in \Omega_A$,

$$A(t) D_- V(t, \phi(0), \phi) + V(t, \phi(0)) D_- A(t)$$
$$\leqslant g(t, V(t, \phi(0)) A(t)).$$

Then, the trivial solution of (7.1.1) is equi-asymptotically stable if the trivial solution of (6.1.2) is equistable.

Proof. If $x(t_0, \phi_0)$ is any solution of (7.1.1) such that

$$A(t_0) a(t_0, \|\phi_0\|_0) = u_0,$$

we have, by Theorem 8.1.3,

$$A(t) V(t, x(t_0, \phi_0)(t)) \leqslant r(t, t_0, u_0), \quad t \geqslant t_0. \tag{8.2.6}$$

Let $0 < \epsilon < \rho$ and $t_0 \in J$ be given. Let $\alpha = \min_{-\tau \leqslant t < \infty} A(t)$. By assumption on $A(t)$, it is clear that $\alpha > 0$. Set $\eta = \alpha b(\epsilon)$. Then, proceeding as in the proof of Theorem 8.2.1 with this η instead of $b(\epsilon)$, it is easy to prove that the trivial solution of (7.1.1) is equistable.

To prove equi-asymptotic stability, let $\eta^* = \alpha b(\rho)$. Let $\delta_1(t_0, \rho)$ be such that $\|\phi_0\|_0 \leqslant \delta_1$ implies $\|x(t_0, \phi_0)(t)\| \leqslant \rho, t \geqslant t_0$. This is possible by equistability. Designate $\delta_0(t_0) = \delta_1(t_0, \rho)$, and suppose that $\|\phi_0\|_0 \leqslant \delta_0$. Since $A(t) \to \infty$ as $t \to \infty$, there exists a positive number $T = T(t_0, \epsilon)$ such that

$$A(t) b(\epsilon) > \alpha b(\rho), \quad t \geqslant t_0 + T. \tag{8.2.7}$$

We then have, using (8.2.3), (8.2.6), and the fact that $u(t, t_0, u_0) < \eta^*$ if $u_0 \leqslant \delta(t_0, \rho)$,

$$A(t) b(\|x(t_0, \phi_0)(t)\|) \leqslant A(t) V(t, x(t_0, \phi_0)(t))$$
$$\leqslant r(t, t_0, u_0)$$
$$< \eta^* = \alpha b(\rho), \quad t \geqslant t_0.$$

If $t \geqslant t_0 + T$, it follows, from the foregoing inequality and (8.2.7), that

$$\|x(t_0, \phi_0)(t)\| < \epsilon, \quad t \geqslant t_0 + T,$$

provided $\|\phi_0\|_0 \leqslant \delta_0$. This concludes the proof of the theorem.

COROLLARY 8.2.2. *The functions $g(t, u) \equiv 0$ and $A(t) \equiv e^{\alpha t}, \alpha > 0$, are admissible in Theorem 8.2.2.*

THEOREM 8.2.3. Assume that there exists a function $V(t, x)$ satisfying the following conditions:

(i) $V \in C[[-\tau, \infty) \times S_\rho, R_+]$, $V(t, x)$ is positive definite, decrescent, and locally Lipschitzian in x;

(ii) for $t > t_0$, $\phi \in \Omega_0$,

$$D_- V(t, \phi(0), \phi) \leqslant -C(\|\phi(0)\|), \qquad C \in \mathscr{K}.$$

Then, the trivial solution of (7.1.1) is uniformly asymptotically stable.

Proof. Since V is positive definite and decrescent, there exist functions $a, b \in \mathscr{K}$ satisfying

$$b(\|x\|) \leqslant V(t, x) \leqslant a(\|x\|), \qquad (t, x) \in J \times S_\rho. \tag{8.2.8}$$

Let $0 < \epsilon < \rho$, $t_0 \in J$ be given. Choose $\delta = \delta(\epsilon) > 0$ such that

$$a(\delta) < b(\epsilon). \tag{8.2.9}$$

We claim that, if $\|\phi_0\|_0 \leqslant \delta$, then $\|x(t_0, \phi_0)(t)\| < \epsilon$, $t \geqslant t_0$. Suppose that this is not true. Then, there exists a solution $x(t_0, \phi_0)$ of (7.1.1) with $\|\phi_0\|_0 \leqslant \delta$ such that

$$\|x(t_0, \phi_0)(t_2)\| = \epsilon$$

and

$$\|x(t_0, \phi_0)(t)\| \leqslant \epsilon, \qquad t \in [t_0, t_2],$$

so that

$$V(t_2, x(t_0, \phi_0)(t_2)) \geqslant b(\epsilon), \tag{8.2.10}$$

because of (8.2.8). Furthermore, this means that $x(t_0, \phi_0)(t) \in S_\rho$, $t \in [t_0, t_2]$. Hence, the choice $u_0 = a(\|\phi_0\|_0)$ and the condition

$$D_- V(t, \phi(0), \phi) \leqslant 0, \qquad t \in (t_0, t_2], \quad \phi \in \Omega_0,$$

give the estimate

$$V(t, x(t_0, \phi_0)(t)) \leqslant a(\|\phi_0\|_0), \qquad t \in [t_0, t_2], \tag{8.2.11}$$

because of Corollary 8.1.1. Now the relations (8.2.10), (8.2.11), and (8.2.9) lead to the contradiction

$$b(\epsilon) \leqslant V(t_2, x(t_0, \phi_0)(t_2)) \leqslant a(\|\phi_0\|_0) \leqslant a(\delta) < b(\epsilon).$$

This proves that the trivial solution of (7.1.1) is uniformly stable.

8.2. STABILITY CRITERIA

To prove uniform asymptotic stability, we have yet to show that the null solution of (7.1.1) is quasi-uniform asymptotically stable. For this purpose, let $x(t_0, \phi_0)$ be any solution of (7.1.1) such that $\|\phi_0\|_0 \leq \delta_0$, where $\delta_0 = \delta(\rho)$. It then follows from uniform stability that

$$x(t_0, \phi_0)(t) \in S_\rho, \quad t \geq t_0.$$

Let now $0 < \epsilon < \delta_0$ be given. Clearly, we have

$$b(\epsilon) \leq a(\delta_0).$$

In view of the assumptions on $L(u)$, which occurs in the definition of Ω_0, it is possible to find a $\beta = \beta(\epsilon) > 0$ such that

$$L(u) > u + \beta \quad \text{if} \quad b(\epsilon) \leq u \leq a(\delta_0). \tag{8.2.12}$$

Moreover, there exists a positive integer $N = N(\epsilon)$ satisfying the inequality

$$b(\epsilon) + N\beta > a(\delta_0). \tag{8.2.13}$$

If, for some $t \geq t_0$, we have

$$V(t, x(t_0, \phi_0)(t)) \geq b(\epsilon),$$

it follows that there exists a $\delta_1 = \delta_1(\epsilon) > 0$ such that $\|x(t_0, \phi_0)(t)\| \geq \delta_1$, because of (8.2.8). This, in turn, implies that

$$C(\|x(t_0, \phi_0)(t)\|) \geq C(\delta_1) \equiv \delta_2. \tag{8.2.14}$$

Obviously, δ_2 depends on ϵ.

With the positive integer N chosen previously, let us construct $N + 1$ numbers $t_k = t_k(t_0, \epsilon)$, $k = 0, 1, 2, ..., N$, such that

$$t_0(t_0, \epsilon) = t_0, \qquad t_{k+1}(t_0, \epsilon) = t_k(t_0, \epsilon) + (\beta/\delta_2) + \tau.$$

It then turns out that

$$t_k(t_0, \epsilon) = t_0 + k[(\beta/\delta_2) + \tau],$$

and, consequently, letting $T(\epsilon) = N[(\beta/\delta_2) + \tau]$, we have

$$t_N(t_0, \epsilon) = t_0 + T(\epsilon).$$

Now, to prove quasi-uniform asymptotic stability, we have to show that

$$\|x(t_0, \phi_0)(t)\| < \epsilon, \qquad t \geq t_0 + T(\epsilon),$$

whenever $\|\phi_0\|_0 \leqslant \delta_0$. It is therefore sufficient to show that

$$V(t, x(t_0, \phi_0)(t)) < b(\epsilon) + (N - k)\beta, \qquad t \geqslant t_k, \qquad (8.2.15)$$

for $k = 0, 1, 2, ..., N$. For $k = 0$, (8.2.15) follows from the first part of the proof and (8.2.13). We wish to prove the desired inequality (8.2.15) by induction. Suppose that, for some $k < N$, we have

$$V(\xi, x(t_0, \phi_0)(\xi)) < b(\epsilon) + (N - k)\beta, \qquad \xi \geqslant t_k,$$

and, for some $t_* \geqslant t_{k+1} - (\beta/\delta_2)$,

$$V(t_*, x(t_0, \phi_0)(t_*)) \geqslant b(\epsilon) + (N - k - 1)\beta.$$

It then follows that

$$b(\epsilon) \leqslant V(t_*, x(t_0, \phi_0)(t_*)) \leqslant a(\delta_0),$$

and, consequently, we derive from (8.2.12) the inequality

$$L(V(t_*, x(t_0, \phi_0)(t_*))) > b(\epsilon) + (N - k)\beta$$
$$> V(\xi, x(t_0, \phi_0)(\xi)),$$

for $t_k \leqslant \xi \leqslant t_*$. This implies that

$$L(V(t_*, \phi(0))) > V(t_* + s, \phi(s)), \qquad -\tau \leqslant s \leqslant 0,$$

where $\phi = x_{t_*}(t_0, \phi_0)$, so that $\phi(0) = x(t_0, \phi_0)(t_*)$. Hence, $\phi \in \Omega_0$. It therefore follows from condition (ii) and the relation (8.2.14) that

$$D_-V(t, \phi(0), \phi) \leqslant -\delta_2 < 0.$$

This shows that, if ever

$$V(t, x(t_0, \phi_0)(t)) < b(\epsilon) + (N - k - 1)\beta,$$

for some $t \geqslant t_{k+1} - (\beta/\delta_2)$, then the same inequality will hold henceforth.

Now, supposing that

$$V(t, x(t_0, \phi_0)(t)) \geqslant b(\epsilon) + (N - k - 1)\beta$$

for $t_{k+1} - (\beta/\delta_2) \leqslant t \leqslant t^*$, we deduce, by arguing as before,

$$V(t^*, x(t_0, \phi_0)(t^*)) \leqslant b(\epsilon) + (N - k)\beta - \delta_2(t^* - t_{k+1} + (\beta/\delta_2))$$
$$\leqslant V(t^*, x(t_0, \phi_0)(t^*)) - \delta_2(t^* - t_{k+1}),$$

which proves that $t^* \leq t_{k+1}$. In other words, we have verified the truth of (8.2.15) for all k.

The proof of the theorem is therefore complete.

The next theorem may be useful in some situations.

THEOREM 8.2.4. Assume that there exist functions $V(t, x)$ and $g(t, u)$ satisfying the following conditions:

(i) $V \in C[[-\tau, \infty) \times S_\rho, R_+]$, $V(t, x)$ is positive definite, decrescent, and locally Lipschitzian in x;

(ii) $g \in C[J \times R_+, R_+]$, $g(t, 0) \equiv 0$, $g(t, u)$ is monotone nondecreasing in u for each t, and, for $t \geq t_0$, $\phi \in C_\rho$,

$$D^+V(t, \phi(0), \phi) + C(\|\phi(0)\|) \leq g(t, V(t, \phi(0))),$$

where $C \in \mathcal{K}$.

Then, the trivial solution of (7.1.1) is uniformly asymptotically stable if the trivial solution of (6.1.2) is uniform stable.

Proof. By Theorem 8.1.2, we have

$$V(t, x(t_0, \phi_0)(t)) \leq -\int_{t_0}^{t} C(\|x(t_0, \phi_0(s))\|)\, ds + r(t, t_0, u_0), \quad t \geq t_0.$$

As this implies,

$$V(t, x(t_0, \phi_0)(t)) \leq r(t, t_0, u_0), \quad t \geq t_0,$$

on the basis of Theorem 8.2.1, it is clear that the trivial solution is uniform-stable. Let $\delta_0 = \delta_1(\rho)$, where $\delta_1(\rho)$ is the number obtained for uniform stability. Similarly, let $\delta_1(\epsilon)$ be the number corresponding to ϵ, and suppose that $\|x(t_0, \phi_0)(t)\| \geq \delta_1(\epsilon)$ for $t \in [t_0, t_0 + T]$, where T is chosen to satisfy the inequality

$$T \geq \frac{b(\rho)}{C(\delta_1(\epsilon))}. \quad (8.2.16)$$

Since $r(t, t_0, u_0) \leq b(\rho)$, $t \geq t_0$, whenever $u_0 = a(\|\phi_0\|_0) \leq \delta(\rho)$, it follows that

$$b(\|x(t_0, \phi_0)(t_0 + T)\|) \leq -C(\delta_1(\epsilon))\, T + b(\rho).$$

This implies, because of (8.2.8), that $0 < b(\delta_1(\epsilon)) \leq 0$. This contradiction proves that there exists a $t_1 \in [t_0, t_0 + T]$ such that

$$\|x(t_0, \phi_0)(t_1)\| < \delta_1(\epsilon).$$

It therefore follows, from the decrescent nature of $V(t, x)$, that, in any case,

$$\| x(t_0, \phi_0)(t) \| < \epsilon, \quad t \geq t_0 + T,$$

whenever $\| \phi_0 \|_0 \leq \delta_0$. The proof is therefore complete.

COROLLARY 8.2.3. *The function $g(t, u) \equiv 0$ is admissible in Theorem 8.2.4.*

We shall now consider Eq. (7.1.8) and illustrate the practicality of using Lyapunov functions instead of functionals.

Take $L(t, x) = A(t) V(t, x) = e^{\alpha t} x^2$, $\alpha > 0$. Then, the set Ω_A is defined by

$$\Omega_A = \left[\phi \in C_\rho : \sup_{-\tau \leq s \leq 0} \phi^2(s) \, e^{\alpha(t+s)} = \phi^2(0) \, e^{\alpha t} \right],$$

which implies $\phi^2(s) \leq \phi^2(0) \, e^{-\alpha s}$, $-\tau \leq s \leq 0$. Hence, using the relation (7.1.9), we see

$$D\_L(t, \phi(0), \phi) = e^{\alpha t} \Big[\alpha \phi^2(0) - 2(a+b) \, \phi^2(0)$$

$$- 2ab\phi(0) \int_{-\tau}^{0} \phi(s) \, ds - 2b^2 \phi(0) \int_{-\tau}^{0} \phi(s-\tau) \, ds \Big].$$

When $\phi \in \Omega_A$, this reduces to, assuming $b \geq 0$,

$$D\_L(t, \phi(0), \phi) \leq \Big[\alpha - 2(a+b) + 2ab \int_{-\tau}^{0} e^{-\alpha s/2} \, ds$$

$$+ 2b^2 \, e^{\alpha \tau} \int_{-\tau}^{0} e^{-\alpha s/2} \, ds \Big] \phi^2(0) \, e^{\alpha t}$$

$$= \Big[\alpha - 2(a+b) + (4ab + 4b^2 \, e^{\alpha \tau}) \frac{(e^{\alpha \tau/2} - 1)}{\alpha} \Big] L(t, \phi(0)).$$

We wish to apply Theorem 8.2.2 with $g(t, u) \equiv 0$.
This means that a, b, α, and τ must satisfy the condition

$$\alpha - 2(a+b) + (4ab + 4b^2 \, e^{\alpha \tau}) \left[\frac{e^{\alpha \tau/2} - 1}{\alpha} \right] \leq 0. \tag{8.2.17}$$

Clearly, by Theorem 8.2.2, the trivial solution of (7.1.8) is uniformly asymptotically (exponentially) stable, provided (8.2.17) is verified.

Observing that $e^x - 1 \geq x$, $x > 0$, and $e^x \geq 1$, it follows, from (8.2.17), that

$$\alpha + 2b(a+b) \, \tau \leq 2(a+b).$$

8.2. STABILITY CRITERIA

Now, choosing $\alpha = 2\gamma(a+b)$, $0 < \gamma < 1$, we readily deduce the condition

$$0 \leqslant b\tau \leqslant 1 - \gamma < 1,$$

which is the known condition for asymptotic stability.

Since α is arbitrary, letting $\alpha \to 0$ and noting that

$$\lim_{\alpha \to 0} \frac{e^{\alpha\tau/2} - 1}{\alpha} = \frac{\tau}{2},$$

the condition (8.2.17) yields

$$0 \leqslant b\tau \leqslant 1,$$

which is the condition obtained before for uniform stability.

If we wish to obtain the stability region in a different way, we have to compute $D_-L(t, \phi(0), \phi)$ using (7.1.8) directly, instead of (7.1.9) as before. We then get the inequality

$$D_-L(t, \phi(0), \phi) \leqslant [\alpha - 2(a - |b| e^{\alpha\tau})] L(t, \phi(0)).$$

To apply Theorem 8.2.2 with $g \equiv 0$, we must have

$$0 < \alpha/2 \leqslant a - |b| e^{\alpha\tau},$$

which implies that the region for uniform asymptotic stability is $|b| < ae^{-\alpha\tau}$. If $\alpha \to 0$, this condition reduces to $|b| \leqslant a$, which is the region for uniform stability for Eq. (7.1.8).

Notice that the latter condition $|b| \leqslant a$ may also be obtained by taking $V(t, x) = x^2$, computing $D_-V(t, \phi(0), \phi)$ for $\phi \in \Omega_1$, and using Theorem 8.1.1 with $g \equiv 0$.

Finally, take $V(t, x) = x^2$, $L(u) = u/q$, $0 < q < 1$, and

$$C(u) = 2a(1 - q^{1/2}) u^2.$$

Then, one can verify the conditions of Theorem 8.2.3 by a calculation similar to the foregoing. The trivial solution of (7.1.8) is uniformly asymptotically stable if $|b| < aq^{1/2}$.

The following theorem shows that the study of the stability properties of the functional differential system can be reduced to that of a single functional differential equation. Naturally, this technique depends heavily on Theorem 8.1.4, as anticipated. Analogous to the stability definitions with respect to the scalar differential equation (6.1.2), if we say that the trivial solution of (8.1.16) is equistable, we mean the following.

DEFINITION 8.2.2. The trivial solution of (8.1.16) is said to be *equistable* if, for each $\epsilon > 0$, $t_0 \in J$, there exists a $\delta = \delta(t_0, \epsilon)$, which is continuous in t_0 for each ϵ, such that

$$u(t_0, \sigma_0)(t) < \epsilon, \qquad t \geq t_0,$$

provided $|\sigma_0|_0 \leq \delta$, where $u(t_0, \sigma_0)$ is any solution of (8.1.16) with an initial function $\sigma_0 \in \mathscr{C}_+$ at $t = t_0$.

Other definitions may be understood similarly.

THEOREM 8.2.5. Assume that there exist a function $V(t, x)$ and a functional $g(t, u, \sigma)$ satisfying the following conditions:

(i) $V \in C[[-\tau, \infty) \times S_\rho, R_+]$, $V(t, x)$ is locally Lipschitzian in x, and

$$b(\| x \|) \leq V(t, x) \leq a(\| x \|), \qquad (t, x) \in J \times S_\rho,$$

where $a, b \in \mathscr{K}$;

(ii) $g \in C[J \times R_+ \times \mathscr{C}_+, R]$, $g(t, 0, 0) \equiv 0$, $g(t, u, \sigma)$ is nondecreasing in σ for each (t, u), and, for $t \in J$, $\phi \in C_\rho$,

$$D^+V(t, \phi(0), \phi) \leq g(t, V(t, \phi(0)), V_t).$$

Then, the trivial solution of (7.1.1) satisfies one of the stability notions, if the trivial solution of (8.1.16) obeys the corresponding one of the stability definitions.

Proof. Suppose that the trivial solution of (8.1.16) is equistable. Let $0 < \epsilon < \rho$ and $t_0 \in J$. Then, given $b(\epsilon) > 0$, $t_0 \in J$, there exists a $\delta = \delta(t_0, \epsilon) > 0$ such that $|\sigma_0|_0 \leq \delta$ implies

$$u(t_0, \sigma_0)(t) < b(\epsilon), \qquad t \geq t_0.$$

Let $x(t_0, \phi_0)$ be any solution of (7.1.1) existing for $t \geq t_0$ such that $V(t_0 + s, \phi_0(s)) \leq \sigma_0(s)$. We then have, by Theorem 8.1.4, the inequality

$$V(t, x(t_0, \phi_0)(t)) \leq r(t_0, \sigma_0)(t), \qquad t \geq t_0,$$

where $r(t_0, \sigma_0)$ is the maximal solution of the functional differential equation (8.1.16). Choose $\delta_1 > 0$ such that $\delta_1 = a^{-1}(\delta)$, and let $\| \phi_0 \|_0 \leq \delta_1$. It turns out that

$$b(\| x(t_0, \phi_0)(t) \|) \leq V(t, x(t_0, \phi_0)(t))$$
$$\leq r(t_0, \sigma_0)(t) < b(\epsilon), \qquad t \geq t_0,$$

which implies that
$$\| x(t_0, \phi_0)(t) \| < \epsilon, \quad t \geq t_0,$$
provided that $\| \phi_0 \|_0 \leq \delta_1$, showing that the trivial solution of (7.1.1) is equistable.

The proof corresponding to other notions may be given by modifying the arguments suitably. The theorem is therefore proved.

COROLLARY 8.2.4. *The functional $g(t, u, \sigma) \equiv 0$ is admissible in Theorem 8.2.4 to yield uniform stability of the trivial solution of (7.1.1).*

8.3. Perturbed systems

Let us consider the perturbed system (7.4.1) corresponding to the unperturbed system (7.1.1). We then have

THEOREM 8.3.1. *Suppose that*

(i) $V \in C[[-\tau, \infty) \times S_\rho, R_+]$, $V(t, x)$ *is positive definite and satisfies a Lipschitz condition in x for a constant $L = L(\rho) > 0$;*

(ii) *for $t > t_0$, $\phi \in \Omega_1$,*
$$D_- V(t, \phi(0), \phi)_{(7.1.1)} \leq 0;$$

(iii) $w \in C[J \times R_+, R_+]$, $w(t, 0) \equiv 0$, *and, for $t > t_0$, $\phi \in \Omega_1$,*
$$\| R(t, \phi) \| \leq w(t, V(t, \phi(0))).$$

Then, the trivial solution of (7.4.1) is equistable (uniform stable) if the trivial solution of (6.1.2) with $g(t, u) = Lw(t, u)$ is equistable (uniform stable).

Proof. Let $t > t_0$ and $\phi \in \Omega_1$. Then, in view of assumptions (i), (ii), and (iii), we have

$$D_- V(t, \phi(0), \phi)_{(7.4.1)} \leq \liminf_{h \to 0^-} h^{-1}[V(t + h, \phi(0) + h\{f(t, \phi) + R(t, \phi)\}$$
$$- V(t + h, \phi(0) + hf(t, \phi))]$$
$$+ \liminf_{h \to 0^-} h^{-1}[V(t + h, \phi(0) + hf(t, \phi)) - V(t, \phi(0))]$$
$$\leq L \| R(t, \phi) \| + D_- V(t, \phi(0), \phi)_{(7.1.1)}$$
$$\leq Lw(t, V(t, \phi(0))) \equiv g(t, V(t, \phi(0))).$$

The desired result is now a consequence of Theorem 8.2.1.

THEOREM 8.3.2. Suppose that

(i) $V \in C[[-\tau, \infty) \times S_\rho, R_+]$, $V(t, x)$ is positive definite and satisfies a Lipschitz condition in x for a constant $L = L(\rho) > 0$;

(ii) $A(t) > 0$ is continuous on $[-\tau, \infty)$, $A(t) \to \infty$ as $t \to \infty$, and, for $t \geq t_0$, $\phi \in \Omega_A$,

$$A(t) D_- V(t, \phi(0), \phi) + V(t, \phi(0)) D_- A(t) \leq 0;$$

(iii) $w \in C[J \times R_+, R_+]$, $w(t, 0) \equiv 0$, and, for $t \geq t_0$, $\phi \in \Omega_A$,

$$A(t) \| R(t, \phi)\| \leq w(t, V(t, \phi(0)) A(t)).$$

Then, the trivial solution of (7.4.1) is equi-asymptotically stable if the null solution of (6.1.2) with $g(t, u) = Lw(t, u)$ is equistable.

Proof. If $t \geq t_0$, $\phi \in \Omega_A$, it follows that

$$D_- V(t, \phi(0), \phi)_{(7.4.1)} \leq L \| R(t, \phi)\| + D_- V(t, \phi(0), \phi)_{(7.1.1)}$$
$$\leq [A(t)]^{-1} [w(t, V(t, \phi(0)) A(t)) - V(t, \phi(0)) D_- A(t)],$$

which implies the inequality

$$A(t) D_- V(t, \phi(0), \phi)_{(7.4.1)} + V(t, \phi(0)) D_- A(t) \leq Lw(t, V(t, \phi(0)) A(t)).$$

Consequently, the conclusion follows by Theorem 8.2.2.

THEOREM 8.3.3. Suppose that

(i) $V \in C[[-\tau, \infty) \times S_\rho, R_+]$, $V(t, x)$ is positive definite and satisfies a Lipschitz condition in x for a constant $L = L(\rho) > 0$;

(ii) for $t \geq t_0$, $\phi \in C_\rho$, and $C \in \mathcal{K}$,

$$D^+ V(t, \phi(0), \phi)_{(7.1.1)} \leq -C(\|\phi(0)\|);$$

(iii) $w \in C[J \times R_+, R_+]$, $w(t, 0) \equiv 0$, $w(t, u)$ is nondecreasing in u for each $t \in J$, and, for $t \geq t_0$, $\phi \in C_\rho$,

$$\| R(t, \phi)\| \leq w(t, V(t, \phi(0))).$$

Then, the uniform stability of the trivial solution of (6.1.2) with $g(t, u) = Lw(t, u)$ assures the uniform asymptotic stability of the trivial solution of (7.4.1).

Proof. Let $t \geq t_0$ and $\phi \in C_\rho$. Then, as previously,

$$D^+ V(t, \phi(0), \phi)_{(7.4.1)} \leq -C(\|\phi(0)\|) + Lw(t, V(t, \phi(0))),$$

and, therefore, the uniform asymptotic stability of the trivial solution of (7.4.1) follows by Theorem 8.2.4.

8.3. PERTURBED SYSTEMS

THEOREM 8.3.4. Assume that

(i) $V \in C[-\tau, \infty) \times S_\rho, R_+]$, $V(t, x)$ satisfies a Lipschitz condition in x for a constant $L = L(\rho) > 0$, and

$$b(\|x\|) \leqslant V(t, x), \qquad (t, x) \in J \times S_\rho, \qquad b \in \mathcal{K};$$

(ii) $g_1 \in C[J \times R_+ \times \mathscr{C}_+, R]$, $g_1(t, 0, 0) \equiv 0$, $g_1(t, u, \sigma)$ is nondecreasing in σ for each (t, u), and, for $(t, \phi) \in J \times C_\rho$,

$$D^+V(t, \phi(0), \phi)_{(7.1.1)} \leqslant g_1(t, V(t, \phi(0)), V_t);$$

(iii) $g_2 \in C[J \times \mathscr{C}_+, R_+]$, $g_2(t, 0) \equiv 0$, $g_2(t, \sigma)$ is nondecreasing in σ for each $t \in J$, and

$$\|R(t, \phi)\| \leqslant g_2(t, \|\phi\|).$$

Then the stability properties of the trivial solution of (8.1.16) with

$$g(t, u, \sigma) = g_1(t, u, \sigma) + Lg_2(t, b^{-1}(\sigma))$$

imply the corresponding stability properties of the trivial solution of (7.4.1).

Proof. Let $t \in J$ and $\phi \in C_\rho$. Then,

$$\begin{aligned}
D^+V(t, \phi(0), \phi)_{(7.4.1)} &\leqslant D^+V(t, \phi(0), \phi)_{(7.1.1)} + L \|R(t, \phi)\| \\
&\leqslant g_1(t, V(t, \phi(0)), V_t) + Lg_2(t, \|\phi\|) \\
&\leqslant g_1(t, V(t, \phi(0)), V_t) + Lg_2(t, b^{-1}(V_t)) \\
&\equiv g(t, V(t, \phi(0)), V_t),
\end{aligned}$$

because of assumptions (i), (ii), and (iii). Now, Theorem 8.2.5 can be applied to yield the stated results.

COROLLARY 8.3.1. The functions

$$g_1(t, u, \sigma) = -\alpha u + \beta \tau \sup_{-\tau \leqslant s \leqslant 0} \sigma(s),$$

$\alpha > 0$, $\beta > 0$, and $g_2(t, \sigma) = \gamma \sup_{-\tau \leqslant s \leqslant 0} \sigma(s)$, γ being sufficiently small, are admissible in Theorem 8.3.4, provided $b(u) = u$ and $0 < \tau < (\alpha - \gamma)/\beta$, to guarantee that the trivial solution of (7.4.1) is exponentially asymptotically stable.

Proof. Under the assumptions, it is easy to see, as in Theorem 6.10.7, that $u' = -\alpha u + (\beta \tau + \gamma) \sup_{-\tau \leqslant s \leqslant 0} u_t(s)$ admits a solution $r(t_0, \sigma_0)$ that tends to zero exponentially as $t \to \infty$, and, therefore, the conclusion follows from Theorem 8.3.4.

8.4. An estimate of time lag

We wish to estimate the time lag τ in order that the solutions of an ordinary differential system

$$x' = f(t, x) \tag{8.4.1}$$

and a functional differential system

$$y' = F(t, y_t) \tag{8.4.2}$$

may have the same behavior, namely, exponential decay. Since Eq. (8.4.2) may also be written as

$$x' = f(t, x) + R(t, x, x_t), \tag{8.4.3}$$

where

$$R(t, x, x_t) = F(t, x_t) - f(t, x),$$

it is sufficient to consider the perturbed system (8.4.3).

THEOREM 8.4.1.   Suppose that

   (i) $V \in C[S_\rho, R_+]$, $V(x)$ is positive definite and satisfies a Lipschitz condition in x for a constant $L = L(\rho) > 0$;
   (ii) $f \in C[J \times S_\rho, R^n]$, and, for $(t, x) \in J \times S_\rho$,

$$D^+V(x)_{(8.4.1)} \leqslant -\alpha V(x), \qquad \alpha > 0;$$

   (iii) $R \in C[J \times S_\rho \times C_\rho, R^n]$, and for $\phi \in C_\rho$,

$$\| R(t, \phi(0), \phi) \| \leqslant N\tau \sup_{-\tau \leqslant s \leqslant 0} V(\phi(s)).$$

Then, if $0 < \tau < \alpha/LN$, every solution $x(t_0, \phi_0)$ of (8.4.3) with $\phi_0 \in C_\rho$ at $t = t_0$, which is defined in the future, tends to zero exponentially as $t \to \infty$.

Proof.   Let $t \in J$ and $\phi \in C_\rho$. Then, using assumptions (i), (ii), and (iii), we have

$$D^+V(t, \phi(0), \phi) = \limsup_{h \to 0^+} h^{-1}[V(\phi(0) + h\{f(t, \phi(0)) + R(t, \phi(0), \phi)\})$$

$$- V(\phi(0))]$$

$$\leqslant L \| R(t, \phi(0), \phi) \| + D^+V(\phi(0))_{(8.4.1)}$$

$$\leqslant LN\tau \sup_{-\tau \leqslant s \leqslant 0} V(\phi(s)) - \alpha V\phi(0)$$

$$\equiv g(t, V(\phi(0)), V(\phi(s))).$$

It is clear that $g(t, u, \sigma)$ satisfies the conditions of Theorem 6.10.4, and, consequently,

$$V(x(t_0, \phi_0)(t)) \leq r(t_0, \sigma_0)(t), \qquad t \geq t_0,$$

where $x(t_0, \phi_0)$ is any solution of (8.4.3) with the initial function $\phi_0 \in C_\rho$ at $t = t_0$, defined in the future, and $r(t_0, \sigma_0)$ is the maximal solution of (8.1.16) with $\sigma_0 \in \mathscr{C}_+$ at $t = t_0$ such that $V(\phi_0(s)) \leq \sigma_0$. All that remains to be shown is that $r(t_0, \sigma_0)$ exists in the future and tends to zero exponentially as $t \to \infty$, if $0 < \tau < \alpha/LN$. In fact, as we have seen in the proof of Theorem 6.10.7, the equation

$$u' = g(t, u, u_t) = -\alpha u + LN\tau \sup_{-\tau \leq s \leq 0} u_t(s)$$

admits solutions $r(t_0, \sigma_0)$ that tend to zero as $t \to \infty$ exponentially, if $0 < \tau < \alpha/LN$. The theorem is therefore established.

COROLLARY 8.4.1. Assume that

(i) $f \in C[J \times S_\rho, R^n]$, and, for all sufficiently small $h > 0$,

$$\| x + hf(t, x) \| \leq \| x \| (1 - \alpha h), \qquad \alpha > 0, \qquad (t, x) \in J \times S_\rho;$$

(ii) $R \in C[J \times S_\rho \times C_\rho, R^n]$, and, for $(t, \phi) \in J \times C_\rho$,

$$\| R(t, \phi(0), \phi) \| \leq N\tau \sup_{-\tau \leq s \leq 0} \| \phi(s) \|.$$

Then, if $0 < \tau < \alpha/N$, the conclusion of Theorem 8.4.1 remains valid.

The proof follows by Theorem 8.4.1 with the choice of $V(x) = \| x \|$.

8.5. Eventual stability

We shall extend the results on eventual stability discussed in Sect. 3.14 to the functional differential system (7.1.1). Following the definition 3.14.1, we can formulate eventual stability concepts in the present case. For example, the set $\phi = 0$ is said to be [with respect to the system (7.1.1)] (E_1) *eventually uniformly stable* if, for every $\epsilon > 0$, there exists a $\delta = \delta(\epsilon) > 0$ and $\tau_0 = \tau_0(\epsilon) > 0$ such that

$$\| x(t_0, \phi_0)(t) \| < \epsilon, \qquad t \geq t_0 \geq \tau_0,$$

provided $\| \phi_0 \|_0 \leq \delta$.

Analogously, we shall denote by (E_1^*) to (E_4^*) the corresponding notions of the set $\sigma = 0$ with respect to the functional differential equation (8.1.16).

THEOREM 8.5.1. Let the following conditions hold:

(i) $V \in C[[-\tau, \infty) \times S_\rho, R_+]$, $V(t, x)$ is locally Lipschitzian in x, and

$$b(\|x\|) \leqslant V(t, x) \leqslant a(\|x\|)$$

for $0 < \alpha < \|x\| < \rho$ and $t \geqslant \theta(\alpha)$, where $a, b \in \mathcal{K}$ and $\theta(u)$ is continuous and monotonic decreasing in u for $0 < u < \rho$;

(ii) $g \in C[J \times R_+ \times \mathscr{C}_+, R]$, and the set $\sigma = 0$ is eventually uniformly stable with respect to (8.1.16);

(iii) $f \in C[J \times C_\rho, R^n]$, and

$$D^+V(t, \phi(0), \phi) \leqslant g(t, V(t, \phi(0)), V_t)$$

for every $\phi \in C_\rho$ such that $0 < \alpha < \|\phi(0)\| < \rho$ and $t \geqslant \theta(\alpha)$.

Then, the set $\phi = 0$ is eventually uniformly stable relative to the system (7.1.1).

Proof. Let $0 < \epsilon < \rho$. By assumption, given $b(\epsilon) > 0$, there exists a $\delta_1 = \delta_1(\epsilon) > 0$ and a $\tau_1 = \tau_1(\epsilon) > 0$ such that

$$u(t_0, \sigma_0)(t) < b(\epsilon), \qquad t \geqslant t_0 \geqslant \tau_1, \tag{8.5.1}$$

whenever $|\sigma_0|_0 \leqslant \delta_1$. Let us set $\delta = a^{-1}(\delta_1)$ and $\tau_2(\epsilon) = \theta[\delta(\epsilon)]$. Let $\tau_0 = \tau_0(\epsilon) = \max[\tau_1(\epsilon), \tau_2(\epsilon)]$. We claim that (E_1) is satisfied with this choice of $\delta(\epsilon)$ and $\tau_0(\epsilon)$. Suppose that this is not true. Then, there would exist a solution $x(t_0, \phi_0)$ of (7.1.1) and two numbers t_1, t_2 such that $t_2 > t_1 > t_0 \geqslant \tau_0(\epsilon)$,

$$\|x(t_0, \phi_0)(t_1)\| = \delta, \qquad \|x(t_0, \phi_0)(t_2)\| = \epsilon,$$

and

$$\delta < \|x(t_0, \phi_0(t))\| < \epsilon, \qquad t \in (t_1, t_2).$$

Moreover, $\|x_t(t_0, \phi_0)\|_0 \leqslant \epsilon$, $t \in [t_0, t_1]$. Setting $\phi = x_t(t_0, \phi_0)$ for $t \in (t_1, t_2)$, we see that $\delta < \|\phi(0)\| < \epsilon$. We choose $|\sigma_0|_0 = a(\|\phi_1\|_0)$, where $\phi_1 = x_{t_1}(t_0, \phi_0)$. Then, condition (iii) and Theorem 8.1.4 yield the inequality

$$V(t, y(t_1, \phi_1)(t)) \leqslant r(t_1, \sigma_0)(t), \qquad t \in [t_1, t_2], \tag{8.5.2}$$

where $y(t_1, \phi_1)$ is any solution of (7.1.1) through (t_1, ϕ_1), and $r(t_1, \sigma_0)$ is the maximal solution of (8.1.16) through (t_1, σ_0). It turns out that (8.5.2) is also true for $x(t_0, \phi_0)$ on the interval $t_1 \leqslant t \leqslant t_2$. Hence, we get

$$b(\epsilon) \leqslant V(t_2, x(t_0, \phi_0)(t_2)) \leqslant r(t_1, \sigma_0)(t_2) < b(\epsilon),$$

on account of the fact that $t_2 > t_1 > t_0 \geq \tau_0(\epsilon)$ and the uniformity of the relation (8.5.1) with respect to t_0. This contradiction shows that (E_1) is valid, and the theorem is proved.

COROLLARY 8.5.1. The uniform stability of the trivial solution $\sigma \equiv 0$ of (8.1.16) is admissible in Theorem 8.5.1 in place of the eventual uniform stability of the set $\sigma = 0$. In particular, $g(t, u, \sigma) \equiv 0$ is admissible.

THEOREM 8.5.2. Assume that

(i) $V \in C[[-\tau, \infty) \times S_\rho, R_+]$, $V(t, x)$ is Lipschitzian in x for a constant $L = L(\rho) > 0$, and

$$b(\| x \|) \leq V(t, x) \leq a(\| x \|),$$

for $0 < \alpha < \| x \| < \rho$ and $t \geq \theta(\alpha)$, where $a, b \in \mathcal{K}$ and $\theta(u)$ is continuous and monotonic decreasing in u for $0 < u < \rho$;

(ii) $f \in C[J \times C_\rho, R^n]$, and

$$D^+ V(t, \phi(0), \phi) \leq 0,$$

for every $\phi \in C_\rho$ such that $0 < \alpha < \| \phi(0) \| < \rho$ and $t \geq \theta(\alpha)$;

(iii) $R \in C[J \times C_\rho, R^n]$, and, for every $\phi \in C_{\rho^*}$, $\rho^* < \rho$ and $t \geq 0$,

$$\int_0^\infty \| R(s, \phi) \| \, ds < \infty.$$

Then the set $\phi = 0$ is eventually uniformly stable with respect to the perturbed system

$$x' = f(t, x_t) + R(t, x_t). \tag{8.5.3}$$

Proof. Let $0 < \epsilon < \rho^*$ be given. Choose the numbers $\delta = \delta(\epsilon)$ and $\tau_1(\epsilon)$ such that

$$2a(\delta) < b(\epsilon) \quad \text{and} \quad \tau_1(\epsilon) = \theta(\delta(\epsilon)). \tag{8.5.4}$$

Define $\lambda(t) = \max_{\| \phi \|_0 \leq \rho^*} \| R(t, \phi) \|$. Since $\lambda(t)$ is integrable, it is possible to find a $\tau_2(\epsilon) > 0$ such that

$$\int_{t_0}^\infty \lambda(s) \, ds < a(\delta)/L, \tag{8.5.5}$$

provided $t_0 \geq \tau_2(\epsilon)$, where L is the Lipschitz constant for $V(t, x)$. Let $\tau_0(\epsilon) = \max[\tau_1(\epsilon), \tau_2(\epsilon)]$.

Suppose that there exists a solution $x(t_0, \phi_0)$ of the perturbed system (8.5.3) and two numbers t_1, t_2 such that $t_2 > t_1 \geq t_0 \geq \tau_0(\epsilon)$,

$$\| x(t_0, \phi_0)(t_1) \| = \delta, \qquad \| x(t_0, \phi_0)(t_2) \| = \epsilon,$$

and

$$\delta < \| x(t_0, \phi_0)(t) \| < \epsilon, \qquad t \in (t_1, t_2).$$

Also,

$$\| x_t(t_0, \phi_0) \| \leq \epsilon \leq \rho^* \qquad \text{for} \qquad t \in [t_0, t_2].$$

Setting $\phi = x_t(t_0, \phi_0)$, $t \in [t_1, t_2]$, it follows that

$$\delta < \| \phi(0) \| < \rho.$$

Hence, as before, we get

$$V(t, x(t_0, \phi_0)(t)) \leq a(\| \phi_1 \|_0) + L \int_{t_1}^{t} \lambda(s)\, ds, \qquad t \in [t_1, t_2],$$

where $\phi_1 = x_{t_1}(t_0, \phi_0)$. At $t = t_2$, we therefore obtain, in view of (8.5.5) and the fact that $\| \phi_1 \|_0 \leq \delta$,

$$b(\epsilon) \leq V(t_2, x(t_0, \phi_0)(t_2)) \leq a(\delta) + a(\delta) = 2a(\delta),$$

which is incompatible with (8.5.4). This shows that (E_1) holds, and the theorem is established.

THEOREM 8.5.3. *Let assumption* (i) *of Theorem* 8.5.1 *hold. Suppose further that* $f \in C[J \times C_\rho, R^n]$ *and*

$$D^+V(t, \phi(0), \phi) \leq -C(\| \phi(0) \|),$$

for every $\phi \in C_\rho$ *such that* $0 < \alpha < \| \phi(0) \| < \rho$ *and* $t \geq \theta(\alpha)$ *and* $C \in \mathscr{K}$. *Then, the set* $\phi = 0$ *is eventually uniformly asymptotically stable.*

Proof. The eventual uniform stability of the set $\phi = 0$ follows by Corollary 8.5.1.

Let $0 < \epsilon < \rho$ be given. Choose $\delta_0 = \delta(\rho)$, $\tau_0 = \tau(\rho)$, and $T(\epsilon) = \tau(\epsilon) + \{a(\rho)/C[\delta(\epsilon)]\}$. Assume that $t_0 \geq \tau_0$ and $\| \phi_0 \|_0 \leq \delta_0$. It is sufficient to show that there is a $t_1 \in [t_0 + \tau(\epsilon), t_0 + T(\epsilon)]$ such that

$$\| x(t_0, \phi_0)(t_1) \| < \delta(\epsilon),$$

in order to complete the proof. Suppose, if possible, that

$$\delta(\epsilon) \leq \| x(t_0, \phi_0)(t) \| < \rho, \qquad t \in [t_0 + \tau(\epsilon), t_0 + T(\epsilon)].$$

Letting $\phi = x_t(t_0, \phi_0)$, we see that

$$\delta(\epsilon) \leqslant \|\phi(0)\| < \rho$$

and $\|\phi\|_0 < \rho$. Hence, it follows that

$$D^+V(t, x(t_0, \phi_0)(t)) \leqslant -C[\|x(t_0, \phi_0)(t)\|] \leqslant -C[\delta(\epsilon)],$$

for $t \in [t_0 + \tau(\epsilon), t_0 + T(\epsilon)]$, which yields

$$V(t_0 + T(\epsilon), x(t_0, \phi_0)(t_0 + T(\epsilon)))$$
$$\leqslant V(t_0 + \tau(\epsilon), x(t_0, \phi_0)(t_0 + \tau(\epsilon)) - C\delta(\epsilon)[T(\epsilon) - \tau(\epsilon)].$$

It then turns out that

$$0 < b[\delta(\epsilon)] \leqslant V(t_0 + T(\epsilon), x(t_0, \phi_0)(t_0 + T(\epsilon)))$$
$$\leqslant a(\|x(t_0, \phi_0)(t_0 + \tau(\epsilon))\|) - C[\delta(\epsilon)] \frac{a(\rho)}{C[\delta(\epsilon)]}$$
$$\leqslant a(\rho) - a(\rho) = 0,$$

which is an absurdity. This means that there exists a

$$t_1 \in [t_0 + \tau(\epsilon), t_0 + T(\epsilon)]$$

such that $\|x(t_0, \phi_0)(t_1)\| < \delta(\epsilon)$, and therefore, in any case,

$$\|x(t_0, \phi_0)(t)\| < \epsilon, \qquad t \geqslant t_0 + T(\epsilon), \qquad t_0 \geqslant \tau(\epsilon),$$

provided $\|\phi_0\|_0 \leqslant \delta_0$. The proof is complete.

8.6. Asymptotic behavior

In this section, some of the results of the Sect. 3.15 will be extended to the functional differential systems.

THEOREM 8.6.1. Assume that

(i) $V \in C[[-\tau, \infty) \times R^n, R_+]$, $V(t, 0) \equiv 0$, $V(t, x)$ satisfies a Lipschitz condition in x for a continuous function $K(t) \geqslant 0$, and

$$b(\|x\|) \leqslant V(t, x),$$

where $b \in \mathscr{K}$ is such that $b(u) \to \infty$ as $u \to \infty$;

(ii) $g \in C[J \times R_+ \times \mathscr{C}_+, R]$, $g(t, u, \sigma)$ is nondecreasing in σ for each (t, u), and, for $t \in J, \phi, \psi \in \mathscr{C}^n$,

$$D^+V(t, \phi(0) - \psi(0), \phi - \psi)$$
$$= \limsup_{h \to 0^+} h^{-1}[V(t + h, \phi(0) - \psi(0) + h\{f(t, \phi) - f(t, \psi)\}$$
$$\quad - V(t, \phi(0) - \psi(0))]$$
$$\leqslant g(t, V(t, \phi(0) - \psi(0)), V_t),$$

where $V_t = V(t + s, \phi(s) - \psi(s))$;

(iii) every solution $u(t_0, \sigma_0)$ of

$$u' = g(t, u, u_t) + K(t)\|f(t, 0)\| \tag{8.6.1}$$

tends to zero as $t \to \infty$.

Then, every solution $x(t_0, \phi_0)$ of the system (7.1.1) tends to zero as $t \to \infty$.

Proof. Let $t \in J$ and $\phi \in \mathscr{C}^n$. Then,

$$D^+V(t, \phi(0), \phi) = \limsup_{h \to 0^+} h^{-1}[V(t + h, \phi(0) + hf(t, \phi)) - V(t, \phi(0))]$$
$$\leqslant K(t)\|f(t, 0)\| + \limsup_{h \to 0^+} h^{-1}[V(t + h, \phi(0)$$
$$\quad + h\{f(t, \phi) - f(t, 0)\}) - V(t, \phi(0))]$$
$$\leqslant K(t)\|f(t, 0)\| + g(t, V(t, \phi(0)), V_t),$$

in view of assumption (ii). Let $x(t_0, \phi_0)$ be any solution of (7.1.1) such that $V(t_0 + s, \phi_0(s)) \leqslant \sigma_0(s)$. It then follows, by Theorem 6.9.4, that

$$V(t, x(t_0, \phi_0)(t)) \leqslant r(t_0, \sigma_0)(t), \qquad t \geqslant t_0, \tag{8.6.2}$$

where $r(t_0, \sigma_0)$ is the maximal solution of the functional differential equation (8.6.1). The stated result is now a direct consequence of the hypotheses (i) and (iii).

The next theorem is very useful in applications, since it does not demand $V(t, x)$ to be positive definite.

THEOREM 8.6.2. *Assume that*

(i) $f \in C[J \times C_\rho, R^n]$, *and*

$$\|f(t, \phi)\| \leqslant M, \qquad t \in J, \qquad \|\phi\|_0 \leqslant \rho^* < \rho;$$

8.6. ASYMPTOTIC BEHAVIOR

(ii) $V \in C[[-\tau, \infty) \times S_\rho, R_+]$, $V(t, x)$ is locally Lipschitzian in x, and

$$D^+ V(t, \phi(0), \phi) \leqslant -C[\phi(0)],$$

for $t \in J$ and $\phi \in \Omega$, where $C(x)$ is positive definite with respect to a closed set Ω in S_ρ;

(iii) all the solutions $x(t_0, \phi_0)$ of (7.1.1) are bounded for $t \geqslant t_0$. Then, every solution of (7.1.1) approaches the set Ω as $t \to \infty$.

Proof. Let $x(t_0, \phi_0)$ be any solution of (7.1.1). By assumption (iii), it is bounded, and, hence, there exists a compact set Q in S_ρ, such that

$$x(t_0, \phi_0)(t) \in Q, \quad t \geqslant t_0.$$

Moreover, it also follows that $\| x_t(t_0, \phi_0)\|_0 \leqslant \rho^* < \rho, t \geqslant t_0$, and therefore, by assumption (i), we have

$$\| f(t, x_t(t_0, \phi_0))\| \leqslant M.$$

Suppose that this solution does not approach Ω as $t \to \infty$. Then, for some $\epsilon > 0$, there exists a sequence $\{t_k\}$, $t_k \to \infty$ as $k \to \infty$, such that

$$x(t_0, \phi_0)(t_k) \in S(\Omega, \epsilon)^C \cap Q,$$

where $S(\Omega, \epsilon)^C$ is the complement of the set $S(\Omega, \epsilon) = [x : d(x, \Omega) < \epsilon]$. We may assume that t_1 is sufficiently large so that, on the intervals $t_k \leqslant t \leqslant t_k + (\epsilon/2M)$, we have

$$x(t_0, \phi_0)(t) \in S(\Omega, \epsilon/2)^C \cap Q. \tag{8.6.3}$$

These intervals may be supposed to be disjoint, by taking a subsequence of $\{t_k\}$, if necessary. By Theorem 8.1.2 and assumption (ii), we get

$$V(t, x(t_0, \phi_0)(t)) \leqslant \sup_{-\tau \leqslant s \leqslant 0} V(t_0, \phi_0(s)) - \int_{t_0}^t C[x(t_0, \phi_0)(s)] \, ds, \tag{8.6.4}$$

for $t \geqslant t_0$. Since $C(x)$ is positive definite with respect to Ω, the relation (8.6.3) shows that there exists a $\delta = \delta(\epsilon/2) > 0$ such that

$$C[x(t_0, \phi_0)(t)] \geqslant \delta, \quad t_k \leqslant t \leqslant t_k + (\epsilon/2M). \tag{8.6.5}$$

It therefore turns out that

$$V\left(t_k + \frac{\epsilon}{2M}, x(t_0, \phi_0)\left(t_k + \frac{\epsilon}{2M}\right)\right) \leqslant \sup_{-\tau \leqslant s \leqslant 0} V(t_0, \phi_0(s)) - \frac{\delta \epsilon k}{2M},$$

on account of (8.6.4) and (8.6.5). The foregoing inequality leads to an absurdity as $k \to \infty$, since, by assumption, $V(t, x) \geqslant 0$. As a consequence, any solution $x(t_0, \phi_0)$ tends to the set Ω as $t \to \infty$, and the theorem is proved.

Making use of two Lyapunov functions, we can extend Theorem 4.2.1 to functional differential system (7.1.1).

THEOREM 8.6.3. Let the following assumptions hold:

(i) $f \in C[J \times C_\rho, R^n]$, $f(t, 0) \equiv 0$, and $f(t, \phi)$ is bounded on $J \times C_\rho$;

(ii) $V_1 \in C[[-\tau, \infty) \times S_\rho, R_+]$, $V_1(t, x)$ is positive definite, decrescent, locally Lipschitzian in x, and, for $t \in J$, $\phi \in C_\rho$,

$$D_+V_1(t, \phi(0), \phi) \leqslant \omega(\phi(0)) \leqslant 0,$$

where $\omega(x)$ is continuous for $x \in S_\rho$;

(iii) $V_2 \in C[[-\tau, \infty) \times S_\rho, R_+]$, $V_2(t, x)$ is bounded on $J \times S_\rho$ and is locally Lipschitzian in x. Furthermore, given any number α, $0 < \alpha < \rho$, there exist positive numbers $\xi = \xi(\alpha) > 0$, $\eta = \eta(\alpha) > 0$, $\eta < \alpha$ such that

$$D^+V_2(t, \phi(0), \phi) > \xi$$

for every $t \geqslant 0$, $\phi \in C_\rho$ such that $\alpha < \| \phi(0) \| < \rho$ and $d(\phi(0), E) < \eta$, where

$$E = [x \in S_\rho : \omega(x) = 0],$$

and $d(x, E)$ is the distance between the point x and the set E.

Then, the trivial solution of (7.1.1) is uniformly asymptotically stable.

Proof. As the proof requires appropriate changes in the proof of Theorem 4.2.1, we shall indicate only the modifications. Let $0 < \epsilon < \rho$ and $t_0 \in J$. Since $V_1(t, x)$ is positive definite and decrescent, there exist functions $a, b \in \mathscr{K}$, satisfying

$$b(\| x \|) \leqslant V_1(t, x) \leqslant a(\| x \|), \qquad (t, x) \in J \times S_\rho.$$

Let us choose $\delta = \delta(\epsilon) > 0$ such that

$$b(\epsilon) > a(\delta).$$

Then, by Corollary 8.2.4, the uniform stability of the trivial solution of (7.1.1) results.

Let us designate $\delta_0 = \delta(\rho)$. Assume that $\| \phi_0 \|_0 < \delta_0$. To prove the theorem, it is sufficient to show that there exists a $T = T(\epsilon)$ such that, for some $t_1 \in [t_0, t_0 + T]$, $\| x(t_0, \phi_0)(t_1) \| < \delta(\epsilon)$. As in the proof of Theorem 4.2.1, this will be achieved in a number of steps:

8.6. ASYMPTOTIC BEHAVIOR

(1°) If $d[x(t_1), x(t_2)] > r > 0$, $t_2 > t_1$, then

$$r \leqslant Mn^{1/2}(t_2 - t_1),$$

where $x(t) = x(t_0, \phi_0)(t)$ and $\|f(t, \phi)\| \leqslant M$, $(t, \phi) \in J \times C_\rho$. There is no change in the proof of this statement.

(2°) By assumption (iii), given $\delta = \delta(\epsilon)$, $0 < \delta(\epsilon) < \rho$, there exist two positive numbers $\xi = \xi(\epsilon)$, $\eta = \eta(\epsilon)$, $\eta < \delta$, such that

$$D^+V_2(t, \phi(0), \phi) > \xi$$

for every $t \in J$, $\phi \in C_\rho$ such that $\delta < \|\phi(0)\| < \rho$, $d(\phi(0), E) < \eta$. Consider the set

$$U = [x \in S_\rho : \delta < \|x\| < \rho \quad \text{and} \quad d(x, E) < \eta],$$

and suppose that $\sup_{\|x\| < \rho} V_2(t, x) = L$. Assume that, at $t = t_1$, we have $x(t_1) = x(t_0, \phi_0)(t_1) \in U$. Then, for $t > t_1$, it follows, by letting $m(t) = V_2(t, x(t))$, that

$$D^+m(t) \geqslant D^+V_2(t, \phi(0), \phi) > \xi,$$

because of condition (iii) and the fact that $V_2(t, x)$ satisfies a Lipschitz condition locally in x. In obtaining the foregoing inequality, we have set $\phi = x_t(t_0, \phi_0)$ so that $\phi(0) = x(t)$ and used the inequality $\|\phi\|_0 = \|x_t(t_0, \phi_0)\|_0 < \rho$, which is a consequence of uniform stability. Since

$$m(t) - m(t_1) = \int_{t_1}^{t} D^+m(s)\, ds,$$

as long as $x(t)$ remains in U, we should have

$$m(t) + m(t_1) \geqslant \int_{t_1}^{t} D^+m(s)\, ds \geqslant \int_{t_1}^{t} D^+V_2(s, \phi(0), \phi)\, ds$$

$$> \xi(t - t_1).$$

This inequality can be realized simultaneously with $m(t) \leqslant L$ only if

$$t < t_1 + (2L/\xi).$$

Hence, it follows that there exists a t_2, $t_1 < t_2 \leqslant t_1 + (2L/\xi)$ with the property that $x(t_2)$ is on the boundary of the set U.

(3°) Consider the sequence $\{t_k\}$ such that

$$t_k = t_0 + k(2L/\xi) \qquad (k = 0, 1, 2, \ldots,).$$

Defining $n(t) = V_1(t, x(t))$ and using assumption (ii), we obtain

$$D^+n(t) \leqslant D^+V_1(t, \phi(0), \phi) \leqslant 0,$$

where $\phi = x_t(t_0, \phi_0)$, as before. Let $\lambda_1 = \lambda\eta/2Mn^{1/2}$, where $\lambda = \inf[|(w(x)|, \delta < \|x\| < \rho,$ and $d(x, E) \geqslant \eta/2]$. Suppose that $x(t)$ satisfies, for $t_k \leqslant t \leqslant t_{k+2}$, the inequality $\delta < \|x(t)\| < \rho$. Then, arguing as in the proof of Theorem 4.2.1, with obvious changes, we can show that

$$V_1(t_{k+2}, x(t_{k+2})) \leqslant V_1(t_k, x(t_k)) - \lambda_1.$$

We now choose an integer K^* such that $\lambda_1 K^* > a(\delta_0)$ and let $T = T(\epsilon) = 4K^*L/\xi$. Assuming that, for $t_0 \leqslant t \leqslant t_0 + T$, we have

$$\|x(t)\| \geqslant \delta(\epsilon),$$

we arrive at the inequality, as in Theorem 4.2.1,

$$V_1(t_0 + T, x(t_0 + T)) \leqslant V_1(t_0, \phi_0) - K^*\lambda_1$$
$$\leqslant a(\delta_0) - K^*\lambda_1 \leqslant 0,$$

which is absurd, since $V_1(t, x)$ is positive definite. It therefore turns out that there exists a $t^* \in [t_0, t_0 + T]$ such that $\|x(t^*)\| < \delta$, and this proves the uniform asymptotic stability of the trivial solution of (7.1.1).

8.7. Notes

The comparison theorems 8.1.1 and 8.1.3 are due to Lakshmikantham [1, 6]. See also Driver [3]. Theorem 8.1.4 is new. Theorems 8.2.1 and 8.2.2 are adapted from the work of Lakshmikantham [1, 6], whereas Theorem 8.2.3 is based on the result of Driver [3]. See also Krasovskii [2, 5]. Theorems 8.2.4 and 8.2.5 are new. The examples in Sect. 8.2 are taken from Lakshmikantham [6] and Driver [3].

All the results of Sect. 8.3 are based on the work of Lakshmikantham [6], whereas Theorem 8.3.4 is new. Section 8.4 contains new results. See also Halanay [22] for particular cases. The results of Sects. 8.5 and 8.6 are new. For many similar results for delay-differential equations, see Oğuztöreli [1].

For related work, see Driver [3], El'sgol'ts [4], Krasovskii [1–5], Lakshmikantham [1], Oğuztöreli [1], and Razumikhin [2, 6]. For the use of vector Lyapunov functions in studying the conditional stability criteria of invariant sets, see Lakshmikantham and Leela [2].

PARTIAL DIFFERENTIAL EQUATIONS

Chapter 9

9.0. Introduction

This chapter is devoted to the study of partial differential inequalities of first order. We consider some basic theorems on partial differential inequalities, discuss a variety of comparison results, and obtain *a priori* bounds of solutions of partial differential equations of first order in terms of solutions of ordinary differential equations as well as solutions of auxiliary partial differential equations. We also treat the uniqueness problem, error estimation of approximate solutions, and simple stability criteria. We make use of Lyapunov-like functions to derive sufficient conditions for stability behavior. For systems of partial differential inequalities of more general type, we merely indicate certain analogous results.

9.1. Partial differential inequalities of first order

We shall use the well-known notation

$$m_t = \frac{\partial m}{\partial t}, \quad m_x = \frac{\partial m}{\partial x}, \quad m_{xy} = \frac{\partial^2 m}{\partial x \, \partial y},$$

whenever convenient. It is, however, necessary to caution the reader not to confuse the symbol m_t with the one used while considering functional differential systems.

Let $\alpha, \beta \in C[J, R]$, and suppose that $\alpha(t) < \beta(t)$, $t \in J$. Assume that $\alpha'(t)$, $\beta'(t)$ exist and are continuous on J. For $t_0 \in J$, we define the following sets:

$$E = [(t, x): t_0 \leq t < \infty, \alpha(t) \leq x \leq \beta(t)],$$
$$E_0 = [(t, x): t_0 < t < \infty, \alpha(t) < x < \beta(t)],$$
$$\partial E_1 = [(t, x): t = t_0, \alpha(t_0) \leq x \leq \beta(t_0)],$$
$$\partial E_2 = [(t, x): t_0 \leq t < \infty, x = \alpha(t)],$$
$$\partial E_3 = [(t, x): t_0 \leq t < \infty, x = \beta(t)].$$

Let us first prove a lemma that is very useful in the subsequent discussion.

LEMMA 9.1.1. Assume that

(i) $m \in C[E, R]$, and $m(t, x)$ possesses partial derivatives (not necessarily continuous) on E and total derivative on $\partial E_2 \cup \partial E_3$;

(ii) if $(t_1, x_1) \in E_0$, $m(t_1, x_1) = m_x(t_1, x_1) = 0$, then $m_t(t_1, x_1) < 0$;

(iii) if $(t_1, x_1) \in \partial E_2$, $m(t_1, x_1) = 0$, and $m_x(t_1, x_1) \leq 0$, then

$$m_t(t_1, x_1) + \alpha'(t_1) m_x(t_1, x_1) < 0;$$

(iv) if $(t_1, x_1) \in \partial E_3$, $m(t_1, x_1) = 0$, and $m_x(t_1, x_1) \geq 0$, then

$$m_t(t_1, x_1) + \beta'(t_1) m_x(t_1, x_1) < 0;$$

(v) $m(t, x) < 0$ on ∂E_1.

Under these assumptions,

$$m(t, x) < 0 \quad \text{on} \quad E. \tag{9.1.1}$$

Proof. Suppose that the inequality (9.1.1) is false. Then, by assumption (v) and the continuity of $m(t, x)$ on E, it follows that there exists a point $(t_1, x_1) \in E$, $t_1 > t_0$, such that

$$m(t_1, x_1) = 0 \quad \text{and} \quad m(t, x) < 0, \quad t < t_1, \quad (t, x) \in E. \tag{9.1.2}$$

Evidently, $m(t_1, x) \leq 0$ for $(t_1, x) \in E$, and, hence, one of the following situations arises:

(a) $(t_1, x_1) \in \partial E_2$ and $m_x(t_1, x_1) \leq 0$;
(b) $(t_1, x_1) \in \partial E_3$ and $m_x(t_1, x_1) \geq 0$;
(c) $(t_1, x_1) \notin \partial E_2 \cup \partial E_3$ and $m_x(t_1, x_1) = 0$.

Consider the case (a). Because of condition (iii), we have

$$\left(\frac{d}{dt} m(t, \alpha(t))\right)_{t=t_1} = m_t(t_1, \alpha(t_1)) + \alpha'(t_1) m_x(t_1, \alpha(t_1))$$
$$= m_t(t_1, x_1) + \alpha'(t_1) m_x(t_1, x_1) < 0. \tag{9.1.3}$$

9.1. PARTIAL DIFFERENTIAL INEQUALITIES OF FIRST ORDER

On the other hand, it turns out, by the relation (9.1.2), that

$$\frac{d}{dt} m(t, \alpha(t))_{t=t_1} \geq 0,$$

which is absurd in view of (9.1.3).

Similarly, in the case (b), we obtain

$$\frac{d}{dt} m(t, \beta(t))_{t=t_1} = m_t(t_1, x_1) + \beta'(t) m_x(t_1, x_1) < 0 \qquad (9.1.4)$$

because of condition (iv). Again, we get a contradiction to (9.1.4), since the relation (9.1.2) also yields

$$\left(\frac{d}{dt} m(t, \beta(t))\right)_{t=t_1} \geq 0.$$

In the last case (c), the contradiction results from assumption (ii). This proves the truth of the inequality (9.1.1).

REMARK 9.1.1. Notice that it is necessary for the proof only that the derivatives m_t, m_x or the total derivative exist at such points (t, x) for which $m(t, x) = 0$.

We shall now prove some basic results in partial differential inequalities of the first order.

THEOREM 9.1.1. Let $f \in C[E \times R \times R, R]$, $u, v \in C[E, R]$, and suppose that the following conditions hold:

(a_0) u, v posses derivatives u_t, v_t, u_x, v_x on E (not necessarily continuous), and the total derivative on $\partial E_2 \cup \partial E_3$;
(a_1)
$$u_t \leq f(t, x, u, u_x),$$
$$v_t > f(t, x, v, v_x) \quad \text{on} \quad E;$$

(a_2) $f(t, x, z, p) - f(t, x, z, q) \geq -\alpha'(t)(p - q)$, $p \geq q$, on ∂E_2;
(a_3) $f(t, x, z, p) - f(t, x, z, q) \leq -\beta'(t)(p - q)$, $p \geq q$, on ∂E_3;
(a_4) $u(t, x) < v(t, x)$ on ∂E_1.

Then,

$$u(t, x) < v(t, x) \quad \text{on} \quad E. \qquad (9.1.5)$$

Proof. It will be shown that the function $m(t, x)$ defined by

$$m(t, x) = u(t, x) - v(t, x)$$

verifies the hypotheses of Lemma 9.1.1. Clearly, assumptions (i) and (v) are satisfied in view of (a_0) and (a_4) of the theorem. Furthermore, $m(t, x)$ fulfills condition (iv) of Lemma 9.1.1. For, let $(t_1, x_1) \in \partial E_3$, $m(t_1, x_1) = 0$, and $m_x(t_1, x_1) \geqslant 0$. Then, we have

$$u(t_1, x_1) = v(t_1, x_1) \quad \text{and} \quad u_x(t_1, x_1) \geqslant v_x(t_1, x_1).$$

Consequently, by the condition (a_3), we derive

$$f(t_1, x_1, u(t_1, x_1), u_x(t_1, x_1)) - f(t_1, x_1, u(t_1, x_1), v_x(t_1, x_1))$$
$$\leqslant -\beta'(t_1)[u_x(t_1, x_1) - v_x(t_1, x_1)].$$

It now results, using the inequalities (a_1), that

$$u_t(t_1, x_1) - v_t(t_1, x_1) < -\beta'(t_1)[u_x(t_1, x_1) - v_x(t_1, x_1)],$$

which implies that

$$m_t(t_1, x_1) + \beta'(t_1) m_x(t_1, x_1) < 0.$$

This verifies condition (iv) of Lemma 9.1.1.

The remaining assumptions may be similarly checked, and the conclusion (9.1.5) is immediate by the application of Lemma 9.1.1. The proof is complete.

It is clear, from Theorem 9.1.1, that one of the inequalities (a_1) must be strict for the validity of the claim (9.1.5). This can, however, be dispensed with, if the function f satisfies a uniqueness condition. To that effect, we present the following.

THEOREM 9.1.2. *Let the assumptions of Theorem* 9.1.1 *hold, except that the differential inequalities* (a_1) *are replaced by*

$$(a_1*) \begin{cases} u_t \leqslant f(t, x, u, u_x), \\ v_t \geqslant f(t, x, v, v_x) \end{cases} \quad \text{on} \quad E.$$

Suppose further that

$$f(t, x, u_1, p) - f(t, x, u_2, p) \leqslant g(t, u_1 - u_2), \quad u_1 > u_2, \quad (9.1.6)$$

where $g \in C[J \times R_+; R]$, *and the maximal solution* $r(t, t_0, 0)$ *of*

$$y' = g(t, y), \quad y(t_0) = 0 \quad (9.1.7)$$

is identically zero. Then,

$$u(t, x) \leqslant v(t, x) \quad \text{on} \quad \partial E_1 \quad (9.1.8)$$

9.1. PARTIAL DIFFERENTIAL INEQUALITIES OF FIRST ORDER

implies
$$u(t, x) \leq v(t, x) \quad \text{on} \quad E. \tag{9.1.9}$$

Proof. Consider the solutions $y(t, \epsilon) = y(t, t_0, 0, \epsilon)$ of the ordinary differential equation
$$y' = g(t, y) + \epsilon, \quad y(t_0) = \epsilon, \tag{9.1.10}$$
for all sufficiently small $\epsilon > 0$, which exist for $t \geq t_0$. Define
$$u(t, x, \epsilon) = u(t, x) - y(t, \epsilon). \tag{9.1.11}$$

Then, because of the relation
$$y'(t, \epsilon) > g(t, y(t, \epsilon)),$$
we obtain
$$\begin{aligned} u_t(t, x, \epsilon) &= u_t(t, x) - y'(t, \epsilon) \\ &< f(t, x, u(t, x), u_x(t, x)) - g(t, y(t, \epsilon)). \end{aligned} \tag{9.1.12}$$

Since $y(t, \epsilon) > 0$, it follows, from (9.1.11), that
$$u(t, x, \epsilon) < u(t, x),$$
and hence, by (9.1.6), there results the inequality
$$f(t, x, u(t, x), u_x(t, x)) - f(t, x, u(t, x, \epsilon), u_x(t, x, \epsilon)) \leq g(t, y(t, \epsilon)),$$
in view of the fact that $u_x(t, x, \epsilon) = u_x(t, x)$. This, together with (9.1.12), yields
$$u_t(t, x, \epsilon) < f(t, x, u(t, x, \epsilon), u_x(t, x, \epsilon)).$$

Moreover, we have also the inequality
$$u(t, x, \epsilon) < v(t, x) \quad \text{on} \quad \partial E_1.$$

Hence, an application of Theorem 9.1.1 shows that
$$u(t, x, \epsilon) < v(t, x) \quad \text{on} \quad E.$$

Since, by assumption, $\lim_{\epsilon \to 0} y(t, \epsilon) \equiv 0$, the stated result (9.1.9) is an immediate consequence of (9.1.11). This completes the proof.

Consider the special case
$$f(t, x, u, u_x) = F(t, x, u) - \lambda(t, x) u_x.$$

Let $\alpha(t), \beta(t)$ be any functions satisfying the ordinary differential inequalities
$$\alpha'(t) \geq \lambda(t, \alpha(t)), \quad \beta'(t) \leq \lambda(t, \beta(t)),$$

such that $\alpha(t) < \beta(t)$. Suppose that we define the set E by means of these functions $\alpha(t)$, $\beta(t)$. Then, we have the following corollaries.

COROLLARY 9.1.1. Let $F \in C[E \times R, R]$, $\lambda \in C[J \times R, R]$, $u, v \in C[E, R]$, and the assumptions (a_0), (a_1), and (a_4) of Theorem 9.1.1 hold with $f = F(t, x, u) - \lambda(t, x) u_x$. Then, (9.1.5) is valid.

COROLLARY 9.1.2. Suppose that

$$F(t, x, u_1) - F(t, x, u_2) \leqslant g(t, u_1 - u_2), \quad u_1 > u_2,$$

where $g(t, u)$ satisfies the same assumptions of Theorem 9.1.2. Assume further that the hypotheses of Corollary 9.1.1 hold except that the inequalities (a_1) are replaced by (a_1^*) of Theorem 9.1.2. Then, (9.1.8) implies (9.1.9).

For the proof of the foregoing corollaries, it is sufficient to show that the conditions (a_2), (a_3) of Theorem 9.1.1 are fulfilled. Verification of this fact is trivial in view of the choice of functions $\alpha(t)$ and $\beta(t)$.

Suppose, on the other hand, we assume that

$$|f(t, x, u, p) - f(t, x, u, q)| \leqslant L |p - q|. \tag{9.1.13}$$

Choose $\alpha(t) = a - L(t - t_0)$, $\beta(t) = -a + L(t - t_0)$, and $\gamma < aL$. Let $t_0 \leqslant t < t_0 + \gamma$. Then, we may define the set E, as before, by means of these functions and the time interval. When (9.1.13) is verified, it can be easily checked that f satisfies the conditions (a_2), (a_3) of Theorem 9.1.1. From this observation stem the following corollaries.

COROLLARY 9.1.3. Let the hypotheses of Theorem 9.1.1 hold except the conditions (a_2) and (a_3). If f verifies the inequality (9.1.13), the conclusion (9.1.5) is valid.

COROLLARY 9.1.4. Let the hypotheses of Theorem 9.1.2 hold except the conditions (a_2) and (a_3). If f verifies the condition (9.1.13), then (9.1.8) implies (9.1.9).

9.2. Comparison theorems

In dealing with the applications of ordinary differential inequalities to partial differential equations, we have to estimate the solutions of such equations, which are functions of several variables, by functions of one variable. While doing so, the following lemmas prove to be useful.

9.2. COMPARISON THEOREMS

LEMMA 9.2.1. Let G be a bounded open set $G \subset R^n$. Let $u \in C[J \times \bar{G}, R]$, and let $M(t) = \max_{x \in \bar{G}} u(t, x)$. If $M(t_0) = u(t_0, x_0)$ and $\partial u(t_0, x_0)/\partial t$ exists, for $(t_0, x_0) \in (0, \infty) \times G$, then

$$D_- m(t_0) \leq \partial u(t_0, x_0)/\partial t. \qquad (9.2.1)$$

Proof. Choose a sequence $\{t_k\}$, so that $t_k < t_0$, $t_k \to t_0$ as $k \to \infty$, and

$$\lim_{k \to \infty} \frac{M(t_k) - M(t_0)}{t_k - t_0} = D_- M(t_0). \qquad (9.2.2)$$

For k sufficiently large, we have

$$\lim_{k \to \infty} \frac{u(t_k, x_0) - u(t_0, x_0)}{t_k - t_0} = \frac{\partial u(t_0, x_0)}{\partial t}. \qquad (9.2.3)$$

On the other hand, by the definition of $M(t)$ and the fact that $M(t_0) = u(t_0, x_0)$, for k sufficiently large, it follows that

$$\frac{M(t_k) - M(t_0)}{t_k - t_0} \leq \frac{u(t_k, x_0) - u(t_0, x_0)}{t_k - t_0},$$

which, on account of the relations (9.2.2) and (9.2.3), yields (9.2.1).

LEMMA 9.2.2. Let G be a bounded open set such that $G \subset R^n$ and $u \in C[J \times \bar{G}, R]$. Let

$$w(t) = \max_{x \in \bar{G}} |u(t, x)|,$$

$$M(t) = \max_{x \in \bar{G}} u(t, x),$$

and

$$N(t) = \max_{x \in \bar{G}} [-u(t, x)].$$

Let $(t_0, x_0) \in (0, \infty) \times G$. Then,

(1°) $w(t_0) > 0$ implies either $w(t_0) = M(t_0)$ or $w(t_0) = N(t_0)$;
(2°) $w(t_0) > 0$ and $w(t_0) = M(t_0)$ implies $D_- w(t_0) \leq D_- M(t_0)$;
(3°) $w(t_0) > 0$ and $w(t_0) = N(t_0)$ implies $D_- w(t_0) \leq D_- N(t_0)$.

Proof. Suppose that $w(t_0) > 0$ and $w(t_0) = u(t_0, x_0)$. It follows from the inequality

$$u(t_0, x) \leq |u(t_0, x)| \leq w(t_0) = u(t_0, x_0), \qquad x \in \bar{G},$$

that $M(t_0) = u(t_0, x_0)$. If, on the other hand, we suppose that $w(t_0) = -u(t_0, x_0) > 0$, the inequality

$$-u(t_0, x) \leqslant |u(t_0, x)| \leqslant w(t_0) = -u(t_0, x_0), \qquad x \in G,$$

shows that $N(t_0) = -u(t_0, x_0)$.

Assume now that $w(t_0) = M(t_0)$, and choose a sequence $\{t_k\}$ such that $t_k < t_0$, $t_k \to t_0$ as $k \to \infty$, and

$$D_- M(t_0) = \lim_{k \to \infty} \frac{M(t_k) - M(t_0)}{t_k - t_0}.$$

Since we have

$$M(t_k) \leqslant w(t_k) \qquad \text{and} \qquad M(t_0) = w(t_0),$$

there results the inequality

$$\frac{w(t_k) - w(t_0)}{t_k - t_0} \leqslant \frac{M(t_k) - M(t_0)}{t_k - t_0},$$

and, consequently, it follows that

$$D_- w(t_0) \leqslant D_- M(t_0).$$

The proof for the case (3°) is analogous.

We shall prove the following simple comparison theorem.

THEOREM 9.2.1. Suppose that $u \in C[\Omega, R]$, $\partial u/\partial t$, $\partial u/\partial x$ exist and are continuous on Ω, where

$$\Omega = [(t, x): t_0 \leqslant t < t_0 + a, |x| \leqslant \beta - M(t - t_0)],$$

and $\beta > Ma$. Let

$$|\partial u/\partial t| \leqslant M |\partial u/\partial x| + g(t, |u|) \qquad \text{on} \qquad \Omega, \qquad (9.2.4)$$

where $g \in C[[t_0, t_0 + a) \times R_+, R_+]$ and $g(t, 0) \equiv 0$. Assume that $r(t, t_0, y_0)$ is the maximal solution of

$$y' = g(t, y), \qquad y(t_0) = y_0 \geqslant 0 \qquad (9.2.5)$$

existing on $t_0 \leqslant t < t_0 + a$. Then,

$$|u(t_0, x)| \leqslant y_0 \qquad \text{for} \qquad |x| \leqslant \beta \qquad (9.2.6)$$

implies that

$$|u(t, x)| \leqslant r(t, t_0, y_0) \qquad \text{on} \qquad \Omega. \qquad (9.2.7)$$

9.2. COMPARISON THEOREMS

Proof. Consider the function

$$m(t) = \max_{|x| \leq \beta - M(t-t_0)} |u(t, x)|,$$

and assume that

$$r(t, t_0, y_0) < m(t).$$

This means that $t > t_0$ and $m(t) > 0$. Moreover, $m(t) = u(t, x_0)$. The point (t, x_0) may be an interior point of Ω or $x_0 = \pm[\beta - M(t - t_0)]$. Suppose that $(t, x_0) \in \text{int } \Omega$. Then $\partial u(t, x_0)/\partial x = 0$, and, as a consequence, by (9.2.4) and Lemmas 9.2.1 and 9.2.2, we get

$$D_- m(t) \leq \left| \frac{\partial u}{\partial t} \right| \leq g(t, m(t)).$$

Suppose now that $x_0 = \beta - M(t - t_0)$. Then, there are two possibilities:

(a) $m(t) = u(t, \beta - M(t - t_0))$,
(b) $m(t) = -u(t, \beta - M(t - t_0))$.

Consider the case (a). If x is sufficiently close to $x_0 = \beta - M(t - t_0)$, $u(t, x) \leq u(t_0, x_0)$, and, therefore, we get

$$\frac{\partial u(t, \beta - M(t - t_0))}{\partial x} \geq 0.$$

As a result, we obtain the inequality

$$\frac{\partial u}{\partial t} - M \frac{\partial u}{\partial x} \leq g(t, m(t)), \tag{9.2.8}$$

at (t, x_0). Furthermore, $u(s, \beta - M(s - t_0)) < m(s)$, $s < t$, and

$$|u(t, \beta - M(t - t_0))| = m(t).$$

Hence, using (9.2.8), we deduce that

$$D_- m(t) \leq D_- |u(t, \beta - M(t - t_0))|$$

$$\leq \frac{\partial u}{\partial t} - M \frac{\partial u}{\partial x}$$

$$\leq \left| \frac{\partial u}{\partial t} \right| - M \frac{\partial u}{\partial x}$$

$$\leq g(t, m(t)).$$

Suppose that the possibility (b) holds. Notice that
$$-u(t, \beta - M(t - t_0)) = \max | u(t, x)|.$$
Hence, $-u(t, x) \leqslant -u(t, \beta - M(t - t_0))$, and, thus,
$$\frac{\partial u(t, x_0)}{\partial x} \leqslant 0.$$
Since
$$\left|\frac{\partial u}{\partial t}\right| \leqslant M \left|\frac{\partial u}{\partial x}\right| + g(t, m(t)),$$
we get
$$\left|\frac{\partial u}{\partial t}\right| + M \frac{\partial u}{\partial x} \leqslant g(t, m(t)) \tag{9.2.9}$$
at (t, x_0). On the other hand,
$$-u(s, \beta - M(s - t_0)) < m(s), \quad s < t,$$
and
$$-u(t, \beta - M(t - t_0)) = m(t),$$
which implies that
$$D_- m(t) \leqslant -\frac{\partial u}{\partial t} + M \frac{\partial u}{\partial x} \tag{9.2.10}$$
at (t, x_0). By the relations (9.2.9) and (9.2.10) and the fact
$$-\frac{\partial u}{\partial t} + M \frac{\partial u}{\partial x} \leqslant \left|\frac{\partial u}{\partial t}\right| + M \frac{\partial u}{\partial x}$$
results the inequality
$$D_- m(t) \leqslant g(t, m(t)).$$
The proof for the case $x_0 = -[\beta - M(t - t_0)]$ is similar. Thus, we have shown that the inequality $r(t, t_0, y_0) < m(t)$ implies that
$$D_- m(t) \leqslant g(t, m(t)).$$
This yields, by Theorem 1.4.1, that $m(t) \leqslant r(t, t_0, y_0)$, contradicting our supposition. Hence (9.2.7) is true, and the theorem is proved.

We shall next prove a more general result whose proof is similar to the one just presented.

9.2. COMPARISON THEOREMS

THEOREM 9.2.2. Let $f \in C[E \times R_+ \times R_+, R_+]$, $u \in C[E, R]$, u possess partial derivatives on E and total derivative on $\partial E_2 \cup \partial E_3$ and on E,

$$\left| \frac{\partial u}{\partial t} \right| \leqslant f(t, x, |u|, |u_x|). \tag{9.2.11}$$

Assume further that

$$f(t, x, z, p) - f(t, x, z, 0) \geqslant -\alpha'(t)p, \quad p \geqslant 0, \quad \text{on} \quad \partial E_2, \tag{9.2.12}$$

$$f(t, x, z, p) - f(t, x, z, 0) \leqslant -\beta'(t)p, \quad p \geqslant 0, \quad \text{on} \quad \partial E_3, \tag{9.2.13}$$

and

$$f(t, x, u, 0) \leqslant g(t, u), \quad u \geqslant 0, \tag{9.2.14}$$

where $g \in C[J \times R_+, R]$, and $g(t, 0) \equiv 0$. Then, if

$$|u(t, x)| \leqslant y_0 \quad \text{on} \quad \partial E_1,$$

we have

$$|u(t, x)| \leqslant r(t, t_0, y_0) \quad \text{on} \quad E, \tag{9.2.15}$$

where $r(t, t_0, y_0)$ is the maximal solution of (9.2.5), existing for $t \geqslant t_0$.

Proof. Let us define $m(t) = \max_{\alpha(t) \leqslant x \leqslant \beta(t)} |u(t, x)|$ and suppose that $r(t, t_0, y_0) < m(t)$. This implies, as before, $t > t_0$, $m(t) > 0$ for $t > t_0$, and there is an x_0 for each t such that $m(t) = |u(t, x_0)|$.

Suppose that the point $(t, x_0) \in E_0$. Then, $\partial u(t, x_0)/\partial x = 0$, and, as a result,

$$D_- m(t) \leqslant \left| \frac{\partial u}{\partial t} \right| \leqslant f(t, x_0, |u(t, x_0)|, |u_x(t, x_0)|)$$

$$\leqslant f(t, x_0, |u(t, x_0)|, 0)$$

$$\leqslant g(t, |u(t, x_0)|)$$

$$= g(t, m(t)),$$

by (9.2.11), (9.2.14), and Lemmas 9.2.1 and 9.2.2.

Suppose that $(t, x_0) \in \partial E_3$, $t > t_0$. Then, there are two possibilities:

(a) $m(t) = u(t, \beta(t))$,
(b) $m(t) = -u(t, \beta(t))$.

Let us consider the case (a). We have

$$u(t, x) \leqslant u(t, x_0),$$

if x is sufficiently close to $x_0 = \beta(t)$, and, therefore,

$$\frac{\partial u(t, \beta(t))}{\partial x} \geq 0.$$

As a result, it follows that

$$\left|\frac{\partial u}{\partial t}\right| \leq f(t, x_0, |u(t, x_0)|, |u_x(t, \beta(t))|).$$

Also, for $s < t$,

$$u(s, \beta(s)) < m(s) \quad \text{and} \quad |u(t, \beta(t))| = m(t).$$

Hence, we deduce, using (9.2.12), (9.2.13), and (9.2.14), that

$$D_- m(t) \leq D_- |u(t, \beta(t))|$$

$$= \frac{\partial u}{\partial t} + \beta'(t)\frac{\partial u}{\partial x}$$

$$\leq f(t, x_0, m(t), u_x(t, x_0)) + \beta'(t)\frac{\partial u}{\partial x}$$

$$\leq f(t, x_0, m(t), 0)$$

$$\leq g(t, m(t)).$$

If (b) holds, we notice that

$$-u(t, x) \leq -u(t, \beta(t)),$$

and hence $\partial u(t, x_0)/\partial x \leq 0$. Consequently,

$$\left|\frac{\partial u}{\partial t}\right| \leq f(t, x_0, m(t), -u_x(t, \beta(t))).$$

On the other hand,

$$-u(s, \beta(s)) < m(s), \quad s < t, \quad \text{and} \quad -u(t, \alpha(t)) = m(t).$$

This implies, as before, using (9.2.11), (9.2.12), and (9.2.14), that

$$D_- m(t) \leq -\frac{\partial u}{\partial t} - \beta'(t)\frac{\partial u}{\partial x} \leq \left|\frac{\partial u}{\partial t}\right| - \beta'(t)\frac{\partial u}{\partial x}$$

$$\leq f(t, x_0, m(t), -u_x(t, \alpha(t))) - \beta'(t)\frac{\partial u}{\partial x}$$

$$\leq f(t, x_0, m(t), 0)$$

$$\leq g(t, m(t)).$$

9.2. COMPARISON THEOREMS

It therefore follows that, in any case, the assumption $r(t, t_0, y_0) < m(t)$ implies the differential inequality

$$D_- m(t) \leqslant g(t, m(t)).$$

The proof for the case $(t, x_0) \in \partial E_2$, $t > t_0$, is similar. By Theorem 1.4.1, we have $m(t) \leqslant r(t, t_0, y_0)$, which contradicts the assumption $m(t) > r(t, t_0, y_0)$. Hence, the conclusion (9.2.15) is valid, and the proof is complete.

Theorem 9.2.2 may be proved by reducing it to Lemma 9.1.1 as follows.

Second proof of Theorem 9.2.2. Let $y(t, \epsilon) = y(t, t_0, y_0, \epsilon)$ be a solution of

$$y' = g(t, y) + \epsilon, \qquad y(t_0) = y_0 + \epsilon, \tag{9.2.16}$$

for sufficiently small $\epsilon > 0$. It is then enough to show that

$$|u(t, x)| < y(t, \epsilon) \quad \text{on} \quad E,$$

since $\lim_{\epsilon \to 0} y(t, \epsilon) \equiv r(t, t_0, y_0)$. Let us first show that

$$u(t, x) < y(t, \epsilon) \quad \text{on} \quad E.$$

Set

$$m(t, x) = u(t, x) - y(t, \epsilon)$$

so that

$$m(t, x) < 0 \quad \text{on} \quad \partial E_1.$$

Suppose that $(t_1, x_1) \in \partial E_3$, $m(t_1, x_1) = 0$, and $m_x(t_1, x_1) \geqslant 0$, which implies

$$u(t_1, x_1) = y(t_1, \epsilon) > 0 \quad \text{and} \quad u_x(t_1, x_1) \geqslant 0.$$

Hence, using (9.2.11), (9.2.13), and (9.2.14), we arrive at the inequality

$$\begin{aligned}
m_t(t_1, x_1) &= u_t(t_1, x_1) - y'(t_1, \epsilon) \\
&\leqslant f(t_1, x_1, u(t_1, x_1), u_x(t_1, x_1)) - g(t_1, y(t_1, \epsilon)) - \epsilon \\
&\leqslant -\beta'(t_1) u_x(t_1, x_1) - \epsilon \\
&= -\beta'(t_1) m_x(t_1, x_1) - \epsilon,
\end{aligned}$$

which leads to

$$m_t(t_1, x_1) + \beta'(t_1) m_x(t_1, x_1) < 0.$$

It turns out that we have so far verified conditions (i), (v), and (iv) of Lemma 9.1.1. The remaining assumptions of Lemma 9.1.1 may easily be checked. This implies, by Lemma 9.1.1, that

$$u(t, x) < y(t, \epsilon) \quad \text{on} \quad E. \tag{9.2.17}$$

Defining $n(t, x) = -[y(t, \epsilon) + u(t, x)]$ and proceeding in a similar way, we can show that $n(t, x)$ satisfies the hypotheses of Lemma 9.1.1. Consequently, we obtain

$$-u(t, x) < y(t, \epsilon) \quad \text{on} \quad E. \tag{9.2.18}$$

The relations (9.2.17) and (9.2.18) together imply that

$$|u(t, x)| < y(t, \epsilon) \quad \text{on} \quad E,$$

and this proves the theorem, as observed earlier.

Finally, we state the following useful comparison theorem.

THEOREM 9.2.3. Let $m \in C[E, R_+], f \in C[E \times R_+ \times R, R]$, satisfying the following assumptions:

(i) $m(t, x)$ possesses continuous partial derivatives u_t, u_x on E and total derivative on $\partial E_2 \cup \partial E_3$;

(ii) for $(t, x) \in E$,

$$\frac{\partial m(t, x)}{\partial t} \leqslant f(t, x, m(t, x), m_x(t, x)); \tag{9.2.19}$$

(iii) the relations (9.2.12), (9.2.13), and (9.2.14) hold.

Then, the inequality $m(t, x) \leqslant y_0$ on ∂E_1 implies

$$m(t, x) \leqslant r(t, t_0, y_0) \quad \text{on} \quad E,$$

where $r(t, t_0, y_0)$ is the maximal solution of (9.2.5).

Proof. Consider the function

$$v(t, x) = m(t, x) - y(t, \epsilon),$$

where $y(t, \epsilon)$ is any solution of (9.2.16) for small $\epsilon > 0$.

We wish to show that $v(t, x)$ satisfies the conditions of Lemma 9.1.1. Clearly, assumptions (i) and (v) of Lemma 9.1.1 hold. Suppose that $(t_1, x_1) \in \partial E_3$, $v(t_1, x_1) = 0$, and $v_x(t_1, x_1) \geqslant 0$, which assures that

$m_x(t_1, x_1) \geq 0$. Hence it follows, in view of the assumptions (9.2.13), (9.2.14), and (9.2.19), that

$$v_t(t_1, x_1) = m_t(t_1, x_1) - y'(t_1, \epsilon)$$
$$\leq f(t_1, x_1, m(t_1, x_1), m_x(t_1, x_1)) - g(t_1, y(t_1, \epsilon)) - \epsilon$$
$$\leq -\beta'(t_1) m_x(t_1, x_1) - \epsilon$$
$$= -\beta'(t_1) v_x(t_1, x_1) - \epsilon,$$

from which we infer the inequality

$$v_t(t_1, x_1) + \beta'(t_1) v_x(t_1, x_1) < 0.$$

This verifies condition (iv) of Lemma 9.1.1. Since it is easy to verify the other conditions of Lemma 9.1.1, the proof of the theorem is complete by the application of Lemma 9.1.1.

COROLLARY 9.2.1. *Under the assumptions of Theorem 9.2.3, if $m(t, x) \leq 0$ on ∂E_1 and if the maximal solution $r(t, t_0, 0)$ is assumed to be the trivial solution of (9.2.5), then $m(t, x) \leq 0$ on E.*

9.3. Upper bounds

In this section, we shall obtain upper bounds of solutions of partial differential equations of first order:

$$u_t = f(t, x, u, u_x). \quad (9.3.1)$$

DEFINITION 9.3.1. Any function $u(t, x)$ is said to be a solution of (9.3.1) if the following conditions hold:

(i) $u \in C[E, R]$, $u(t, x)$ possesses continuous partial derivatives on E_0 and total derivative on $\partial E_2 \cup \partial E_3$;
(ii) $u(t_0, x) = \phi(x)$, where $\phi(x)$ is continuous on $\alpha(t_0) \leq x \leq \beta(t_0)$;
(iii) $u(t, x)$ satisfies the equation (9.3.1) for $(t, x) \in E_0$.

The following theorem gives an estimate on the growth of solutions of Eq. (9.3.1).

THEOREM 9.3.1. *Let $f \in C[E \times R \times R, R]$,*

$$f(t, x, z, p) - f(t, x, z, 0) \geq -\alpha'(t)p, \quad p \geq 0, \quad \text{on} \quad \partial E_2, \quad (9.3.2)$$

and

$$f(t, x, z, p) - f(t, x, z, 0) \leq -\beta'(t)p, \quad p \geq 0, \quad \text{on} \quad \partial E_3. \quad (9.3.3)$$

Assume that
$$f(t, x, u, 0) \leq g(t, u), \quad u \geq 0, \tag{9.3.4}$$
where $g \in C[J \times R_+, R]$ and $g(t, 0) \equiv 0$. Suppose that $r(t) = r(t, t_0, y_0)$ is the maximal solution of (9.2.5) existing for $t \geq t_0$. If $u(t, x)$ is any solution of (9.3.1) such that
$$|\phi(x)| \leq y_0 \quad \text{on} \quad \partial E_1, \tag{9.3.5}$$
we have
$$|u(t, x)| \leq r(t) \quad \text{on} \quad E. \tag{9.3.6}$$

Proof. Let $u(t, x)$ be any solution of (9.3.1) satisfying (9.3.5), and let $y(t, \epsilon) = y(t, t_0, y_0, \epsilon)$ be a solution of (9.2.16) for small $\epsilon > 0$. Since $\lim_{\epsilon \to 0} y(t, \epsilon) = r(t)$, it turns out that it is enough to show that
$$|u(t, x)| < y(t, \epsilon) \quad \text{on} \quad E, \tag{9.3.7}$$
to prove (9.3.6). In fact, we prove
$$-u(t, x) < y(t, \epsilon) \quad \text{and} \quad u(t, x) < y(t, \epsilon) \quad \text{on} \quad E,$$
so that (9.3.7) holds. First of all, consider the function
$$m(t, x) = u(t, x) - y(t, \epsilon).$$
It is clear that conditions (i) and (v) of Lemma 9.1.1 are satisfied. Let us suppose that
$$(t_1, x_1) \in \partial E_3, \quad m(t_1, x_1) = 0, \quad \text{and} \quad m_x(t_1, x_1) \geq 0. \tag{9.3.8}$$
Hence, it follows, on account of (9.3.3) and (9.3.4), that
$$\begin{aligned} m_t(t_1, x_1) &= u_t(t_1, x_1) - y'(t_1, \epsilon) \\ &= f(t_1, x_1, u(t_1, x_1), u_x(t_1, x_1)) - g(t_1, y(t_1, \epsilon)) - \epsilon \\ &\leq -\beta'(t_1) u_x(t_1, x_1) - \epsilon \\ &= -\beta'(t_1) m_x(t_1, x_1) - \epsilon. \end{aligned}$$
Consequently, we have
$$m_t(t_1, x_1) + \beta'(t_1) m_x(t_1, x_1) < 0,$$
verifying condition (iv) of Lemma 9.1.1. The verification of the remaining assumptions is similar. Thus, by Lemma 9.1.1,
$$u(t, x) < y(t, \epsilon) \quad \text{on} \quad E.$$

We consider next the function

$$m(t, x) = -[y(t, \epsilon) + u(t, x)],$$

and, proceeding as before, it can be shown that $m(t, x)$ satisfies the hypotheses of Lemma 9.1.1, so that

$$-u(t, x) < y(t, \epsilon) \quad \text{on} \quad E.$$

The proof is therefore complete, as observed earlier.

COROLLARY 9.3.1. *The conclusion of Theorem 9.3.1 remains true even if we replace (9.3.4) by*

$$|f(t, x, u, 0)| \leqslant g(t, |u|).$$

THEOREM 9.3.2. *Suppose that, instead of (9.3.4), we have the weaker assumption*

$$f(t, x, u, 0) \leqslant g(t, u), \tag{9.3.9}$$

for $u \in \Omega$, where Ω is defined by

$$\Omega = [u: r(t) < u < r(t) + \epsilon_0, t \geqslant t_0]$$

for some $\epsilon_0 > 0$. Then, the assertion of Theorem 9.3.1 is valid.

Proof. Let $[t_0, T]$ be any given compact interval. Then, by Lemma 1.3.1, the maximal solution $r(t, \epsilon) = r(t, t_0, y_0, \epsilon)$ of (9.2.16) exists on $[t_0, T]$, for all sufficiently small $\epsilon > 0$ and

$$\lim_{\epsilon \to 0} r(t, \epsilon) = r(t),$$

uniformly on $[t_0, T]$. As a result, there exists an $\epsilon_0 > 0$ such that $r(t, \epsilon) < r(t) + \epsilon_0$ for $t \in [t_0, T]$. Furthermore, we have, by Theorem 1.2.1,

$$r(t) < r(t, \epsilon), \quad t \in [t_0, T],$$

which implies the inequality

$$r(t) < r(t, \epsilon) < r(t) + \epsilon_0, \quad t \in [t_0, T]. \tag{9.3.10}$$

To prove (9.3.6), it is sufficient to show that

$$|u(t, x)| < r(t, \epsilon) \quad \text{on} \quad E.$$

We proceed to show, as in Theorem 9.3.1, that

$$u(t, x) < r(t, \epsilon) \quad \text{and} \quad -u(t, x) < r(t, \epsilon) \quad \text{on} \quad E.$$

Following the proof of Theorem 9.3.1, we wish to verify the assumptions of Lemma 9.1.1. In verifying assumption (iv) for

$$m(t, x) = u(t, x) - r(t, \epsilon),$$

we arrive at the step

$$u(t_1, x_1) = r(t_1, \epsilon) \quad \text{and} \quad u_x(t_1, x_1) \geq 0,$$

analogous to (9.3.8). This, together with (9.3.10), leads to the fact $u = u(t_1, x_1) \in \Omega$, and, hence, we use the weaker assumption (9.3.9) to show that

$$m_t(t_1, x_1) + \beta'(t_1) m_x(t_1, x_1) < 0,$$

as before. With this change in the argument, the rest of the proof is similar.

Instead of the maximal solution of (9.2.5), if we know a positive solution of the differential inequality

$$y' > g(t, y), \quad y(t_0) = y_0 > 0, \tag{9.3.11}$$

it is easy to prove the following result.

THEOREM 9.3.3. Let $f \in C[E \times R \times R, R]$, and let the boundary conditions (9.3.2) and (9.3.3) hold. Assume that $y(t) > 0$ is a solution of (9.3.11). Suppose that

$$f(t, x, u, 0) \leq g(t, u), \tag{9.3.12}$$

for $u \in \Omega^*$, where

$$\Omega^* = [u : u = y(t), t \geq t_0].$$

Then, (9.3.5) implies

$$|u(t, x)| < y(t) \quad \text{on} \quad E.$$

Proof. All that is necessary to show is that

$$u(t, x) < y(t) \quad \text{and} \quad -u(t, x) < y(t) \quad \text{on} \quad E.$$

While verifying the assumptions of Lemma 9.1.1, with

$$m(t, x) = u(t, x) - y(t),$$

we arrive at

$$u(t_1, x_1) = y(t_1) \quad \text{and} \quad u_x(t_1, x_1) \geq 0$$

This implies that $u = u(t_1, x_1) \in \Omega^*$, and, hence, we can use (9.3.12) to verify the hypotheses of Lemma 9.1.1. With this observation, the rest of the proof runs similar to the proof of Theorem 9.3.1.

The next theorem allows us to estimate the difference of any solution of (9.3.1) and any solution of

$$v_t = F(t, x, v, v_x). \tag{9.3.13}$$

THEOREM 9.3.4. Let $f, F \in C[E \times R \times R, R]$, and

$$|f(t, x, u, p) - F(t, x, v, p)| \leq g(t, |u - v|), \tag{9.3.14}$$

where $g \in C[E \times R_+, R_+]$ and $g(t, 0) \equiv 0$. Assume that $r(t) = r(t, t_0, y_0)$ is the maximal solution of (9.2.5) existing for $t \geq t_0$. Suppose that

$$f(t, x, z, p) - f(t, x, z, q) \geq -\alpha'(t)(p - q), \quad p \geq q, \quad \text{on} \quad \partial E_2,$$

and

$$F(t, x, z, p) - F(t, x, z, q) \leq -\beta'(t)(p - q), \quad p \geq q, \quad \text{on} \quad \partial E_3.$$

Let $u(t, x)$, $v(t, x)$ be any two solutions of (9.3.1) and (9.3.13), respectively, such that $u(t_0, x) = \phi(x)$, $v(t_0, x) = \psi(x)$, on ∂E_1, and

$$|\phi(x) - \psi(x)| \leq y_0 \quad \text{on} \quad \partial E_1.$$

Under these assumptions,

$$|u(t, x) - v(t, x)| \leq r(t) \quad \text{on} \quad E.$$

The proof is very much the same as the proof of Theorem 9.3.1.

COROLLARY 9.3.2. The function $F(t, x, u, u_x) \equiv f(t, x, u, u_x)$ is admissible in Theorem 9.3.4 to yield an estimate for the difference of any two solutions of (9.3.1). In fact, instead of (9.3.14),

$$f(t, x, u, p) - f(t, x, v, p) \leq g(t, u - v), \quad u \geq v,$$

is sufficient.

We shall next consider *a priori* bounds for solutions of (9.3.1) in terms of solutions of another partial differential equation. Sometimes, this has an advantage of obtaining sharper bounds.

THEOREM 9.3.5. Let $f \in C[E \times R \times R, R]$, and

$$|f(t, x, u, u_x)| \leqslant G(t, x, |u|, |u_x|),$$

where $G \in C[E \times R_+ \times R_+, R_+]$. Suppose that

$$G(t, x, z, p) - G(t, x, z, q) \geqslant -\alpha'(t)(p - q), \qquad p \geqslant q, \qquad \text{on} \qquad \partial E_2,$$
(9.3.15)

and

$$G(t, x, z, p) - G(t, x, z, q) \leqslant -\beta'(t)(p - q), \qquad p \geqslant q, \qquad \text{on} \qquad \partial E_3.$$
(9.3.16)

Assume that

$$G(t, x, z_1, p) - G(t, x, z_2, p) \leqslant g(t, z_1 - z_2), \qquad z_1 \geqslant z_2, \quad (9.3.17)$$

where $g \in C[J \times R_+, R]$, and the maximal solution $r(t, t_0, 0)$ of (9.1.7) is identically zero. Let $u(t, x)$ be a solution of (9.3.1) and $z(t, x) \geqslant 0$ be a solution of

$$z_t = G(t, x, z, z_x), \tag{9.3.18}$$

such that $z_x(t, x) \geqslant 0$ and

$$|u(t, x)| \leqslant z(t, x) \qquad \text{on} \qquad \partial E_1.$$

Under these assumptions, we have

$$|u(t, x)| \leqslant z(t, x) \qquad \text{on} \qquad E.$$

Proof. Let $u(t, x)$ be any solution of (9.3.1). We shall show that

$$|u(t, x)| < z(t, x) + y(t, \epsilon) \qquad \text{on} \qquad E,$$

where $y(t, \epsilon) = y(t, t_0, 0, \epsilon)$ is any solution of (9.1.10), for sufficiently small $\epsilon > 0$.

First of all, define the function

$$m(t, x) = u(t, x) - z(t, x) - y(t, \epsilon).$$

Evidently, $m(t, x) < 0$ on ∂E_1. Suppose that $(t_1, x_1) \in \partial E_3$, $m(t_1, x_1) = 0$, and $m_x(t_1, x_1) \geqslant 0$. We then derive that

$$u(t_1, x_1) - z(t_1, x_1) = y(t_1, \epsilon) > 0$$

and

$$u_x(t_1, x_1) \geqslant z_x(t_1, x_1).$$

9.3. UPPER BOUNDS

It therefore follows that

$$m_t(t_1, x_1) \leq |f(t_1, x_1, u(t_1, x_1), u_x(t_1, x_1))|$$
$$- G(t_1, x_1, z(t_1, x_1), z_x(t_1, x_1)) - g(t_1, y(t_1, \epsilon)) - \epsilon$$
$$\leq G(t_1, x_1, u(t_1, x_1), u_x(t_1, x_1)) - G(t_1, x_1, z(t_1, x_1), u_x(t_1, x_1))$$
$$+ G(t_1, x_1, z(t_1, x_1), u_x(t_1, x_1)) - G(t_1, x_1, z(t_1, x_1), z_x(t_1, x_1))$$
$$- g(t_1, y(t_1, \epsilon)) - \epsilon$$
$$\leq -\beta'(t_1) m_x(t_1, x_1) - \epsilon.$$

Thus, we have

$$m_t(t_1, x_1) + \beta'(t_1) m_x(t_1, x_1) < 0,$$

verifying condition (iv) of Lemma 9.1.1. The other assumptions of Lemma 9.1.1 may be checked similarly. Thus, by Lemma 9.1.1, we have

$$u(t, x) < z(t, x) + y(t, \epsilon) \quad \text{on} \quad E.$$

On the other hand, defining

$$n(t, x) = -[z(t, x) + y(t, \epsilon) + u(t, x)],$$

it is easy to show that $n(t, x)$ satisfies the hypotheses of Lemma 9.1.1, and, consequently,

$$-u(t, x) \leq z(t, x) + y(t, \epsilon) \quad \text{on} \quad E.$$

It therefore turns out that

$$|u(t, x)| \leq z(t, x) \quad \text{on} \quad E,$$

since $\lim_{\epsilon \to 0} y(t, \epsilon) \equiv 0$. The proof is complete.

The following theorem may be proved, using the arguments similar to the preceding one with necessary changes.

THEOREM 9.3.6. *Let* $f, F \in C[E \times R \times R, R]$, *and*

$$|f(t, x, u, u_x) - F(t, x, v, v_x)| \leq G(t, x, |u - v|, |u_x - v_x|),$$

where $G \in C[E \times R_+ \times R_+, R_+]$ *satisfies the conditions* (9.3.15), (9.3.16), *and* (9.3.17) *of Theorem* 9.3.5. *Suppose that the maximal solution* $r(t, t_0, 0)$ *of* (9.1.7) *is identically zero. Let* $u(t, x)$, $v(t, x)$ *be any* solutions of (9.3.1) and (9.3.13), respectively. Then, if $z(t, x) \geqslant 0$ is the solution of (9.3.18) such that

$$|u(t, x) - v(t, x)| \leqslant z(t, x) \quad \text{on} \quad \partial E_1,$$

we have

$$|u(t, x) - v(t, x)| \leqslant z(t, x) \quad \text{on} \quad E.$$

9.4. Approximate solutions and uniqueness

We shall consider the partial differential inequality

$$|u_t - f(t, x, u, u_x)| \leqslant \delta. \tag{9.4.1}$$

DEFINITION 9.4.1. A function $u(t, x, \delta)$ is aid to be a δ-approximate solution of (9.3.1) if

(i) $u \in C[E, R]$, and $u(t, x, \delta)$ possesses continuous partial derivatives on E_0 and total derivative on $\partial E_2 \cup \partial E_3$;
(ii) $u(t_0, x, \delta) = \phi(x)$, where $\phi(x)$ is continuous on $\alpha(t_0) \leqslant x \leqslant \beta(t_0)$;
(iii) $u(t, x, \delta)$ satisfies the inequality (9.4.1) on E_0.

The following theorem estimates the error between a solution and a δ-approximate solution of (9.3.1).

THEOREM 9.4.1. Let $f \in C[E \times R \times R, R]$, and satisfy

$$f(t, x, u, p) - f(t, x, v, p) \leqslant g(t, u - v), \quad u > v, \tag{9.4.2}$$

where $g \in C[J \times R_+, R]$ and $g(t, 0) \equiv 0$. Assume that

$$r(t, \delta) = r(t, t_0, 0, \delta)$$

is the maximal solution of

$$y' = g(t, y) + \delta, \quad y(t_0) = 0,$$

existing for $t \geqslant t_0$, and the inequalities

$$f(t, x, z, p) - f(t, x, z, q) \geqslant -\alpha'(t)(p - q), \quad p \geqslant q, \quad \text{on} \quad \partial E_2, \tag{9.4.3}$$

$$f(t, x, z, p) - f(t, x, z, q) \leqslant -\beta'(t)(p - q), \quad p \geqslant q, \quad \text{on} \quad \partial E_3, \tag{9.4.4}$$

hold. Let $u(t, x)$, $u(t, x, \delta)$ be a solution and a δ-approximate solution of (9.3.1). Then,

$$|u(t, x) - u(t, x, \delta)| \leqslant r(t, \delta) \quad \text{on} \quad E. \tag{9.4.5}$$

9.4. APPROXIMATE SOLUTIONS AND UNIQUENESS

Proof. Consider the function
$$m(t, x) = u(t, x) - u(t, x, \delta) - y(t, \epsilon),$$
where $y(t, \epsilon) = y(t, t_0, 0, \delta, \epsilon)$ is a solution of
$$y' = g(t, y) + \delta + \epsilon, \quad y(t_0) = \epsilon,$$
for small $\epsilon > 0$. Suppose that, for $(t_1, x_1) \in \partial E_3$, $m(t_1, x_1) = 0$ and $m_x(t_1, x_1) \geq 0$. This shows that
$$u(t_1, x_1) > u(t_1, x_1, \delta)$$
and
$$u_x(t_1, x_1) \geq u_x(t_1, x_1, \delta).$$

Thus, using (9.4.2) and (9.4.4), we have

$$m_t(t_1, x_1) = f(t_1, x_1, u(t_1, x_1), u_x(t_1, x_1))$$
$$- f(t_1, x_1, u(t_1, x_1, \delta), u_x(t_1, x_1))$$
$$+ f(t_1, x_1, u(t_1, x_1, \delta), u_x(t_1, x_1))$$
$$- f(t_1, x_1, u(t_1, x_1, \delta), u_x(t_1, x_1, \delta))$$
$$- [u_t(t_1, x_1, \delta) - f(t_1, x_1, u(t_1, x_1, \delta), u_x(t_1, x_1, \delta))]$$
$$- g(t_1, y(t_1, \epsilon)) - \epsilon - \delta$$
$$\leq -\beta'(t_1) m_x(t_1, x_1) - \epsilon,$$

so that
$$m_t(t_1, x_1) + \beta'(t_1) m_x(t_1, x_1) < 0.$$

This proves condition (iv) of Lemma 9.1.1. It is easy to show that the other assumptions of Lemma 9.1.1 also hold. Hence, by Lemma 9.1.1,
$$u(t, x) - u(t, x, \delta) < y(t, \epsilon) \quad \text{on} \quad E.$$

Proceeding similarly, we can show that
$$u(t, x, \delta) - u(t, x) < y(t, \epsilon) \quad \text{on} \quad E.$$

The estimate (9.4.5) results immediately, noting that
$$\lim_{\epsilon \to 0} y(t, \epsilon) = r(t, t_0, 0, \delta).$$

The proof is complete.

COROLLARY 9.4.1. If the function $g(t, u) = Ku$, $K > 0$, then (9.4.5) takes the form

$$|u(t, x) - u(t, x, \delta)| \leq (\delta/K)[\exp K(t - t_0) - 1] \quad \text{on} \quad E.$$

We next state a uniqueness theorem of Perron type whose proof is an immediate consequence of Theorem 9.4.1 or Corollary 9.3.2.

THEOREM 9.4.2. Let $f \in C[E \times R \times R, R]$ and the condition (9.4.2) hold. Assume further that the boundary conditions (9.4.3) and (9.4.4) are satisfied. If $y(t) \equiv 0$ is the maximal solution of

$$y' = g(t, y), \quad y(t_0) = 0,$$

for $t \geq t_0$, then the partial differential equation (9.3.1) admits atmost one solution.

THEOREM 9.4.3. Under the assumptions of Theorem 9.4.2, given $\epsilon > 0$, there exists a $\delta(\epsilon) > 0$ such that

$$|\phi(x) - \psi(x)| < \delta \quad \text{on} \quad \partial E_1 \quad \text{implies} \quad |u(t, x) - v(t, x)| < \epsilon \quad \text{on} \quad E,$$

where $u(t, x)$, $v(t, x)$ are any two solutions of (9.3.1) satisfying $u(t_0, x) = \phi(x)$, $v(t_0, x) = \psi(x)$, respectively, provided the solutions $y(t, t_0, y_0)$ of (9.2.5) exist for $t \geq t_0$ and are continuous with respect to the initial values (t_0, y_0).

Proof. By Corollary 9.3.2, we have, choosing

$$y_0 = \max_{\partial E_1} |\phi(x) - \psi(x)|,$$

the inequality

$$|u(t, x) - v(t, x)| \leq r(t, t_0, y_0) \quad \text{on} \quad E.$$

By assumption, given $\epsilon > 0$, there exists $\delta = \delta(\epsilon) > 0$ such that, if $y_0 < \delta$, then $r(t, t_0, y_0) < \epsilon$, $t \geq t_0$, where $r(t, t_0, y_0)$ is the maximal solution of (9.2.5). The stated result follows at once.

9.5. Systems of partial differential inequalities of first order

Many of the results obtained in the earlier sections of this chapter can be extended to more general systems. In this section, we shall indicate some basic theorems only.

9.5. SYSTEMS OF INEQUALITIES

Let $\alpha, \beta \in C[J, R^n]$ such that $\alpha(t) < \beta(t)$, $t \in J$. Assume that $\alpha'(t), \beta'(t)$ exist and are continuous on J. For $t_0 \in J$, we may now define the sets $E, E_0, \partial E_1, \partial E_2$, and ∂E_3, following Sect. 9.1. First of all, we shall extend the Lemma 9.1.1.

LEMMA 9.5.1. Assume that

(i) $m \in C[E, R^N]$ and $m(t, x)$ possesses partial derivatives on E and total derivative on $\partial E_2 \cup \partial E_3$;

(ii) for some index j, $1 \leqslant j \leqslant N$, if $(t_1, x_1) \in E_0$, $m^j(t_1, x_1) = 0$, $m^i(t_1, x_1) \leqslant 0$, $i \neq j$, and $m_x^j(t_1, x_1) = 0$, then $m_t^j(t_1, x_1) < 0$;

(iii) for some index j, $1 \leqslant j \leqslant N$, if $(t_1, x_1) \in \partial E_2$, $m^j(t_1, x_1) = 0$, $m^i(t_1, x_1) \leqslant 0$, $i \neq j$, and $m_x^j(t_1, x_1) \leqslant 0$, then

$$m_t^j(t_1, x_1) + \sum_{k=1}^n \alpha_k'(t_1) m_{x_k}^j(t_1, x_1) < 0;$$

(iv) for some index j, $1 \leqslant j \leqslant N$, if $(t_1, x_1) \in \partial E_3$, $m^j(t_1, x_1) = 0$, $m^i(t_1, x_1) \leqslant 0$, $i \neq j$, and $m_x^j(t_1, x_1) \geqslant 0$, then

$$m_t^j(t_1, x_1) + \sum_{k=1}^n \beta_k'(t_1) m_{x_k}^j(t_1, x_1) < 0;$$

(v) $m(t, x) < 0$ on ∂E_1.

Then, we have

$$m(t, x) < 0 \quad \text{on} \quad E. \tag{9.5.1}$$

Proof. Suppose that the inequality (9.5.1) is false. Then, the set Z given by

$$Z = \bigcup_{i=1}^N [(t, x) \in E : m^i(t, x) \geqslant 0]$$

is nonempty. Take the projection of Z on t-axis and denote by t_1 its greatest lower bound. It follows from assumption (v) that $t_1 > t_0$. The set Z is closed, and, hence, we conclude that there is a point $(t_1, x_1) \in E$ and an index j, $1 \leqslant j \leqslant N$, such that

$$\begin{aligned} m^j(t_1, x_1) &= 0, \\ m^j(t, x) &< 0, \quad t < t_1, \quad (t, x) \in E, \end{aligned} \tag{9.5.2}$$

and, for $i \neq j$,

$$m^i(t_1, x) \leqslant 0.$$

Now, one of the following situations arises:

(a) $(t_1, x_1) \in \partial E_2$ and $m_x^j(t_1, x_1) \leq 0$;
(b) $(t_1, x_1) \in \partial E_3$ and $m_x^j(t_1, x_1) \geq 0$;
(c) $(t_1, x_1) \notin \partial E_2 \cup \partial E_3$ and $m_x^j(t_1, x_1) = 0$.

Suppose that (a) holds. Then,

$$\frac{d}{dt} m^j(t_1, \alpha(t_1)) = m_t^j(t_1, \alpha(t_1)) + \sum_{k=1}^{n} \alpha_k'(t_1) m_{x_k}^j(t_1, \alpha(t_1))$$

$$= m_t^j(t_1, x_1) + \sum_{k=1}^{n} \alpha_k'(t_1) m_{x_k}^j(t_1, x_1) < 0, \quad (9.5.3)$$

in view of condition (iii). On the other hand, it follows by (9.5.2) that

$$\frac{d}{dt} m^j(t_1, \alpha(t_1)) \geq 0,$$

which is incompatible with (9.5.3). Following now the proof of Lemma 9.1.1 and that just given, it is easy to obtain a contradiction for the cases (b) and (c) proving that the inequality (9.5.1) is true.

An extension of Theorem 9.1.1 to systems of inequalities is the following.

THEOREM 9.5.1. Let $f \in C[E \times R^N \times R^n, R^N]$, $u, v \in C[E, R^N]$ satisfy the following assumptions:

(i) u, v possess continuous partial derivatives on E and total derivatives on ∂E_2, ∂E_3;

(ii)
$$u_t^i \leq f^i(t, x, u, u_x^i),$$
$$v_t^i > f^i(t, x, v, v_x^i)$$

on E, $i = 1, 2, \ldots, N$;

(iii) for each i,

$$f^i(t, x, z, p) - f^i(t, x, z, q) \geq - \sum_{k=1}^{n} \alpha_k'(t)[p_k - q_k], \quad p \geq q, \quad \text{on} \quad \partial E_2;$$

(iv) for each i,

$$f^i(t, x, z, p) - f^i(t, x, z, q) \leq - \sum_{k=1}^{n} \beta_k'(t)[p_k - q_k], \quad p \geq q, \quad \text{on} \quad \partial E_3;$$

9.5. SYSTEMS OF INEQUALITIES

(v) $u(t, x) < v(t, x)$ on ∂E_1 ;

(vi) $f(t, x, u, p)$ is quasi-monotone in u for each fixed (t, x, p). These conditions imply that

$$u(t, x) < v(t, x) \quad \text{on} \quad E. \tag{9.5.4}$$

Proof. Consider the function

$$m(t, x) = u(t, x) - v(t, x).$$

Evidently, $m(t, x)$ verifies assumptions (i) and (v) of Lemma 9.5.1, on account of assumptions (i) and (v). Suppose that, for some index j, $1 \leqslant j \leqslant N$, $(t_1, x_1) \in \partial E_3$, we have $m^j(t_1, x_1) = 0$, $m^i(t_1, x_1) \leqslant 0, i \neq j$, and $m_x^j(t_1, x_1) \geqslant 0$. This implies

$$\begin{aligned} u^j(t_1, x_1) &= v^j(t_1, x_1), \\ u^i(t_1, x_1) &\leqslant v^i(t_1, x_1), \quad i \neq j, \end{aligned} \tag{9.5.5}$$

and

$$u_x^j(t_1, x_1) \geqslant v_x^j(t_1, x_1).$$

Consequently,

$$\begin{aligned} u_t^j(t_1, x_1) - v_t^j(t_1, x_1) &< f^j(t_1, x_1, u(t_1, x_1), u_x^j(t_1, x_1)) \\ &\quad - f^j(t_1, x_1, v(t_1, x_1), v_x^j(t_1, x_1)). \end{aligned}$$

Since $f(t, x, u, p)$ is quasi-monotone in u, we have, because of the relations (9.5.5), the inequality

$$\begin{aligned} f^j(t_1, x_1, u^1(t_1, x_1), &..., u^j(t_1, x_1), ..., u^N(t_1, x_1), u_x^j(t_1, x_1)) \\ &\leqslant f^j(t_1, x_1, v^1(t_1, x_1), ..., v^j(t_1, x_1), ..., v^N(t_1, x_1), v_x^j(t_1, x_1)). \end{aligned}$$

Hence, using condition (iv),

$$\begin{aligned} u_t^j(t_1, x_1) - v_t^j(t_1, x_1) &< f^j(t_1, x_1, u(t_1, x_1), u_x^j(t_1, x_1)) \\ &\quad - f^j(t_1, x_1, v(t_1, x_1), v_x^j(t_1, x_1)) \\ &\leqslant - \sum_{k=1}^{n} \beta_k'(t_1)[u_{x_k}^j(t_1, x_1) - v_{x_k}^j(t_1, x_1)], \end{aligned}$$

which implies that

$$m_t^j(t_1, x_1) + \sum_{k=1}^{n} \beta_k'(t_1) \, m_{x_k}^j(t_1, x_1) < 0.$$

This verifies assumption (iv) of Lemma 9.5.1. The checking of the remaining conditions is similar, and the stated result follows, by Lemma 9.5.1.

Making use of a similar argument, we can prove the following theorem, which generalizes Theorem 9.1.2.

THEOREM 9.5.2. Let the assumptions of Theorem 9.5.1 hold, except that condition (ii) is replaced by

$$u_t^i \leqslant f^i(t, x, u, u_x^i),$$
$$v_t^i \geqslant f^i(t, x, v, v_x^i) \quad \text{on} \quad E \quad (i = 1, 2, ..., N).$$

Assume further that, for each i,

$$f^i(t, x, u, p) - f^i(t, x, v, p) \leqslant g(t, u^i - v^i), \quad u^i > v^i,$$

where $g \in C[J \times R_+, R]$, and the maximal solution $r(t, t_0, 0)$ of (9.1.7) is identically zero. Then,

$$u(t, x) \leqslant v(t, x) \quad \text{on} \quad E$$

whenever

$$u(t, x) \leqslant v(t, x) \quad \text{on} \quad \partial E_1.$$

We shall next prove a comparison theorem analogous to Theorem 9.2.3.

THEOREM 9.5.3. Let $m \in C[E, R_+^N]$, $f \in C[E \times R^N \times R^n, R^N]$ such that the following conditions hold:

(i) $m(t, x)$ possesses continuous partial derivatives on E and total derivative on $\partial E_2 \cup \partial E_3$;
(ii) for $(t, x) \in E$ and $i = 1, 2, ..., N$,

$$m_t^i(t, x) \leqslant f^i(t, x, m(t, x), m_x^i(t, x)); \tag{9.5.6}$$

(iii)
$$f(t, x, u, 0) \leqslant g(t, u), \tag{9.5.7}$$

where $g \in C[J \times R_+^N, R^N]$, $g(t, 0) \equiv 0$, and $g(t, u)$ is quasi-monotone in u for each $t \in J$;
(iv) for each i,

$$f^i(t, x, u, p) - f^i(t, x, u, 0) \geqslant -\sum_{k=1}^{n} \alpha_k'(t) p_k, \quad p \geqslant 0, \quad \text{on} \quad \partial E_2, \tag{9.5.8}$$

$$f^i(t, x, u, p) - f^i(t, x, u, 0) \leqslant -\sum_{k=1}^{n} \beta_k'(t) p_k, \quad p \geqslant 0, \quad \text{on} \quad \partial E_3. \tag{9.5.9}$$

Then, $m(t, x) \leq y_0$, on ∂E_1, implies
$$m(t, x) \leq r(t, t_0, y_0) \quad \text{on} \quad E,$$
where $r(t, t_0, y_0)$ is the maximal solution of the differential system
$$y' = g(t, y), \quad y(t_0) = y_0,$$
existing for $t \geq t_0$.

Proof. Let $y(t, \epsilon) = y(t, t_0, y_0, \epsilon)$ be any solution of the ordinary differential system
$$y' = g(t, y) + \epsilon, \quad y(t_0) = y_0 + \epsilon,$$
for sufficiently small vector $\epsilon > 0$. Since it is known that
$$\lim_{\epsilon \to 0} y(t, \epsilon) = r(t, t_0, y_0),$$
it is sufficient to show that
$$m(t, x) < y(t, \epsilon) \quad \text{on} \quad E. \tag{9.5.10}$$
It is easy to verify that $v(t, x) = m(t, x) - y(t, \epsilon)$ satisfies the hypotheses of Lemma 9.5.1. Clearly, assumptions (i) and (v) hold. Moreover, for an index j, $1 \leq j \leq N$, $(t_1, x_1) \in \partial E_3$, suppose that $v^j(t_1, x_1) = 0$, $v^i(t_1, x_1) \leq 0$, $i \neq j$, and $v_x{}^j(t_1, x_1) \geq 0$.
  Then,
$$m^j(t_1, x_1) = y^j(t_1, \epsilon) > 0, \quad m^i(t_1, x_1) \leq y^i(t_1, \epsilon), \quad i \neq j,$$
and
$$m_x{}^j(t_1, x_1) \geq 0.$$
Thus, using the relations (9.5.6), (9.5.7), (9.5.8), and the quasi-monotone character of $g(t, u)$, we get

$$\begin{aligned}
m_t{}^j(t_1, x_1) &- dy^j/dt(t_1, \epsilon) \\
&\leq f^j(t_1, x_1, m(t_1, x_1), m_x{}^j(t_1, x_1)) - g^j(t_1, y(t_1, \epsilon)) - \epsilon^j \\
&\leq f^j(t_1, x_1, m(t_1, x_1), m_x{}^j(t_1, x_1)) \\
&\quad - g^j(t_1, m^1(t_1, x_1),..., y^j(t_1, \epsilon),..., m^N(t_1, x_1)) - \epsilon^j \\
&\leq f^j(t_1, x_1, m(t_1, x_1), m_x{}^j(t_1, x_1)) \\
&\quad - f^j(t_1, x_1, m(t_1, x_1), 0) + f^j(t_1, x_1, m(t_1, x_1), 0) \\
&\quad - g^j(t_1, m(t_1, x_1)) - \epsilon^j \\
&\leq - \sum_{k=1}^{n} \beta_k{'}(t_1) m_{x_k}^j(t_1, x_1) - \epsilon^j \\
&= - \sum_{k=1}^{n} \beta_k{'}(t_1) v_{x_k}^j(t_1, x_1) - \epsilon^j.
\end{aligned}$$

This assures that

$$v_t^j(t_1, x_1) + \sum_{k=1}^{n} \beta_k'(t_1) v_{x_k}^j(t_1, x_1) < 0,$$

proving that hypothesis (iv) of Lemma 9.5.1 is valid. Arguing similarly, it is easy to verify the assumptions of Lemma 9.5.1, and consequently (9.5.10) is true. This proves the theorem.

Following the proofs just presented, the growth estimates of solutions and approximate solutions and a uniqueness result for the systems of the type

$$u_t^i = f^i(t, x, u, u_x^i) \quad (i = 1, 2, ..., N),$$

where $f \in C[E \times R^N \times R^n, R^N]$, may be proved. We leave it as an exercise to the reader.

Finally, we shall give below a theorem on differential inequalities for overdetermined systems of inequalities. For such systems, the time variable t also may be a vector, such that $t_\gamma^0 \leq t_\gamma < \infty$, $\gamma = 1, 2, ..., k$. The sets E, E_0, ∂E_1, ∂E_2, and ∂E_3 may be understood with this change. Then, we have

THEOREM 9.5.4. Let $f \in C[E \times R^N \times R^n, R^N]$, $u, v \in C[E, R^N]$ satisfy the following conditions:

(i) $u(t, x)$, $v(t, x)$ possess continuous partial derivatives on E and total derivative on $\partial E_2 \cup \partial E_3$;

(ii)
$$u_{t_\gamma}^i \leq f_\gamma^i(t, x, u, u_x^i),$$
$$v_{t_\gamma}^i > f_\gamma^i(t, x, v, v_x^i),$$

on E, $i = 1, 2, ..., N$, $\gamma = 1, 2, ..., k$;

(iii) for each i and γ,

$$f_\gamma^i(t, x, z, p) - f_\gamma^i(t, x, z, q) \geq -\sum_{j=1}^{n} \alpha_j'(t)[p_j - q_j], \quad p \geq q, \quad \text{on} \quad \partial E_2,$$

$$f_\gamma^i(t, x, z, p) - f_\gamma^i(t, x, z, q) \leq -\sum_{j=1}^{n} \beta_j'(t)[p_j - q_j], \quad p \geq q, \quad \text{on} \quad \partial E_3;$$

(iv) the function $f(t, x, u, p)$ satisfies the quasi-monotone property in u for each (t, x, p);

(v) $u(t, x) < v(t, x)$ on ∂E_1.

9.5. SYSTEMS OF INEQUALITIES

Under these assumptions,
$$u(t, x) < v(t, x) \quad \text{on} \quad E.$$

Proof. Introduce Mayer's transformation
$$t_\gamma = t_\gamma^0 + \lambda_\gamma s,$$
where $\lambda_\gamma \geqslant 0$ ($\gamma = 1, 2,..., k$). Define the functions
$$\tilde{u}(s, x, \lambda) = u(t_0 + \lambda s, x),$$
$$\tilde{v}(s, x, \lambda) = v(t_0 + \lambda s, x).$$

By assumption (v), it follows that
$$\tilde{u}(0, x, \lambda) < \tilde{v}(0, x, \lambda),$$
and the functions $\tilde{u}$, $\tilde{v}$ are defined on
$$E = [(s, x): 0 \leqslant s < \infty, \alpha(s) \leqslant x \leqslant \beta(s)],$$
where s is a single variable. Moreover,
$$\tilde{u}_s = \sum_{\gamma=1}^{k} \lambda_\gamma u_{t_\gamma}(t_0 + \lambda s, x),$$
$$\tilde{v}_s = \sum_{\tilde{\gamma}=1}^{k} \lambda_{\tilde{\gamma}} v_{t_{\tilde{\gamma}}}(t_0 + \lambda s, x).$$

We define also the functions for each i,
$$F^i(s, x, \tilde{u}, p) = \sum_{\gamma=1}^{k} \lambda_\gamma f_\gamma^i(t_0 + \lambda s, x, u, p),$$
so that the inequalities (ii) reduce to
$$\tilde{u}_s \leqslant F^i(s, x, \tilde{u}, \tilde{u}_x^i),$$
$$\tilde{v}_s > F^i(s, x, \tilde{v}, \tilde{v}_x^i) \quad \text{on} \quad E.$$

Evidently, $F(s, x, z, p)$ satisfies quasi-monotone property in z for each (t, x, p). Furthermore, the inequalities (iii) reduce to
$$F^i(s, x, z, p) - F^i(s, x, z, q) \geqslant -k \sum_{j=1}^{n} \alpha_j'(t)[p_j - q_j], \quad p \geqslant q, \quad \text{on} \quad \partial E_2$$
and
$$F^i(s, x, z, p) - F^i(s, x, z, q) \leqslant -k \sum_{j=1}^{n} \beta_j'(t)[p_j - q_j], \quad p \geqslant q, \quad \text{on} \quad \partial E_3.$$

It then follows that the functions $\tilde{u}$, $\tilde{v}$, and F satisfy the assumptions of Theorem 9.5.1, and, consequently,

$$\tilde{u}(s, x, \lambda) < \tilde{v}(s, x, \lambda) \quad \text{on} \quad E,$$

and, in particular, for $s = 1$,

$$\tilde{u}(1, x, \lambda) < \tilde{v}(1, x, \lambda).$$

Let, now, (t, x) be an arbitrary point, and let

$$\lambda = (t - t_0) = (t_1 - t_1^0, ..., t_k - t_k^0)$$

be scuh that $\lambda_k \geqslant 0$. Then, on E, we have

$$u(t, x) = \tilde{u}(1, x, t - t_0) < \tilde{v}(1, x, t - t_0) = v(t, x),$$

which is exactly the relation we have to prove. This completes the proof.

One could formulate and prove analogous results for systems of the type

$$u_{t_j}^i = F_j^i(t, x, u, u_x^i),$$

on the strength of Theorem 9.5.4. We shall not attempt such a formulation of the results.

9.6. Lyapunov-like function

Let us consider a first-order partial differential system of the form

$$u_t = f(t, x, u, u_x^i), \tag{9.6.1}$$

where $f \in C[E \times R^N \times R^n, R^N]$. We wish to estimate the growth of solutions of (9.6.1) by means of a Lyapunov-like function. To this end, we have

THEOREM 9.6.1.   Assume that

(i) $V \in C[J \times R^N, R_+]$, $V(t, u)$ possesses continuous partial derivatives with respect to t and the components of u, and

$$\frac{\partial V(t, u)}{\partial t} + \frac{\partial V(t, u)}{\partial u} \cdot f(t, x, u, u_x^i) \leqslant G(t, x, V(t, u), V_x(t, u)),$$

where $G \in C[E \times R_+ \times R, R]$ and $V_x(t, u) = (\partial V/\partial u) \cdot (\partial u/\partial x)$;

9.6. LYAPUNOV-LIKE FUNCTION

(ii) $G(t, x, z, p) - G(t, x, z, 0) \geq -\alpha'(t)p$, $p \geq 0$, on ∂E_2;
(iii) $G(t, x, z, p) - G(t, x, z, 0) \leq -\beta'(t)p$, $p \geq 0$, on ∂E_3;
(iv) $G(t, x, z, 0) \leq g(t, z)$, $z > 0$, where $g \in C[J \times R_+, R]$;
(v) the maximal solution $r(t, t_0, y_0)$ of (9.2.5) exists for $t \geq t_0$.

Then, any solution $u(t, x)$ of (9.6.1) satisfying

$$V(t_0, \phi(x)) \leq y_0 \quad \text{on} \quad \partial E_1 \quad (9.6.2)$$

allows the estimate

$$V(t, u(t, x)) \leq r(t, t_0, y_0) \quad \text{on} \quad E. \quad (9.6.3)$$

Proof. Let $u(t, x)$ be any solution of (9.6.1) such that (9.6.2) holds. Consider the function

$$m(t, x) = V(t, u(t, x)).$$

By assumption (i), we have

$$\frac{\partial m(t, x)}{\partial t} = \frac{\partial V(t, u(t, x))}{\partial t} + \frac{\partial V(t, u(t, x))}{\partial u} \cdot f(t, x, u(t, x), u_x{}^i(t, x))$$

$$\leq G(t, x, m(t, x), m_x(t, x)),$$

and

$$m(t_0, x) \leq y_0 \quad \text{on} \quad \partial E_1.$$

It is evident that the hypotheses of Theorem 9.2.3 are fulfilled, and, as a result,

$$V(t, u(t, x)) = m(t, x) \leq r(t, t_0, y_0) \quad \text{on} \quad E,$$

where $r(t, t_0, y_0)$ is the maximal solution of (9.2.5). The proof is therefore complete.

THEOREM 9.6.2. *Assume that*

(i) the assumption (i) of Theorem 9.6.1 is satisfied;
(ii) $G(t, x, z, p) - G(t, x, z, q) \geq -\alpha'(t)(p - q)$, $p \geq q$, on ∂E_2;
(iii) $G(t, x, z, p) - G(t, x, z, q) \leq -\beta'(t)(p - q)$, $p \geq q$, on ∂E_3;
(iv) $G(t, x, z_1, p) - G(t, x, z_2, p) \leq g(t, z_1 - z_2)$, $z_1 \geq z_2$;
(v) the maximal solution $r(t, t_0, 0)$ of (9.1.7) is identically zero.

Then, if $z(t, x) \geq 0$ is the solution of (9.3.18) such that $z(t_0, x) = \psi(x) \geq 0$ on ∂E_1 and

$$V(t_0, \phi(x)) \leq \psi(x) \quad \text{on} \quad \partial E_1,$$

we have
$$V(t, u(t, x)) \leqslant z(t, x) \quad \text{on} \quad E.$$

Proof. If $u(t, x)$ is any solution of (9.6.1), we obtain, as in Theorem 9.6.1, the inequality
$$\frac{\partial m(t, x)}{\partial t} \leqslant G(t, x, m(t, x), m_x(t, x)).$$

Since $z(t, x)$ is the solution of (9.3.18), we have
$$\frac{\partial z(t, x)}{\partial t} = G(t, x, z(t, x), z_x(t, x)).$$

As the hypotheses of Theorem 9.1.2 are satisfied, the conclusion follows immediately, and the theorem is proved.

Let us now assume the existence of solutions of (9.6.1). Suppose also that the system (9.6.1) has the trivial solution $u \equiv 0$. We may then formulate the definition of stability of the trivial solution of (9.6.1).

DEFINITION 9.6.1. The trivial solution $u \equiv 0$ of (9.6.1) is said to be *stable* if, for every $\epsilon > 0$ and $t_0 \in J$, there exists a $\delta > 0$ such that $\| \phi(x) \| < \delta$ on ∂E_1 implies
$$\| u(t, x) \| < \epsilon \quad \text{on} \quad E,$$
where $u(t, x)$ is any solution of (9.6.1) with $u(t_0, x) = \phi(x)$ on ∂E_1.

The trivial solution $u \equiv 0$ of (9.6.1) is said to be *asymptotically stable* if it is stable and, for every $\epsilon > 0$, $t_0 \in J$, there exist positive numbers δ_0 and T such that $\| \phi(x) \| < \delta_0$ on ∂E_1 implies
$$\| u(t, x) \| < \epsilon, \quad t \geqslant t_0 + T, \quad \alpha(t) \leqslant x \leqslant \beta(t).$$

On the strength of Theorem 9.6.1, it is easy to state the sufficient conditions for the stability behavior of the trivial solution of (9.6.1).

THEOREM 9.6.3. Let the assumptions of Theorem 9.6.1 hold. Suppose further that
$$b(\| u \|) \leqslant V(t, u) \leqslant a(\| u \|), \tag{9.6.4}$$

where $a, b \in \mathscr{K}$. Then, the stability or asymptotic stability of the trivial solution of the ordinary differential equation (9.2.5) implies the stability or asymptotic stability of the trivial solution of the system (9.6.1).

9.6. LYAPUNOV-LIKE FUNCTION

Proof. Suppose that the trivial solution of (9.2.5) is stable. Let $\epsilon > 0$ and $t_0 \in J$. Then, given $b(\epsilon) > 0$ and $t_0 \in J$, there exists a $\delta > 0$ such that $y_0 \leqslant \delta$ implies

$$y(t, t_0, y_0) < \epsilon, \qquad t \geqslant t_0, \tag{9.6.5}$$

where $y(t, t_0, y_0)$ is any solution of (9.2.5). By Theorem 9.6.1,

$$V(t, u(t, x)) \leqslant r(t, t_0, y_0) \quad \text{on} \quad E, \tag{9.6.6}$$

for any solution $u(t, x)$ of (9.6.1), $r(t, t_0, y_0)$ being the maximal solution of (9.2.5). Choose a positive number δ_1 such that $a(\delta_1) = \delta$, and assume that $\| \phi(x) \| \leqslant \delta_1$. This implies that

$$V(t_0, \phi(x)) \leqslant a(\| \phi(x) \|) \leqslant a(\delta_1) = \delta.$$

Choose $y_0 = \sup_{x \in \partial E_1} V(t_0, \phi(x))$. It then follows, by the relations (9.6.4), (9.6.5), and (9.6.6), that

$$b(\| u(t, x) \|) \leqslant V(t, u(t, x)) \leqslant r(t, t_0, y_0) < b(\epsilon) \quad \text{on} \quad E,$$

which leads to a further inequality

$$\| u(t, x) \| < \epsilon \quad \text{on} \quad E,$$

provided $\| \phi(x) \| \leqslant \delta_1$. This proves the stability of the trivial solution of (9.6.1).

Now, suppose the trivial solution of (9.2.5) is asymptotically stable. Let $\epsilon > 0$ and $t_0 \in J$. Then, given $b(\epsilon) > 0$ and $t_0 \in J$, there exist two positive numbers δ_0 and T such that $y_0 \leqslant \delta_0$ implies

$$y(t, t_0, y_0) < b(\epsilon), \qquad t \geqslant t_0 + T.$$

As before, we choose $y_0 = \sup_{x \in \partial E_1} V(t_0, \phi(x))$. Furthermore, let $a(\tilde{\delta}_0) = \delta_0$ and assume that $\| \phi(x) \| \leqslant \tilde{\delta}_0$. These considerations show that, as previously,

$$b(\| u(t, x) \|) \leqslant V(t, u(t, x)) \leqslant r(t, t_0, y_0) < b(\epsilon),$$

for $t \geqslant t_0 + T$ and $\alpha(t) \leqslant x \leqslant \beta(t)$. From this follows the inequality

$$\| u(t, x) \| < \epsilon, \qquad t \geqslant t_0 + T, \qquad \alpha(t) \leqslant x \leqslant \beta(t),$$

provided $\| \phi(x) \| \leqslant \tilde{\delta}_0$. It is easy to see that this assures the asymptotic stability of the trivial solution of (9.6.1), in view of the foregoing proof. The theorem is completely proved.

Theorem 9.6.2 may also be used to discuss stability properties of the trivial solution of (9.6.1). For this purpose, let us assume that Eq. (9.3.18) possesses the trivial solution and that all the solutions $z(t, x)$ with $z(t_0, x) = \psi(x) \geqslant 0$ are nonnegative on E. Then, we can define stability notions with respect to the trivial solution of (9.3.18), noting that all the solutions are nonnegative.

DEFINITION 9.6.2. The trivial solution of (9.3.18) is stable if, for every $\epsilon > 0$ and $t_0 \in J$, there exists a $\delta > 0$ such that $z(t, x) < \epsilon$ on E, provided $\psi(x) \leqslant \delta$ on ∂E_1.

The definition for asymptotic stability may be similarly formed.

THEOREM 9.6.4. Let the assumptions of Theorem 9.6.2 hold, and let $V(t, u)$ satisfy the inequality (9.6.4). Then, the stability or asymptotic stability of the partial differential equation (9.3.18) implies the stability or asymptotic stability of the trivial solution of the partial differential system (9.6.1).

9.7. Notes

The results of Sect. 9.1 are due to Plis [6]. Lemmas 9.2.1 and 9.2.2 and Theorem 9.2.1 are adapted from Szarski [8]. Theorems 9.2.2 and 9.2.3 are new. The contents of Sects. 9.3 and 9.4 are modeled on the basis of the work of Plis [6] and are new. For the result of the type given in Theorem 9.5.4, see Szarski [8]. The other results of Sect. 9.5 are new. Section 9.6 contains new results. For further related work, see Plis [1–5] and Szarski [1–3, 6, 8].

Chapter 10

10.0. Introduction

In this chapter, we shall investigate partial differential equations of parabolic type. First of all, we shall concentrate in obtaining certain results concerning parabolic differential inequalities in bounded domains and comparison theorems connected with such inequalities. We consider different kinds of initial boundary-value problems, obtain bounds and error estimates, and prove uniqueness of solutions. Stability criteria of the steady-state solutions is discussed. Many of the results have been extended to systems of parabolic differential equations and inequalities in bounded domains. Introducing the concept of Lyapunov functions, we give sufficient conditions for stability and boundedness of various types. Criteria for conditional stability and boundedness are discussed in terms of several Lyapunov functions. Regarding the parabolic differential equations in unbounded domains, we have basic results concerning parabolic differential inequalities and uniqueness of solutions. Finally, we treat the exterior boundary-value problem. We have given uniqueness criteria only.

10.1. Parabolic differential inequalities in bounded domains

Let H be a region of (t, x) space in R^{n+1} satisfying the following conditions:

(i) H is open, contained in the zone $t_0 < t < \infty$, $t_0 \geqslant 0$, and the intersection of $\bar{H}$, the closure of H, with any zone $t_0 \leqslant t \leqslant T$ is bounded;

(ii) for any $t_1 \in [t_0, \infty)$, the projection P_{t_1} on R^n of the intersection of $\bar{H}$ with the plane $t = t_1$ is nonempty;

(iii) for every $(t_1, x_1) \in \bar{H}$ and for every sequence $\{t_k\}$ such that

$t_k \in [t_0, \infty)$, $t_k \to t_1$ as $k \to \infty$, there is a sequence $\{x_k\}$ satisfying $x_k \in P_{t_k}$ and $x_k \to x_1$ as $k \to \infty$.

We shall denote by ∂H that part of the boundary of H which is contained in the zone $t_0 < t < \infty$.

It is easy to see that the topological product $H = (t_0, \infty) \times D$, where $D \subset R^n$ is an open, bounded region, satisfies the preceding conditions. Also, the pyramid defined by the inequalities

$$t_0 < t < t_0 + T, \qquad |x_i - x_i^0| \leqslant a_i - L(t - t_0), \qquad i = 1, 2, \ldots, n,$$

where $L \geqslant 0$, $a_i > 0$, and $T = \min(a_i/L)$, verifies the requirements of H. On the other hand, $H = H_1 \cup H_2$, where

$$H_1 = [(t, x) \colon 0 < t < 1, 0 < x < 2],$$
$$H_2 = [(t, x) \colon 1 \leqslant t < 2, 0 < x < 1],$$

does not satisfy condition (iii) at the point $(1, \tfrac{3}{2})$.

Let us prove some basic lemmas, which we shall use frequently.

LEMMA 10.1.1. Assume that

(i) $m \in C[\bar{H}, R]$, the partial derivatives m_t, m_x, m_{xx} exist and are continuous in H;

(ii) $m(t, x) < 0$ on $P_{t_0} \cup \partial H$;

(iii) if $(t_1, x_1) \in P_{t_1}$, $m(t_1, x_1) = 0$, $m_x(t_1, x_1) = 0$, and the quadratic form

$$\sum_{i,k=1}^{n} \frac{\partial^2 m(t_1, x_1)}{\partial x_i \, \partial x_k} \lambda_i \lambda_k \leqslant 0$$

for arbitrary vector λ, then $m_t(t_1, x_1) < 0$.

These assumptions imply that

$$m(t, x) < 0 \qquad \text{on} \qquad \bar{H}. \tag{10.1.1}$$

Proof. Suppose that the inequality (10.1.1) is not true. Then, the set

$$Z = [(t, x) \in \bar{H} \colon m(t, x) \geqslant 0]$$

is nonempty. Let Z_t be the projection of Z on t axis and $t_1 = \inf Z_t$. We then have

$$m(t, x) \leqslant 0 \qquad \text{on} \qquad \bar{H} \cap [t_0, t_1].$$

10.1. PARABOLIC DIFFERENTIAL INEQUALITIES

We assert that $m(t_1, x)$ has a maximum equal to zero for some $x_1 \in P_{t_1}$. If this is false, on account of assumption (ii), we must have $m(t, x) < 0$ for all $x \in P_{t_1}$. This contradicts the definition of t_1. Hence, there is an $x_1 \in \text{int } P_{t_1}$ such that $m(t_1, x_1) = 0$. It therefore follows that

$$m_t(t_1, x_1) \geq 0. \tag{10.1.2}$$

Since $m(t, x)$ attains its interior maximum at (t_1, x_1), we obtain $m_x(t_1, x_1) = 0$ and the quadratic form

$$\sum_{i,k=1}^{n} \frac{\partial^2 m(t_1, x_1)}{\partial x_i \, \partial x_k} \lambda_i \lambda_k \leq 0$$

for an arbitrary vector λ. We then deduce by condition (iii) that $m_t(t_1, x_1) < 0$, which is incompatible with (10.1.2). Thus, the set Z is empty, and the inequality (10.1.1) is true.

Let $\alpha(t, x)$ be continuous on ∂H. We shall denote by ∂H_α that part of the boundary ∂H on which $\alpha(t, x) > 0$, that is,

$$\partial H_\alpha = [(t, x) \in \partial H : \alpha(t, x) > 0]. \tag{10.1.3}$$

Let τ be a direction defined at each point of ∂H_α in a continuous manner. We shall say that a direction τ at $(t_1, x_1) \in \partial H_\alpha$ points into the interior of H, if there exists a finite ray with (t_1, x_1) as its origin, all of whose interior points lie in H. For any continuous function $u(t, x)$, $\partial u/\partial \tau$ is defined as follows. If τ points into the interior of H, $\partial u/\partial \tau$ is a directional derivative, that is,

$$\frac{\partial u(t, x)}{\partial \tau} = \lim_{\Delta \tau \to 0} \frac{u(t, x) - u(t_1, x_1)}{-\Delta \tau}, \tag{10.1.4}$$

where (t_1, x_1) lies on the ray issuing from (t, x) in the direction of τ, the distance from (t_1, x_1) to (t, x) being $\Delta \tau$. We shall always assume that the direction τ points into the interior of H, and we shall take $\partial u/\partial \tau$ to be defined in the sense of (10.1.4).

If, in addition, the direction τ is orthogonal to the t axis, then it is denoted by τ_0, and the corresponding directional derivative will be given by

$$\frac{\partial u(t, x)}{\partial \tau_0} = \lim_{\Delta \tau_0 \to 0} \frac{u(t, x) - u(t, x_1)}{-\Delta \tau_0},$$

where (t, x_1) lies on the ray issuing from (t, x) in the direction of τ_0, the distance between (t, x_1) and (t, x) being $\Delta \tau_0$. We shall use both types of directional derivative in the sequel.

We are now in a position to prove a variant of Lemma 10.1.1.

LEMMA 10.1.2. Let the hypotheses of Lemma 10.1.1 remain the same except that condition (ii) is replaced by

(iia) $m(t, x) < 0$ on P_{t_0} and $\partial H - \partial H_\alpha$;

(iib) if $(t_1, x_1) \in \partial H_\alpha$ and $m(t_1, x_1) = 0$, then $\partial m(t_1, x_1)/\partial \tau < 0$, where, for each $(t, x) \in \partial H_\alpha$, $\partial m/\partial \tau$ is assumed to exist.

Then, the conclusion (10.1.1) is valid.

Proof. The proof is similar to the proof of Lemma 10.1.1, the only difference being that (t_1, x_1) cannot lie on ∂H_α, in view of condition (iib). For, if $(t_1, x_1) \in \partial H_\alpha$, since $m(t_1, x_1) = 0$, we have, by condition (iib),

$$\frac{\partial m(t_1, x_1)}{\partial \tau} < 0. \qquad (10.1.5)$$

On the other hand, from the fact that $m(t, x) \leqslant 0$ on $\bar{H} \cap [t_0, t_1]$, we deduce that

$$\frac{\partial m(t_1, x_1)}{\partial \tau} \geqslant 0,$$

contradicting (10.1.5).

Let us now consider the partial differential operator given by

$$T[u] = u_t - f(t, x, u, u_x, u_{xx}), \qquad (10.1.6)$$

where $f \in C[\bar{H} \times R \times R^n \times R^{n^2}, R]$, $u_x = (\partial u/\partial x_1, ..., \partial u/\partial x_n)$, and $u_{xx} = (\partial^2 u/\partial x_1^2, \partial^2 u/\partial x_1 \partial x_2, ..., \partial^2 u/\partial x_n^2)$.

DEFINITION 10.1.1. The function $f(t, x, u, P, R)$ is said to be *elliptic* at a point (t_1, x_1) if, for any u, P_i, Q_{ik}, R_{ik} ($i, k = 1, 2, ..., n$), the quadratic form

$$\sum_{i,k=1}^{n} (Q_{ik} - R_{ik}) \lambda_i \lambda_k \leqslant 0,$$

for arbitrary vector λ, implies

$$f(t_1, x_1, u, P, Q) \leqslant f(t_1, x_1, u, P, R).$$

If the foregoing property holds for every $(t, x) \in H$, then $f(t, x, u, P, R)$ is said to be elliptic in H. Moreover, we shall say that the differential operator T is *parabolic* if $f(t, x, u, P, R)$ is elliptic.

10.1. PARABOLIC DIFFERENTIAL INEQUALITIES

A fundamental result in parabolic differential inequalities may now be proved.

THEOREM 10.1.1. Assume that

(i) $u, v \in C[\bar{H}, R]$, the partial derivatives u_t, u_x, u_{xx}, v_t, v_x, v_{xx} exist and are continuous in H;

(ii) $f \in C[\bar{H} \times R \times R^n \times R^{n^2}, R]$, the differential operator T is parabolic, and
$$T[u] < T[v] \quad \text{on} \quad H;$$

(iii) $u(t, x) < v(t, x)$ on $P_{t_0} \cup \partial H$.

Under these assumptions, we have
$$u(t, x) < v(t, x) \quad \text{on} \quad \bar{H}. \tag{10.1.7}$$

Proof. We consider the function
$$m(t, x) = u(t, x) - v(t, x).$$

Clearly, $m(t, x)$ satisfies assumptions (i), (ii) of Lemma 10.1.1. We shall show that condition (iii) of Lemma 10.1.1 is also true. Suppose that $(t_1, x_1) \in P_{t_1}$, $m(t_1, x_1) = 0$, $m_x(t_1, x_1) = 0$, and the quadratic form
$$\sum_{i,k=1}^{n} m_{x_i x_k}(t_1, x_1) \lambda_i \lambda_k \leqslant 0$$

for arbitrary vector λ. This implies that
$$\begin{aligned} u(t_1, x_1) &= v(t_1, x_1), \\ u_x(t_1, x_1) &= v_x(t_1, x_1), \end{aligned} \tag{10.1.8}$$

and
$$\sum_{i,k=1}^{n} \frac{\partial^2 [u(t_1, x_1) - v(t_1, x_1)]}{\partial x_i \, \partial x_k} \lambda_i \lambda_k \leqslant 0.$$

Consequently, using the ellipticity of the function f, we obtain
$$\begin{aligned} &f(t_1, x_1, u(t_1, x_1), u_x(t_1, x_1), u_{xx}(t_1, x_1)) \\ &\leqslant f(t_1, x_1, u(t_1, x_1), u_x(t_1, x_1), v_{xx}(t_1, x_1)). \end{aligned}$$

This, together with the inequality $T[u] < T[v]$, yields
$$\begin{aligned} m_t(t_1, x_1) &< f(t_1, x_1, u(t_1, x_1), u_x(t_1, x_1), v_{xx}(t_1, x_1)) \\ &\quad - f(t_1, x_1, v(t_1, x_1), v_x(t_1, x_1), v_{xx}(t_1, x_1)) = 0, \end{aligned}$$

in view of (10.1.8). By Lemma 10.1.1, the validity of (10.1.7) follows.

The next theorem allows nonlinear boundary conditions and is, as such, more general than Theorem 10.1.1.

THEOREM 10.1.2. Let the assumptions of Theorem 10.1.1 remain the same except that the boundary condition is replaced by

(iiia) $u(t, x) < v(t, x)$ on P_{t_0} and $\partial H - \partial H_\alpha$;
(iiib) for each $(t, x) \in \partial H_\alpha$, $\partial u/\partial \tau$, $\partial v/\partial \tau$ exist, and

$$\alpha(t, x) \frac{\partial u(t, x)}{\partial \tau} + Q(t, x, u(t, x)) \leq 0,$$

$$\alpha(t, x) \frac{\partial v(t, x)}{\partial \tau} + Q(t, x, v(t, x)) > 0,$$

where $Q \in C[\partial H_\alpha \times R, R]$.

Then the inequality (10.1.7) is true.

Proof. The proof is similar to the proof of Theorem 10.1.1 except that we have to verify conditions (iia), (iib) of Lemma 10.1.2. The definition of $m(t, x)$ and condition (iiia) imply condition (iia). If $(t_1, x_1) \in \partial H_\alpha$ and $m(t_1, x_1) = 0$, then it is easily checked that $\partial m(t_1, x_1)/\partial \tau < 0$ because of condition (iiib) and the fact that $\alpha(t_1, x_1) > 0$ whenever $(t_1, x_1) \in \partial H_\alpha$. This verifies condition (iib), and therefore Lemma 10.1.2 assures the desired result.

COROLLARY 10.1.1. Assume that

(i) $u \in C[\overline{H}, R]$, the partial derivatives u_t, u_x, u_{xx} exist and are continuous in H;
(ii) $f \in C[\overline{H} \times R \times R^n \times R^{n^2}, R]$, $f(t, x, 0, 0, 0) \equiv 0$, the differential operator is parabolic, and $T[u] < 0$ on H;
(iii) $u(t, x) < 0$ on $P_{t_0} \cup \partial H$.

Then, $u(t, x) < 0$ on $\overline{H}$.

COROLLARY 10.1.2. The Corollary 10.1.1 remains true even if condition (iii) is replaced by

(iii*) $u(t, x) < 0$ on P_{t_0} and $\partial H - \partial H_\alpha$;
(iv) for each $(t, x) \in \partial H_\alpha$, $\partial u/\partial \tau$ exists
and

$$\alpha(t, x) \frac{\partial u(t, x)}{\partial \tau} + Q(t, x, u(t, x)) < 0$$

on ∂H_α, where $Q \in C[\partial H_\alpha \times R, R]$ and $Q(t, x, 0) \equiv 0$.

10.2. Comparison theorems

A comparison theorem that plays a prominent part in applications is the following result.

THEOREM 10.2.1. Suppose that

(i) $m \in C[\bar{H}, R_+]$, $m(t, x)$ possesses continuous partial derivatives m_t, m_x, and m_{xx} in H;

(ii) $f \in C[\bar{H} \times R \times R^n \times R^{n^2}, R]$, the differential operator T is parabolic, and $T[m] \leq 0$ on H;

(iii) $g \in C[J \times R_+, R]$, and $r(t) = r(t, t_0, y_0) \geq 0$ is the maximal solution of
$$y' = g(t, y), \quad y(t_0) = y_0 \geq 0, \tag{10.2.1}$$
existing on $[t_0, \infty)$;

(iv) $m(t, x) \leq r(t, t_0, y_0)$ on $P_{t_0} \cup \partial H$.

Then, one of the assumptions

(a) $f(t, x, u, 0, 0) \leq g(t, u), u > 0$,
(b) $f(t, x, u, 0, 0) \leq g(t, u), r(t) < u < r(t) + \epsilon_0$, for some $\epsilon_0 > 0$,

implies that
$$m(t, x) \leq r(t, t_0, y_0) \quad \text{on} \quad \bar{H}. \tag{10.2.2}$$

Proof. We consider the function
$$v(t, x) = m(t, x) - r(t, \epsilon),$$
where $r(t, \epsilon) = r(t, t_0, y_0, \epsilon)$ is the maximal solution of
$$y' = g(t, y) + \epsilon, \quad y(t_0) = y_0 + \epsilon,$$
for sufficiently small $\epsilon > 0$, which exists on any compact interval $[t_0, t_0 + \beta]$, $\beta > 0$, by Theorem 1.3.1. Also,
$$\lim_{\epsilon \to 0} r(t, \epsilon) \equiv r(t, t_0, y_0).$$
Furthermore, this implies that there exists an $\epsilon_0 > 0$ such that
$$r(t) < r(t, \epsilon) < r(t) + \epsilon_0, [t_0, t_0 + \beta]. \tag{10.2.3}$$

We shall show that, under either one of the assumptions (a) or (b), $v(t, x)$ satisfies the hypotheses of Lemma 10.1.1. Clearly, the conditions of Lemma 10.1.1 hold. To verify condition (iii), let $(t_1, x_1) \in P_{t_1}$, $v(t_1, x_1) = 0$, $v_x(t_1, x_1) = 0$, and the quadratic form

$$\sum_{i,k=1}^{n} \frac{\partial^2 v(t_1, x_1)}{\partial x_i \, \partial x_k} \lambda_i \lambda_k \leq 0$$

for arbitrary vector λ. It therefore follows that

$$m(t_1, x_1) = r(t_1, \epsilon), \qquad m_x(t_1, x_1) = 0, \qquad (10.2.4)$$

and

$$\sum_{i,k=1}^{n} \frac{\partial^2 m(t_1, x_1)}{\partial x_i \, \partial x_k} \lambda_i \lambda_k \leq 0.$$

Since the function f is elliptic, we obtain

$$f(t_1, x_1, m(t_1, x_1), m_x(t_1, x_1), m_{xx}(t_1, x_1))$$
$$\leq f(t_1, x_1, m(t_1, x_1), 0, 0). \qquad (10.2.5)$$

Notice that the relations (10.2.3) and (10.2.4) yield

$$r(t_1) < m(t_1, x_1) < r(t_1) + \epsilon_0.$$

Also, $m(t_1, x_1) > 0$, since $r(t_1, \epsilon) > 0$. Thus, in any case, the assumption (a) or (b) may be used. Hence,

$$\begin{aligned} v_t(t_1, x_1) &= m_t(t_1, x_1) - r'(t_1, \epsilon) \\ &\leq f(t_1, x_1, m(t_1, x_1), m_x(t_1, x_1), m_{xx}(t_1, x_1)) - r'(t_1, \epsilon) \\ &\leq f(t_1, x_1, m(t_1, x_1), 0, 0) - g(t_1, r(t_1, \epsilon)) - \epsilon \\ &\leq -\epsilon < 0 \end{aligned}$$

on account of (10.2.5) and the assumption (a) or (b). By Lemma 10.1.1, we then have

$$v(t, x) < 0 \quad \text{on} \quad \bar{H},$$

which implies that

$$m(t, x) \leq r(t, t_0, y_0) \quad \text{on} \quad \bar{H},$$

since $\lim_{\epsilon \to 0} r(t, \epsilon) = r(t, t_0, y_0)$. The proof is complete.

The boundary condition (iv) of Theorem 10.2.1 may be replaced by nonlinear boundary conditions to achieve the same conclusion. This we state as a corollary, the proof of which may be deduced by reducing to Lemma 10.1.2.

10.2. COMPARISON THEOREMS

COROLLARY 10.2.1. The conclusion of Theorem 10.2.1 remains valid, if, in place of boundary condition (iv), we have

(a) $m(t, x) \leqslant r(t, t_0, y_0)$ on P_{t_0} and $\partial H - \partial H_\alpha$;
(b) for each $(t, x) \in \partial H_\alpha$, $\partial m / \partial \tau_0$ exists and

$$\alpha(t, x) \frac{\partial m(t, x)}{\partial \tau_0} + Q(t, x, m(t, x)) \leqslant 0 \quad \text{on} \quad \partial H_\alpha,$$

where $Q \in C[\partial H_\alpha \times R_+, R_+]$, $Q(t, x, u) > 0$ if $u > 0$, for each $(t, x) \in \partial H_\alpha$.

COROLLARY 10.2.2. Let assumptions (i) and (ii) of Theorem 10.2.1 hold. Suppose that $g \in C[J \times R_+, R]$ and that $y(t) > 0$ is a differentiable function satisfying the differential inequality

$$y'(t) > g(t, y(t)), \quad y(t_0) = y_0 > 0.$$

Then, the assumption

$$f(t, x, u, 0, 0) \leqslant g(t, u), \quad u = y(t), \quad t \geqslant t_0,$$

implies that

$$m(t, x) < y(t) \quad \text{on} \quad \bar{H},$$

provided either

(a) $m(t, x) < y(t)$ on $P_{t_0} \cup \partial H$; or
(b) $m(t, x) < y(t)$ on P_{t_0} and $\partial H - \partial H_\alpha$ and

$$\alpha(t, x) \frac{\partial m(t, x)}{\partial \tau_0} + Q(t, x, m(t, x)) \leqslant 0 \quad \text{on} \quad \partial H_\alpha,$$

where $Q \in C[\partial H_\alpha \times R_+, R]$ and $Q(t, x, u) > 0$ if $u > 0$, for each $(t, x) \in \partial H_\alpha$.

In certain situations, the next theorem is more suitable in applications, since it offers a better estimate. Moreover, it shows that the strict inequality $T[u] < T[v]$ in Theorem 10.1.1 may be relaxed if f satisfies certain additional restrictions.

THEOREM 10.2.2. Assume that

(i) $m, v \in C[\bar{H}, R_+]$, the partial derivatives $m_t, m_x, m_{xx}, u_t, u_x, u_{xx}$ exist and are continuous in H;

(ii) $f \in C[\bar{H} \times R_+ \times R^n \times R^{n^2}, R]$, the differential operator T is parabolic, and

$$T[m] \leqslant T[v] \quad \text{on} \quad H;$$

(iii) $g \in C[J \times R_+, R]$, $g(t, 0) \equiv 0$, the maximal solution of

$$y' = g(t, y), \quad y(t_0) = 0 \tag{10.2.6}$$

is identically zero, and

$$f(t, x, z_1, P, R) - f(t, x, z_2, P, R) \leqslant g(t, z_1 - z_2), \quad z_1 > z_2; \tag{10.2.7}$$

(iv) $m(t, x) \leqslant v(t, x)$ on $P_{t_0} \cup \partial H$.

Under these assumptions,

$$m(t, x) \leqslant v(t, x) \quad \text{on} \quad \bar{H}. \tag{10.2.8}$$

Proof. Consider the solutions $y(t, \epsilon) = y(t, t_0, 0, \epsilon)$ of the ordinary differential equation

$$y' = g(t, y) + \epsilon, \quad y(t_0) = \epsilon, \tag{10.2.9}$$

for sufficiently small $\epsilon > 0$. Define the function

$$z(t, x) = v(t, x) + y(t, \epsilon).$$

Clearly, $m(t, x) < z(t, x)$ on $P_{t_0} \cup \partial \bar{H}$. Furthermore, observing that $y(t, \epsilon) > 0$ and using the relation (10.2.7), we have

$$\begin{aligned}
T(z) &= z_t(t, x) - f(t, x, z(t, x), z_x(t, x), z_{xx}(t, x)) = v_t(t, x) + y'(t, \epsilon) \\
&\quad - f(t, x, v(t, x) + y(t, \epsilon), v_x(t, x), v_{xx}(t, x)) \\
&\quad + f(t, x, v(t, x), v_x(t, x), v_{xx}(t, x)) \\
&\quad - f(t, x, v(t, x), v_x(t, x), v_{xx}(t, x)) \\
&\geqslant T[v] + \epsilon.
\end{aligned}$$

This implies, in view of the fact that $T[v] \geqslant T[m]$, the inequality $T[m] < T[z]$. Hence, applying Theorem 10.1.1 to functions $m(t, x)$, $z(t, x)$, we obtain the relation

$$m(t, x) < z(t, x) \quad \text{on} \quad \bar{H}.$$

Since, by assumption, $\lim_{\epsilon \to 0} y(t, \epsilon) \equiv 0$, the desired result (10.2.8) follows immediately. The proof is complete.

THEOREM 10.2.3. *Let the assumptions of Theorem 10.2.2 remain the same except that boundary condition* (iv) *is replaced by*

(a) $m(t, x) \leqslant v(t, x)$ on P_{t_0} and $\partial H - \partial H_\alpha$;

(b) for each $(t, x) \in \partial H_\alpha$, $\partial m/\partial \tau_0$, $\partial v/\partial \tau_0$ exist, and

$$\alpha(t, x) \frac{\partial m(t, x)}{\partial \tau_0} + Q(t, x, m(t, x)) \leq 0,$$

$$\alpha(t, x) \frac{\partial v(t, x)}{\partial \tau_0} + Q(t, x, v(t, x)) \geq 0 \quad \text{on} \quad \partial H_\alpha,$$

where $Q \in C[\partial H_\alpha \times R_+, R]$, and Q is increasing in u for each (t, x).

Proof. The proof is similar to the proof of Theorem 10.2.2. In the present case, we verify the assumptions of Theorem 10.1.2. Evidently,

$$m(t, x) < z(t, x) \quad \text{on} \quad P_{t_0} \quad \text{and} \quad \partial H - \partial H_\alpha.$$

The monotonicity of the function Q in u, together with the fact that $z(t, x) > v(t, x)$, shows that, on ∂H_α,

$$\alpha(t, x) \frac{\partial z(t, x)}{\partial \tau_0} = \alpha(t, x) \frac{\partial v(t, x)}{\partial \tau_0} \geq -Q(t, x, v(t, x))$$
$$> -Q(t, x, z(t, x)).$$

The application of Theorem 10.1.2 now yields the stated result.

As an immediate consequence of Theorems 10.2.2 and 10.2.3, we derive weak maximum and minimum principles.

COROLLARY 10.2.3. Assume that

(i) $u \in C[\bar{H}, R]$, $u(t, x)$ possesses continuous partial derivatives u_t, u_x, and u_{xx} in $\bar{H}$;

(ii) $f \in C[\bar{H} \times R \times R^n \times R^{n^2}, R]$, and the differential operator T is parabolic;

(iii) $f(t, x, z, 0, 0) \leq 0$ if $z > 0$ and $T[u] \leq 0$;

(iv) either

(a) $u(t, x) \leq 0$ on $P_{t_0} \cup \partial H$; or

(b) $u(t, x) \leq 0$ on P_{t_0} and $\partial H - \partial H_\alpha$, and for each $(t, x) \in \partial H_\alpha$, $\partial u/\partial \tau_0$ exists and

$$\alpha(t, x) \frac{\partial u(t, x)}{\partial \tau_0} + Q(t, x, u(t, x)) \leq 0 \quad \text{on} \quad \partial H_\alpha,$$

where $Q \in C[\partial H_\alpha \times R, R]$ and $Q(t, x, z)$ is increasing in z for each (t, x).

Then we have
$$u(t, x) \leq 0 \quad \text{on} \quad \bar{H}.$$

COROLLARY 10.2.4. Let assumptions (i) and (ii) of Corollary 10.2.3 hold. Assume further that

(iii*) $f(t, x, z, 0, 0) \geq 0$ if $z < 0$ and $T(u) \geq 0$;
(iv*) either
    (a) $u(t, x) \geq 0$ on $P_{t_0} \cup \partial H$; or
    (b) $u(t, x) \geq 0$ on P_{t_0} and $\partial H - \partial H_\alpha$, and for each $(t, x) \in \partial H_\alpha$, $\partial u/\partial \tau_0$ exists and

$$\alpha(t, x) \frac{\partial u(t, x)}{\partial \tau_0} + Q(t, x, u(t, x)) \geq 0 \quad \text{on} \quad \partial H_\alpha,$$

where $Q \in C[\partial H_\alpha \times R, R]$ and $Q(t, x, z)$ is increasing in z for each (t, x).

Then, we have
$$u(t, x) \geq 0 \quad \text{on} \quad \bar{H}.$$

Finally, we shall prove a comparison theorem that will be useful in considering the stability of steady-state solutions of nonlinear diffusion equations.

Let G be an open bounded region in R^n and ∂G be its boundary. Denote by H the topological product $[0, \infty) \times G$.

THEOREM 10.2.4. Suppose that

(i) $u, v \in C[\bar{H}, R]$, the partial derivatives u_t, v_t, u_x, v_x, u_{xx}, v_{xx} exist and are continuous in H;

(ii) $f \in C[\bar{H} \times R \times R^n \times R^{n^2}, R]$, the differential operator T is parabolic, $\partial f/\partial u$ exists and is continuous, and $T[v] \leq T[u]$ on H;

(iii) the derivatives $\partial u(t, x)/\partial \tau_0$, $\partial v(t, x)/\partial \tau_0$ in the direction of the outward normal to the hypersurface $(0, \infty) \times \partial G$ exist, and

$$\frac{\partial u(t, x)}{\partial \tau_0} = F_1(u(t, x)), \quad \frac{\partial v(t, x)}{\partial \tau_0} = F_2(v(t, x)) \quad \text{on} \quad (0, \infty) \times \partial G,$$

where F_1, F_2 are continuous functions with closed domain and bounded derivatives, such that
$$F_1(u) < F_2(u),$$

if u belongs to the common domain of definition of F_1 and F_2;

(iv)
$$v(t, 0) < u(t, 0), \quad x \in \bar{G}.$$

Then, we have
$$v(t, x) < u(t, x) \quad \text{on} \quad \bar{H}. \qquad (10.2.10)$$

10.2. COMPARISON THEOREMS

Proof. We divide the proof into two parts. The first is a proof of the theorem if the condition $T[v] \leq T[u]$ is replaced by strict inequality, that is, $T[v] < T[u]$. We consider the function

$$m(t, x) = v(t, x) - u(t, x),$$

and proceed as in the proof of Lemma 10.1.1. The only difference in the proof is in showing that $m(t_1, x_1)$ has a maximum equal to zero for some $x_1 \in G$. Suppose that $x_1 \in \partial G$. Then, by assumption,

$$\frac{\partial m(t_1, x_1)}{\partial \tau_0} = \frac{\partial v(t_1, x_1)}{\partial \tau_0} - \frac{\partial u(t_1, x_1)}{\partial \tau_0}$$

$$= F_2(v(t_1, x_1)) - F_1(u(t_1, x_1))$$

$$> 0,$$

since $u(t_1, x_1) = v(t_1, x_1)$. Let $x^* \in G$ be a point on the normal to the hypersurface $(0, \infty) \times \partial G$ at x_1, sufficiently close to x_1. Then, we obtain, from the fact that $m(t, x)$ is continuously differentiable and by application of mean value theorem, that $m(t_1, x^*) > 0$. Since $m(t, x)$ attains its maximum equal to zero at (t_1, x_1), this is an absurdity. Hence, $x_1 \in G$. The rest of the proof is standard, and (10.2.10) is therefore true.

We now prove the second part, that is, we shall not demand $T[v] < T[u]$. Let us deny the conclusion (10.2.10). Then, there exists a $t_1 > 0$ and an $x_1 \in G$ such that

$$v(t_1, x_1) \geq u(t_1, x_1). \tag{10.2.11}$$

We define a function $w(t, x)$ by

$$w(t, x) = v(t, x) + \frac{\epsilon}{(n-1)(t+1)^{n-1}},$$

where both $\epsilon > 0$ and $n \geq 2$ will be specified later. We have

$$w_t(t, x) = v_t(t, x) - \frac{\epsilon}{(t+1)^n},$$

$$w_x(t, x) = v_x(t, x),$$

$$w_{xx}(t, x) = v_{xx}(t, x),$$

and

$$w(0, x) = v(0, x) + \frac{\epsilon}{n-1}.$$

Since $\partial f/\partial u$ exists, it follows that

$$f(t, x, w, w_x, w_{xx}) - w_t = f(t, x, v, v_x, v_{xx})$$
$$+ \frac{f_v(t, x, v, v_x, v_{xx})}{(n-1)(t+1)^{n-1}} + O(\epsilon^2)$$
$$- v_t + \frac{\epsilon}{(t+1)^n}$$
$$> f(t, x, v, v_x, v_{xx}) - v_t = T[v], \quad (10.2.12)$$

provided

$$\frac{\epsilon}{(t+1)^{n-1}} \left[\frac{f_v(t, x, v, v_x, v_{xx})}{n-1} + \frac{1}{t+1} \right] > 0, \quad (10.2.13)$$

and $\epsilon > 0$ is sufficiently small. Let

$$\mu = \min_{\substack{x \in \bar{G} \\ t \in [0, t_1]}} f_v(t, x, v(t, x), v_x(t, x), v_{xx}(t, x)).$$

We now choose n so large that

$$\frac{\mu}{n-1} + \frac{1}{t_1 + 1} > 0 \quad \text{or} \quad n > -\mu(t_1 + 1) + 1.$$

With this value of n, the inequality (10.2.13) holds for $x \in \bar{G}$ and $t \in [0, t_1]$. There is a number $\rho > 0$ such that

$$F_2(v(t, x_1)) - F_1(v(t, x_1)) > \rho,$$

for $t \in (0, t_1]$ and $x_1 \in \partial G$, since the left side of the foregoing inequality is positive on a closed set. Hence, for all $w(t, x_1)$ belonging to the domain of F_1, there results

$$w_x(t, x_1) = v_x(t, x_1) = F_2(v(t, x_1))$$
$$> F_1(v(t, x_1)) + \rho$$
$$> F_1(v(t, x_1)) + \frac{\epsilon M}{(n-1)(t+1)^{n-1}}$$
$$> F_1\left(v(t, x_1) + \frac{\epsilon}{(n-1)(t+1)^{n-1}}\right)$$
$$= F_1(w(t, x_1)), \quad (10.2.14)$$

if ϵ is sufficiently small. Here M is an upper bound of $|F_1'|$. Let us now choose $\epsilon > 0$ so that (10.2.12), (10.2.14), and

$$\min_{x \in \bar{G}}[u(0, x) - v(0, x)] > \epsilon \quad (10.2.15)$$

hold. Note that there is a positive value of ϵ satisfying (10.2.15), since the left side is the minimum of a continuous positive function on a closed set. Thus, it follows, from (10.2.12) and $T[v] \leqslant T[u]$, that

$$T[w] < T[v] \leqslant T[u] \quad \text{on} \quad x \in \bar{G}, \quad t \in [0, t_1].$$

Also,

$$w(0, x) = v(0, x) + \frac{\epsilon}{n-1} < u(0, x), \quad x \in \bar{G}.$$

Applying the first part of the proof, we obtain

$$w(t, x) < u(t, x) \quad \text{on} \quad [0, t_1] \times \bar{G}.$$

However, this, together with (10.2.11), leads to

$$w(t_1, x_1) < u(t_1, x_1) \leqslant v(t_1, x_1)$$
$$= w(t_1, x_1) - \frac{\epsilon}{(n-1)(t_1+1)^{n-1}}.$$

This contradiction shows that there does not exist a $t_1 > 0$ satisfying (10.2.11), and, hence, the proof is complete.

10.3. Bounds, under and over functions

Consider the partial differential equation

$$u_t = f(t, x, u, u_x, u_{xx}), \qquad (10.3.1)$$

where $f \in C[\bar{H} \times R \times R^n \times R^{n^2}, R]$.

DEFINITION 10.3.1. Given an initial $\phi(t, x)$, which is defined and continuous on $P_{t_0} \cup \partial H$, a *solution* of (10.3.1) is any function $u(t, x)$ satisfying the following properties:

(i) $u(t, x)$ is defined and continuous for $(t, x) \in \bar{H}$;

(ii) $u(t, x)$ possesses continuous partial derivatives u_t, u_x, u_{xx} in H and satisfies (10.3.1) for $(t, x) \in H$;

(iii) $u(t, x) = \phi(t, x)$ for $(t, x) \in P_{t_0} \cup \partial H$.

The problem of finding a solution to the partial differential equation (10.3.1) is called a *first initial-boundary-value problem*.

Let $Q \in C[\partial H_\alpha \times R, R]$, and suppose that

(a) $u(t, x) = \phi(t, x)$ on P_{t_0} and $\partial H - \partial H_\alpha$,

(b) $\alpha(t, x)[\partial u(t, x)/\partial \tau] + Q(t, x, u(t, x)) = \phi(t, x)$ on ∂H_α, where it is assumed that $\partial u/\partial \tau$ exists for $(t, x) \in \partial H_\alpha$.

If, instead of boundary condition (iii) in the Definition 10.3.1, we ask (a) and (b) to be satisfied, we have a *mixed initial-boundary-value problem*.

REMARK 10.3.1. If $Q(t, x, u) = \beta(t, x)u$, where $\beta(t, x) > 0$ on ∂H_α, the problem is said to be a first mixed problem; and if β is not restricted to be positive on ∂H_α, it is called a second mixed problem. If $\alpha(t, x) \equiv 0$, the boundary condition (b) is of Dirichlet type, and the first mixed problem coincides with the first Fourier problem. On the otherhand, if $\alpha(t, x) \equiv 1$ and $\beta(t, x) \equiv 0$, the boundary condition (b) is of Newmann type, in which case the mixed problem reduces to second Fourier problem. If $\alpha(t, x) \equiv 1$, so that $\partial H_\alpha = \partial H$ and $\beta(t, x)$ is continuous on ∂H, the problem is called a third initial-boundary-value problem. If, in addition, H is a cylinder and the directions τ are inward conormals, it is said to be a second initial-boundary-value problem.

In what follows, we shall assume the existence of solutions for the two boundary-value problems just stated, that is, first and mixed initial-boundary-value problems.

THEOREM 10.3.1. Assume that

(i) $f \in C[\bar{H} \times R \times R^n \times R^{n^2}, R]$, $f(t, x, u, P, R)$ is elliptic, and one of the assumptions

(a) $|f(t, x, u, 0, 0)| \leq g(t, |u|)$,

(b) $|f(t, x, u, 0, 0)| \leq g(t, |u|)$, $r(t) < u < r(t) + \epsilon_0$, for some $\epsilon_0 > 0$, holds;

(ii) $g \in C[J \times R_+, R_+]$, and the maximal solution $r(t) = r(t, t_0, y_0)$ of (10.2.1) exists for $t \geq t_0$;

(iii) $u(t, x)$ is a solution of the first initial-boundary-value problem, such that

$$|u(t, x)| \leq r(t, t_0, y_0) \quad \text{on} \quad P_{t_0} \cup \partial H.$$

Then,

$$|u(t, x)| \leq r(t, t_0, y_0) \quad \text{on} \quad \bar{H}.$$

Proof. By Theorem 1.3.1, the maximal solution $r(t, \epsilon) = r(t, t_0, y_0, \epsilon)$ of

$$y' = g(t, y) + \epsilon, \quad y(t_0) = y_0 + \epsilon,$$

for sufficiently small $\epsilon > 0$, exists on any compact interval $[t_0, t_0 + \gamma]$,

10.3. BOUNDS, UNDER AND OVER FUNCTIONS

$\gamma > 0$, and $\lim_{\epsilon \to 0} r(t, \epsilon) \equiv r(t, t_0, y_0)$, uniformly on $[t_0, t_0 + \gamma]$. Furthermore, we have

$$r(t) < r(t, \epsilon) < r(t) + \epsilon_0, \qquad t \in [t_0, t_0 + \gamma], \qquad (10.3.2)$$

for some $\epsilon_0 > 0$. Let us first consider the function

$$m(t, x) = u(t, x) - r(t, \epsilon).$$

It is easy to see that $m(t, x)$ satisfies conditions (i), (ii) of Lemma 10.1.1. The condition (iii) is also verified. For, let $(t_1, x_1) \in P_{t_1}$, $m(t_1, x_1) = 0$, $m_x(t_1, x_1) = 0$, and, for an arbitrary vector λ,

$$\sum_{i,k=1}^{n} \frac{\partial^2 m(t_1, x_1)}{\partial x_i \, \partial x_k} \lambda_i \lambda_k \leqslant 0.$$

Since this implies that $u(t_1, x_1) = r(t_1, \epsilon) > 0$, $u_x(t_1, x_1) = 0$, and

$$\sum_{i,k=1}^{n} \frac{\partial^2 u(t_1, x_1)}{\partial x_i \, \partial x_k} \lambda_i \lambda_k \leqslant 0,$$

we have, on account of the ellipticity of f,

$$\begin{aligned} f(t_1, x_1, u(t_1, x_1), u_x(t_1, x_1), u_{xx}(t_1, x_1)) \\ \leqslant f(t_1, x_1, u(t_1, x_1), 0, 0). \end{aligned} \qquad (10.3.3)$$

Moreover, because of (10.3.2),

$$r(t_1) < |u(t_1, x_1)| < r(t_1) + \epsilon_0. \qquad (10.3.4)$$

Thus, either of the assumptions (a) or (b) of the theorem shows that

$$\begin{aligned} m_t(t_1, x_1) &= u_t(t_1, x_1) - r'(t_1, \epsilon) \\ &= f(t_1, x_1, u(t_1, x_1), u_x(t_1, x_1), u_{xx}(t_1, x_1)) - g(t_1, r(t_1, \epsilon)) - \epsilon \\ &\leqslant f(t_1, x_1, u(t_1, x_1), 0, 0) - g(t_1, r(t_1, \epsilon)) - \epsilon \\ &\leqslant g(t_1, |u(t_1, x_1)|) - g(t_1, r(t_1, \epsilon)) - \epsilon \\ &\leqslant -\epsilon, \end{aligned}$$

using (10.3.3) and the fact that $u(t_1, x_1) = r(t_1, \epsilon)$. We therefore obtain, by Lemma 10.1.1, that

$$u(t, x) < r(t, \epsilon) \qquad \text{on} \qquad \bar{H}. \qquad (10.3.5)$$

Let us next consider the function

$$n(t, x) = -[r(t, \epsilon) + u(t, x)]$$

and show that it also fulfills the conditions of Lemma 10.1.1. It is only required to check condition (iii) of Lemma 10.1.1. Suppose that $(t_1, x_1) \in P_{t_1}$, $n(t_1, x_1) = 0$, $n_x(t_1, x_1) = 0$, and

$$\sum_{i,k=1}^{n} \frac{\partial^2 n(t_1, x_1)}{\partial x_i \, \partial x_k} \lambda_i \lambda_k \leq 0,$$

for some arbitrary vector $\lambda \neq 0$. Then, we have

$$-u(t_1, x_1) = r(t_1, \epsilon) > 0,$$
$$-u_x(t_1, x_1) = 0,$$

and

$$\sum_{i,k=1}^{n} -\frac{\partial^2 u(t_1, x_1)}{\partial x_i \, \partial x_k} \lambda_i \lambda_k \leq 0.$$

By the ellipticity of the function f, we get

$$f(t_1, x_1, u(t_1, x_1), 0, 0) \leq f(t_1, x_1, u(t_1, x_1), u_x(t_1, x_1), u_{xx}(t_1, x_1)). \quad (10.3.6)$$

Also, (10.3.4) holds. Hence, as before, we obtain

$$\begin{aligned} n_t(t_1, x_1) &= -r'(t_1, \epsilon) - u_t(t_1, x_1) \\ &= -g(t_1, r(t_1, \epsilon)) \\ &\quad - \epsilon - f(t_1, x_1, u(t_1, x_1), u_x(t_1, x_1), u_{xx}(t_1, x_1)) \\ &\leq -g(t_1, r(t_1, \epsilon)) - \epsilon + g(t_1, r(t_1, \epsilon)) \\ &\leq -\epsilon \end{aligned}$$

on account of (10.3.6) and either one of the conditions (a) or (b), noting that $|u(t_1, x_1)| = -u(t_1, x_1) = r(t_1, \epsilon)$. Hence, by Lemma 10.1.1,

$$-u(t, x) < r(t, \epsilon) \quad \text{on} \quad \bar{H}. \quad (10.3.7)$$

The two inequalities (10.3.5) and (10.3.7) now yield the desired estimate

$$|u(t, x)| \leq r(t, t_0, y_0) \quad \text{on} \quad \bar{H},$$

and the proof is complete.

COROLLARY 10.3.1. *The assertion of Theorem 10.3.1 remains true if $u(t, x)$ is a solution of the mixed initial-boundary-value problem such that*

(a$_1$) $|u(t, x)| \leq r(t, t_0, y_0)$ on P_{t_0} and $\partial H - \partial H_\alpha$;

(a_2) for some $\beta > 0$,

$$\left| \alpha(t, x) \frac{\partial u(t, x)}{\partial \tau_0} + Q(t, x, u(t, x)) \right| \leq \beta r(t, t_0, y_0) \quad \text{on} \quad \partial H_\alpha,$$

where $Q \in C[\partial H_\alpha \times R, R]$, $Q(t, x, z)$ is increasing in z for each $(t, x) \in \partial H_\alpha$, $Q(t, x, -z) = -Q(t, x, z)$, and $\beta r(t, t_0, y_0) \leq Q(t, x, r(t, t_0, y_0))$.

The next theorem offers a better bound and is a variant of Theorem 10.2.2.

THEOREM 10.3.2. Assume that

(i) $f \in C[\bar{H} \times R \times R^n \times R^{n^2}, R]$, the differential operator T is parabolic, and

$$f(t, x, -u, -P, -R) = -f(t, x, u, P, R); \tag{10.3.8}$$

(ii) $g \in C[J \times R_+, R]$, $g(t, 0) \equiv 0$, the maximal solution $r(t, t_0, 0)$ of (10.2.6) is identically zero, and

$$f(t, x, z_1, P, R) - f(t, x, z_2, P, R) \leq g(t, z_1 - z_2), \quad z_1 > z_2; \tag{10.3.9}$$

(iii) $v \in C[\bar{H}, R_+]$, $v(t, x)$ possesses continuous partial derivatives $v_x \geq 0$, v_t, v_{xx}, and $T[v] \geq 0$;

(iv) $u(t, x)$ is a solution of the first initial-boundary-value problem satisfying

$$|u(t, x)| \leq v(t, x) \quad \text{on} \quad P_{t_0} \cup \partial H.$$

Then, we have

$$|u(t, x)| \leq v(t, x) \quad \text{on} \quad \bar{H}. \tag{10.3.10}$$

Proof. Let us consider the function

$$n(t, x) = -[u(t, x) + v(t, x) + y(t, \epsilon)],$$

where $y(t, \epsilon) = y(t, t_0, 0, \epsilon)$ is any solution of (10.2.9), for sufficiently small $\epsilon > 0$. Let $(t_1, x_1) \in P_{t_1}$, $n(t_1, x_1) = 0$, $n_x(t_1, x_1) = 0$, and the quadratic form

$$\sum_{i,k=1}^{n} \frac{\partial^2 n(t_1, x_1)}{\partial x_i \partial x_k} \lambda_i \lambda_k \leq 0$$

for some vector λ. This means, noting $y(t, \epsilon) > 0$, that

$$-u(t_1, x_1) > v(t_1, x_1),$$
$$-u_x(t_1, x_1) = v_x(t_1, x_1) \geq 0,$$

and
$$\sum_{i,k=1}^{n} \frac{\partial^2 [-u(t_1, x_1) - v(t_1, x_1)]}{\partial x_i \, \partial x_k} \lambda_i \lambda_k \leqslant 0.$$

Because of the ellipticity of f, it results that
$$f(t_1, x_1, -u(t_1, x_1), -u_x(t_1, x_1), -u_{xx}(t_1, x_1))$$
$$\leqslant f(t_1, x_1, -u(t_1, x_1), -u_x(t_1, x_1), v_{xx}(t_1, x_1)).$$

Furthermore, using (10.3.8), (10.3.9), and the preceding inequality, we obtain
$$\begin{aligned} n_t(t_1, x_1) &= -u_t(t_1, x_1) - v_t(t_1, x_1) - y'(t_1, \epsilon) \\ &\leqslant -f(t_1, x_1, u(t_1, x_1), u_x(t_1, x_1), u_{xx}(t_1, x_1)) \\ &\quad - f(t_1, x_1, v(t_1, x_1), v_x(t_1, x_1), v_{xx}(t_1, x_1)) - y'(t_1, \epsilon) \\ &\leqslant f(t_1, x_1, -u(t_1, x_1), -u_x(t_1, x_1), v_{xx}(t_1, x_1)) \\ &\quad - f(t_1, x_1, v(t_1, x_1), v_x(t_1, x_1), v_{xx}(t_1, x_1)) - y'(t_1, \epsilon) \\ &\leqslant g(t_1, y(t_1, \epsilon)) - g(t_1, y(t_1, \epsilon)) - \epsilon \\ &\leqslant -\epsilon, \end{aligned}$$

which implies that $n_t(t_1, x_1) < 0$. Clearly, $n(t, x)$ satisfies all the assumptions of Lemma 10.1.1, and hence
$$-u(t, x) < v(t, x) + y(t, \epsilon) \quad \text{on} \quad \bar{H}.$$

Proceeding similarly, we can show, on the basis of Lemma 10.1.1, that
$$u(t, x) < v(t, x) + y(t, \epsilon) \quad \text{on} \quad \bar{H}.$$

The preceding two inequalities, together with the fact that $\lim_{\epsilon \to 0} y(t, \epsilon) \equiv 0$, yield the estimate (10.3.10). The theorem is proved.

COROLLARY 10.3.2. *Let the hypotheses of Theorem 10.3.2 remain the same except that condition* (iv) *is replaced by*

(iv*) $u(t, x)$ *is a solution of the mixed initial-boundary-value problem satisfying*

(a) $|u(t, x)| \leqslant v(t, x)$ on P_{t_0} and $\partial H - \partial H_\alpha$;

(b)
$$\left| \alpha(t, x) \frac{\partial u(t, x)}{\partial \tau_0} + Q(t, x, u(t, x)) \right|$$
$$\leqslant \alpha(t, x) \frac{\partial v(t, x)}{\partial \tau_0} + Q(t, x, v(t, x)),$$

10.3. BOUNDS, UNDER AND OVER FUNCTIONS

where $Q \in C[\partial H_\alpha \times R, R]$, $Q(t, x, z)$ is increasing in z for each $(t, x) \in \partial H_\alpha$, and $Q(t, x, -z) = -Q(t, x, z)$.

Then, (10.3.10) is valid.

We shall now introduce the notion of under and over functions with respect to the parabolic equation (10.3.1).

DEFINITION 10.3.2. Let $u \in C[\bar{H}, R]$, and let $u(t, x)$ possess continuous partial derivatives u_t, u_x, u_{xx} in H. If $u(t, x)$ satisfies the parabolic differential inequality

$$T[u] < 0 \quad \text{on} \quad H,$$

together with $u(t, x) = \phi(t, x)$ for $(t, x) \in P_{t_0} \cup \partial H$, we shall say that $u(t, x)$ is an *under function* with respect to the first initial-boundary-value problem. On the other hand, if

$$T[u] > 0 \quad \text{on} \quad H,$$

$u(t, x)$ is said to be an *over function*. If $\partial u/\partial \tau$ exists on ∂H_α and $u(t, x)$ satisfies

$$u(t, x) = \phi(t, x) \quad \text{on} \quad P_{t_0} \quad \text{and} \quad \partial H - \partial H_\alpha,$$

$$\alpha(t, x) \frac{\partial u(t, x)}{\partial \tau} + Q(t, x, u(t, x)) = \phi(t, x) \quad \text{on} \quad \partial H_\alpha,$$

we shall say that $u(t, x)$ is an under or over function with respect to the mixed initial-boundary-value problem according as $T[u] < 0$ or $T[u] > 0$, on H.

As a direct consequence of Theorems 10.1.1 and 10.1.2, we have the following.

THEOREM 10.3.3. Let $f \in C[\bar{H} \times R \times R^n \times R^{n^2}, R]$ and the differential operator T be parabolic. Suppose that $u(t, x)$ and $v(t, x)$ are under and over functions with respect to the first initial-boundary-value problem. If $z(t, x)$ is any solution of the same problem such that

$$u(t, x) < z(t, x) < v(t, x) \quad \text{on} \quad P_{t_0} \cup \partial H,$$

then

$$u(t, x) < z(t, x) < v(t, x) \quad \text{on} \quad \bar{H}. \tag{10.3.11}$$

The inequality (10.3.11) remains true, even when $u(t, x)$ and $v(t, x)$

are under and over functions with respect to the mixed initial-boundary-value problem, provided $z(t, x)$ is any solution of the same problem and

$$u(t, x) < z(t, x) < v(t, x) \quad \text{on} \quad P_{t_0} \quad \text{and} \quad \partial H - \partial H_\alpha,$$

$$\alpha(t, x) \frac{\partial u(t, x)}{\partial \tau} + Q(t, x, u(t, x)) \leqslant \alpha(t, x) \frac{\partial z(t, x)}{\partial \tau} + Q(t, x, z(t, x))$$

$$\leqslant \alpha(t, x) \frac{\partial v(t, x)}{\partial \tau} + Q(t, x, v(t, x)).$$

10.4. Approximate solutions and uniqueness

We shall begin with the theorems that estimate the difference between a solution and an approximate solution of (10.3.1).

THEOREM 10.4.1.   Assume that

(i) $f \in C[\bar{H} \times R \times R^n \times R^{n^2}, R]$, the operator T is parabolic, and

$$f(t, x, z_1, P, R) - f(t, x, z_2, P, R) \leqslant g(t, z_1 - z_2), \quad z_1 > z_2; \quad (10.4.1)$$

(ii) $v \in C[\bar{H}, R]$, $v(t, x)$ possesses continuous partial derivatives v_t, v_x, v_{xx} such that

$$|T[v]| \leqslant \delta(t), \quad (10.4.2)$$

where $\delta \in C[J, R_+]$;

(iii) $g \in C[J \times R_+, R]$, $g(t, 0) \equiv 0$, and $r(t, t_0, y_0)$ is the maximal solution of

$$y' = g(t, y) + \delta(t), \quad y(t_0) = y_0 \geqslant 0, \quad (10.4.3)$$

existing for $t \geqslant t_0$;

(iv) $u(t, x)$ is any solution of the first initial-boundary-value problem such that

$$|u(t, x) - v(t, x)| \leqslant r(t, t_0, y_0) \quad \text{on} \quad P_{t_0} \cup \partial H.$$

Then, the estimate

$$|u(t, x) - v(t, x)| \leqslant r(t, t_0, y_0) \quad \text{on} \quad \bar{H} \quad (10.4.4)$$

is valid.

Proof. Define

$$m(t, x) = u(t, x) - v(t, x) - y(t, \epsilon),$$

10.4. APPROXIMATE SOLUTIONS AND UNIQUENESS

where $y(t, \epsilon) = y(t, t_0, y_0, \epsilon)$ is any solution of

$$y' = g(t, y) + \delta(t) + \epsilon, \qquad y(t_0) = y_0 + \epsilon, \tag{10.4.5}$$

for sufficiently small $\epsilon > 0$. Suppose that $(t_1, x_1) \in P_{t_1}$, $m(t_1, x_1) = 0$, $m_x(t_1, x_1) = 0$, and, for some nonzero vector λ,

$$\sum_{i,k=1}^{n} \frac{\partial^2 m(t_1, x_1)}{\partial x_i \, \partial x_k} \lambda_i \lambda_k \leq 0.$$

Since $y(t, \epsilon) > 0$, this implies that

$$u(t_1, x_1) > v(t_1, x_1),$$
$$u_x(t_1, x_1) = v_x(t_1, x_1),$$

and

$$\sum_{i,k=1}^{n} \frac{\partial^2 (u-v)}{\partial x_i \, \partial x_k} \lambda_i \lambda_k \leq 0.$$

The last inequality yields, because of the ellipticity of f,

$$f(t_1, x_1, u(t_1, x_1), u_x(t_1, x_1), u_{xx}(t_1, x_1))$$
$$\leq f(t_1, x_1, u(t_1, x_1), u_x(t_1, x_1), v_{xx}(t_1, x_1)).$$

It follows, in view of the preceding inequality, that

$$u_t(t_1, x_1) - v_t(t_1, x_1) \leq f(t_1, x_1, u(t_1, x_1), u_x(t_1, x_1), v_{xx}(t_1, x_1))$$
$$- f(t_1, x_1, v(t_1, x_1), v_x(t_1, x_1), v_{xx}(t_1, x_1))$$
$$- T[v].$$

Hence, the relations (10.4.1) and (10.4.2) show that

$$u_t(t_1, x_1) - v_t(t_1, x_1) \leq g(t_1, y(t_1, \epsilon)) + \delta(t).$$

We thus have

$$m_t(t_1, x_1) = u_t(t_1, x_1) - v_t(t_1, x_1) - y'(t_1, \epsilon)$$
$$\leq -\epsilon < 0.$$

It is easy to see that $m(t, x)$ satisfies the hypotheses of Lemma 10.1.1, and hence

$$u(t, x) - v(t, x) < y(t, \epsilon) \qquad \text{on} \qquad \bar{H}.$$

In a similar way, considering the function

$$n(t, x) = v(t, x) - u(t, x) - y(t, \epsilon)$$

and verifying the assumptions of Lemma 10.1.1, we can obtain

$$v(t, x) - u(t, x) < y(t, \epsilon) \quad \text{on} \quad \bar{H}.$$

The last two inequalities, together with the fact that

$$\lim_{\epsilon \to 0} y(t, \epsilon) \equiv r(t, t_0, y_0),$$

assure the stated estimate (10.4.4).

THEOREM 10.4.2. Let $u(t, x)$ be any solution of the mixed initial-boundary-value problem such that

(a) $|u(t, x) - v(t, x)| \leqslant r(t, t_0, y_0)$ on P_{t_0} and $\partial H - \partial H_\alpha$;
(b) for each $(t, x) \in \partial H_\alpha$, $\partial(|u - v|)/\partial \tau_0$ exists, and, for some $\beta > 0$,

$$\alpha(t, x) \frac{\partial(|u(t, x) - v(t, x)|)}{\partial \tau_0} + Q(t, x, |u(t, x) - v(t, x)|) \leqslant \beta r(t, t_0, y_0),$$

where $Q \in C[\partial H_\alpha \times R_+, R]$, $Q(t, x, z)$ is increasing in z for each (t, x), and $\beta r(t, t_0, y_0) \leqslant Q(t, x, r(t, t_0, y_0))$, other hypotheses (i), (ii), and (iii) of Theorem 10.4.1 being the same.

Then (10.4.4) is true.

REMARK 10.4.1. If $u(t, x)$, $v(t, x)$ are any two solutions of the boundary-value problem, we can deduce the estimate of the difference between them, as a consequence of Theorems 10.4.1 and 10.4.2. Similar remark holds for the theorem given below.

THEOREM 10.4.3. Assume that

(i) $f, G \in C[\bar{H} \times R \times R^n \times R^{n^2}, R]$, $G(t, x, z, P, R)$ is elliptic, and, if $z_1 > z_2$, $P_1 \geqslant P_2$,

$$f(t, x, z_1, P_1, R_1) - f(t, x, z_2, P_2, R_2)$$
$$\leqslant G(t, x, z_1 - z_2, P_1 - P_2, R_1 - R_2); \qquad (10.4.6)$$

(ii) $v \in C[\bar{H}, R]$, $v(t, x)$ possesses continuous partial derivatives v_t, v_x, v_{xx}, and

$$|T[v]| \leqslant \delta(t, x), \qquad (10.4.7)$$

where $\delta \in C[\bar{H}, R_+]$;

(iii)

$$G(t, x, z_1, P, R) - G(t, x, z_2, P, R) \leqslant g(t, z_1 - z_2), \quad z_1 > z_2, \quad (10.4.8)$$

10.4. APPROXIMATE SOLUTIONS AND UNIQUENESS

where $g \in C[J \times R_+, R]$, $g(t, 0) \equiv 0$, and the maximal solution $r(t, t_0, 0)$ of (10.2.6) is identically zero;

(iv) $z \in C[\bar{H}, R_+]$, $z_x \geqslant 0$, z_t, z_{xx} exist and are continuous in H, and

$$z_t \geqslant G(t, x, z, z_x, z_{xx}) + \delta(t, x). \tag{10.4.9}$$

Then,

$$|u(t, x) - v(t, x)| \leqslant z(t, x) \quad \text{on} \quad P_{t_0} \cup \partial H$$

implies

$$|u(t, x) - v(t, x)| \leqslant z(t, x) \quad \text{on} \quad H,$$

where $u(t, x)$ is any solution of the first initial-boundary-value problem.

Proof. As usual, we shall reduce the theorem to Lemma 10.1.1. We shall first consider the function

$$m(t, x) = u(t, x) - v(t, x) - z(t, x) - y(t, \epsilon),$$

where $y(t, \epsilon) = y(t, t_0, 0, \epsilon)$ is any solution of (10.2.9) for small $\epsilon > 0$. Suppose that $(t_1, x_1) \in P_{t_1}$, $m(t_1, x_1) = 0$, $m_x(t_1, x_1) = 0$, and, for some nonzero vector λ,

$$\sum_{i,k=1}^{n} \frac{\partial^2 m(t_1, x_1)}{\partial x_i \, \partial x_k} \lambda_i \lambda_k \leqslant 0.$$

Since $z(t, x) \geqslant 0$, $y(t, \epsilon) > 0$, the preceding supposition implies that

$$u(t_1, x_1) > v(t_1, x_1),$$
$$u_x(t_1, x_1) - v_x(t_1, x_1) = z_x(t_1, x_1) \geqslant 0,$$
$$\sum_{i,k=1}^{n} \frac{\partial^2 [\{u(t_1, x_1) - v(t_1, x_1)\} - z(t_1, x_1)]}{\partial x_i \, \partial x_k} \lambda_i \lambda_k \leqslant 0.$$

The ellipticity of G shows that

$$G(t_1, x_1, z(t_1, x_1) + y(t_1, \epsilon), z_x(t_1, x_1), u_{xx}(t_1, x_1) - v_{xx}(t_1, x_1))$$
$$\leqslant G(t_1, x_1, z(t_1, x_1) + y(t_1, \epsilon), z_x(t_1, x_1), z_{xx}(t_1, x_1)).$$

In view of this and the relations (10.4.6), (10.4.7), we have

$$u_t(t_1, x_1) - v_t(t_1, x_1)$$
$$\leqslant G(t_1, x_1, z(t_1, x_1) + y(t_1, \epsilon), z_x(t_1, x_1), z_{xx}(t_1, x_1))$$
$$+ \delta(t_1, x_1). \tag{10.4.10}$$

Hence, using (10.4.8) and (10.4.9), we derive

$$m_t(t_1, x_1) \leqslant G(t_1, x_1, z(t_1, x_1) + y(t_1, \epsilon), z_x(t_1, x_1), z_{xx}(t_1, x_1)) + \delta(t_1, x_1)$$
$$- G(t_1, x_1, z(t_1, x_1), z_x(t_1, x_1), z_{xx}(t_1, x_1)) - \delta(t_1, x_1)$$
$$- g(t_1, y(t_1, \epsilon)) - \epsilon$$
$$\leqslant -\epsilon < 0.$$

This shows that $m(t, x)$ verifies the assumptions of Lemma 10.1.1, and, therefore,

$$u(t, x) - v(t, x) < z(t, x) + y(t, \epsilon) \quad \text{on} \quad \bar{H}.$$

Arguing similarly, we can prove

$$v(t, x) - u(t, x) < z(t, x) + y(t, \epsilon) \quad \text{on} \quad \bar{H}.$$

Since $\lim_{\epsilon \to 0} y(t, \epsilon) \equiv 0$, by assumption, it follows from the preceding two inequalities that

$$|u(t, x) - v(t, x)| \leqslant z(t, x) \quad \text{on} \quad \bar{H},$$

proving the theorem.

We shall next consider the uniqueness problem.

THEOREM 10.4.4. Suppose that

(i) $f \in C[\bar{H} \times R \times R^n \times R^{n^2}, R]$, the operator T is parabolic, and

$$f(t, x, z_1, P, R) - f(t, x, z_2, P, R) \leqslant g(t, z_1 - z_2), \qquad z_1 > z_2;$$

(ii) $g \in C[J \times R_+, R]$, $g(t, 0) \equiv 0$, and the maximal solution $r(t, t_0, 0)$ of (10.2.6) is identically zero.

Under these assumptions, there is at most one solution to either one of the initial-boundary-value problems.

The proof is a direct consequence of Theorems 10.4.1 and 10.4.2.

10.5. Stability of steady-state solutions

Let problem D represent the partial differential equation of the form

$$u_t = f(x, u, u_x, u_{xx}), \tag{10.5.1}$$

for $x \in [a, b]$, $t > 0$, together with the boundary conditions

$$u_x(t, a) = f_1(u(t, a)), \qquad u_x(t, b) = f_2(u(t, b)), \qquad t > 0,$$

where f_1, f_2 are continuous functions with bounded derivatives. Let us assume that $\partial f/\partial u$ exists and is continuous, $f(x, u, P, R)$ is nondecreasing in R for each (x, u, P). We use the notation $u(t, x, \phi)$ to denote a solution of problem D such that

$$u(0, x, \phi) = \phi(x), \quad x \in [a, b],$$

where $\phi \in C[[a, b], R]$.

DEFINITION 10.5.1. Let $u(t, x, \phi)$ be a solution of problem D. We shall say that $u(t, x, \phi)$ is a *steady-state solution* if $u(t, x, \phi) = \phi(x)$, $t > 0$.

DEFINITION 10.5.2. The steady-state solution $u(t, x, \phi)$ of problem D is said to be *stable* if, given $\epsilon > 0$, there is a $\delta > 0$ such that

$$\max_{x \in [a,b]} |\phi(x) - \psi(x)| < \delta$$

implies

$$\max_{x \in [a,b]} |u(t, x, \phi) - u(t, x, \psi)| < \epsilon, \quad t \geq 0.$$

DEFINITION 10.5.3. Let

$$A = [(x, u): x \in [a, b] \quad \text{and} \quad \psi_1(x) \leq u \leq \psi_2(x)],$$

where ψ_1, ψ_2 are arbitrary functions, twice continuously differentiable on $[a, b]$. Let B be the set of functions on $[a, b]$ such that $\psi \in B$ implies

$$[(x, \psi(x)): x \in [a, b]] \subset A.$$

The steady-state solution $u(t, x, \phi)$ of problem D is said to be *asymptotically stable* if it is stable and

$$\lim_{t \to \infty} [\max_{x \in [a,b]} |u(t, x, \phi) - u(t, x, \psi)|] = 0,$$

whenever $\psi \in B$. The set A is called the *domain of attraction*.

Sufficient conditions for a steady-state solution $u(t, x, \phi)$ of problem D to be stable are given by the following theorem.

THEOREM 10.5.1. Assume that there exists a one-parameter family $v(x, \lambda)$, $\lambda \in [\lambda_1, \lambda_2]$, of solutions of the equation

$$f(x, v, v_x, v_{xx}) = 0 \tag{10.5.2}$$

fulfilling the following conditions:

(i) there is a $\lambda^* \in (\lambda_1, \lambda_2)$ such that $v_x(a, \lambda^*) = f_1(v(a, \lambda^*))$ and $v_x(b, \lambda^*) = f_2(v(b, \lambda^*))$;
(ii) $v_\lambda(x, \lambda) > 0$, $x \in [a, b]$, $\lambda \in [\lambda_1, \lambda_2]$;
(iii) $v_x(a, \lambda) > f_1(v(a, \lambda))$ and $v_x(b, \lambda) < f_2(v(b, \lambda))$, $\lambda \in [\lambda_1, \lambda^*)$;
(iv) $v_x(a, \lambda) < f_1(v(a, \lambda))$ and $v_x(b, \lambda) > f_2(v(b, \lambda))$, $\lambda \in (\lambda^*, \lambda_2]$.

Then, if $\phi(x) = v(x, \lambda^*)$, the steady-state solution $u(t, x, \phi)$ of problem D is stable.

Proof. Let $\epsilon > 0$ be given. Choose $\lambda_0 \in [\lambda_1, \lambda^*)$, $\lambda^0 \in (\lambda^*, \lambda_2]$ such that

$$\max_{x \in [a,b]} |v(x, \lambda^*) - v(x, \lambda_0)| < \epsilon \tag{10.5.3}$$

and

$$\max_{x \in [a,b]} |v(x, \lambda^0) - v(x, \lambda^*)| < \epsilon. \tag{10.5.4}$$

Then, define the number δ by

$$\delta = \min[\min_{x \in [a,b]} (v(x, \lambda^*) - v(x, \lambda_0)),$$
$$\min_{x \in [a,b]} (v(x, \lambda^0) - v(x, \lambda^*))]. \tag{10.5.5}$$

Since $v_\lambda(x, \lambda) > 0$ for all $x \in [a, b]$, it is clear that $\delta > 0$. Let $\psi(x)$ satisfy the inequality

$$\max_{x \in [a,b]} |\phi(x) - \psi(x)| < \delta,$$

and assume that $\phi(x) = v(x, \lambda^*)$. We then have, by (10.5.5),

$$v(x, \lambda_0) \leq \phi(x) - \delta < \psi(x) < \phi(x) + \delta \leq v(x, \lambda_0), \qquad x \in [a, b].$$

The fact that $v(x, \lambda)$ is a family of solutions of (10.5.2) and $u(t, x, \psi)$ is a solution to the problem D implies that

$$f(x, v(x, \lambda_0), v_x(x, \lambda_0), v_{xx}(x, \lambda_0)) - v_t(x, \lambda_0)$$
$$= f(x, u(t, x, \psi), u_x(t, x, \psi), u_{xx}(t, x, \psi)) - u_t(t, x, \psi)$$
$$= f(x, v(x, \lambda^0), v_x(x, \lambda^0), v_{xx}(x, \lambda^0)) - v_t(x, \lambda^0).$$

By a successive application of Theorem 10.2.4, we deduce that

$$v(x, \lambda_0) < u(t, x, \psi) < v(x, \lambda^0), \qquad x \in [a, b], \qquad t \geq 0.$$

This, because of the relations (10.5.3) and (10.5.4), yields

$$u(t, x, \phi) - \epsilon = v(x, \lambda^*) - \epsilon < v(x, \lambda_0) < u(t, x, \psi)$$

and

$$u(t, x, \phi) + \epsilon = v(x, \lambda^*) + \epsilon > v(x, \lambda^0) > u(t, x, \psi).$$

It is evident from the preceding inequalities that

$$\max_{x \in [a,b]} |u(t, x, \phi) - u(t, x, \psi)| < \epsilon, \qquad t \geqslant 0,$$

whenever $\max_{x \in [a,b]} |\phi(x) - \psi(x)| < \delta$. The proof is complete.

In the situation in which it is difficult or impossible to find a one-parameter family $v(x, \lambda)$, satisfying the conditions of Theorem 10.5.1, it may still be possible to find an upper bound. This we state as a corollary.

COROLLARY 10.5.1. *Suppose that there exists a solution $v(x)$ of (10.5.2) satisfying*

$$v_x(a) < f_1(v(a)) \qquad \text{and} \qquad v_x(b) > f_2(v(b)).$$

Then, if $u(t, x, \psi)$ is a solution to problem D such that

$$\psi(x) < v(x), \qquad x \in [a, b],$$

we have

$$u(t, x, \psi) < v(x), \qquad x \in [a, b], \qquad t \geqslant 0.$$

A similar corollary may be stated establishing a lower bound.

THEOREM 10.5.2. *Let the hypotheses of Theorem 10.5.1 hold. Suppose further that, for $x \in [a, b]$ and $\lambda \in [\lambda_1, \lambda_2]$,*

$$f_v(x, v(x, \lambda), v_x(x, \lambda), v_{xx}(x, \lambda)) \neq 0. \tag{10.5.6}$$

Then, if $\phi(x) = v(x, \lambda^*)$, the steady-state solution $u(t, x, \phi)$ is asymptotically stable, and the set

$$A = [(x, u): x \in [a, b] \qquad \text{and} \qquad v(x, \lambda_1) < u < v(x, \lambda_2)] \tag{10.5.7}$$

is a region of attraction.

Proof. The stability of steady-state solution $u(t, x, \phi)$ follows by Theorem 10.5.1. Let A be the set defined by (10.5.7), and let B be the set of functions such that $\psi \in B$ implies $[(x, \psi(x)) : x \in [a, b]] \subset A$. We

shall first show that, for any $\epsilon > 0$ and any $\psi \in B$, there exists a $T_1 > 0$ such that
$$\max_{\substack{x \in [a,b] \\ t > T_1}} [u(t, x, \psi) - u(t, x, \phi)] < \epsilon. \tag{10.5.8}$$

Let $\epsilon > 0$ be given, and let $\lambda^0 \in (\lambda^*, \lambda_2]$ be such that
$$v(x, \lambda^0) - v(x, \lambda^*) < \epsilon, \quad x \in [a, b]. \tag{10.5.9}$$

Without loss of generality, we may assume that $f_v > 0$, in view of (10.5.6). We then define three positive numbers μ_1, μ_2, μ_3 by
$$\mu_1 = \min_{x \in [a,b]} [v(x, \lambda^0) - v(x, \lambda^*)],$$
$$\mu_2 = \min_{\substack{x \in [a,b] \\ \lambda \in [\lambda^*, \lambda_2]}} [f_v(x, v(x, \lambda), v_x(x, \lambda), v_{xx}(x, \lambda))],$$
and
$$\mu_3 = \max_{\substack{x \in [a,b] \\ \lambda \in [\lambda_1, \lambda_2]}} [v_\lambda(x, \lambda)].$$

Let $\bar{\lambda} \in (\lambda^*, \lambda^0)$ be such that
$$v(x, \bar{\lambda}) - v(x, \lambda^*) \leqslant \mu_1/2, \quad x \in [a, b].$$

Define a positive number μ_4 by
$$\mu_4 = \min_{\lambda \in [\bar{\lambda}, \lambda_2]} [\{f_1(v(a, \lambda)) - v_x(a, \lambda)\}, \{v_x(b, \lambda) - f_2(v(b, \lambda))\}]. \tag{10.5.10}$$

Let $H(\lambda)$ be a function defined for $\lambda \in [\lambda^0, \lambda_2]$, such that, for $h < H(\lambda)$, we have
$$f_1(v(a, \lambda) - h) - f_1(v(a, \lambda)) > \mu_4/2, \tag{10.5.11}$$
$$f_2(v(b, \lambda) - h) - f_2(v(b, \lambda)) < \mu_4/2. \tag{10.5.12}$$

Consider the function
$$w = w(x, \lambda) = v(x, \lambda) - \delta,$$
where $\delta > 0$ will be specified later. Since f_v exists, for sufficiently small $\delta > 0$, we have
$$\begin{aligned} f(x, w, w_x, w_{xx}) &= f(x, v - \delta, v_x, v_{xx}) \\ &= f(x, v, v_x, v_{xx}) - \delta f_v(x, v, v_x, v_{xx}) + O(\delta^2) \\ &\leqslant f(x, v, v_x, v_{xx}) - \delta \mu_2 + O(\delta^2) \\ &< f(x, v, v_x, v_{xx}) \\ &= 0. \end{aligned} \tag{10.5.13}$$

10.5. STABILITY OF STEADY-STATE SOLUTIONS

We let $\delta_2 > 0$ be such that

$$\delta_2 < \min_{x \in [a,b]} [v(x, \lambda_2) - \psi(x)].$$

Let us now choose a positive δ so small that (10.5.13) holds, and also

$$\delta \leqslant \min[\mu_1/3, \delta_1, \delta_2], \qquad (10.5.14)$$

where

$$\delta_1 = \min_{\lambda \in [\lambda^0, \lambda_2]} H(\lambda). \qquad (10.5.15)$$

It follows, from the inequality (10.5.13), that there exists a $\mu_5 > 0$ satisfying

$$f(x, w, w_x, w_{xx}) < -\mu_5, \qquad \lambda \in [\lambda^0, \lambda_2], \qquad x \in [a, b].$$

Let $m = m(t, x) = w(x, \lambda(t))$, where

$$\lambda(t) = \lambda^* + (\lambda_2 - \lambda^*) e^{-pt},$$

$p > 0$, being a number to be specified shortly. Then,

$$\begin{aligned} f(x, m, m_x, m_{xx}) - m_t &= f(x, w, w_x, w_{xx}) \\ &\quad + w_\lambda p(\lambda_2 - \lambda^*) e^{-pt} \\ &\leqslant -\mu_5 + \mu_3 p(\lambda_2 - \lambda^*). \end{aligned} \qquad (10.5.16)$$

Choose $p = \mu_5/\mu_3(\lambda_2 - \lambda^*)$ so that the right-hand side of (10.5.16) is equal to zero. Then the inequality

$$f(x, m, m_x, m_{xx}) - m_t \leqslant f(x, u, u_x, u_{xx}) - u_t = 0$$

is verified for $x \in [a, b]$, $t \in [0, T_1]$, where T_1 is the solution of the equation $\lambda(T_1) = \lambda^0$. The inequality (10.5.11), together with (10.5.14) and (10.5.15), shows that

$$f_1(v(a, \lambda) - \delta) - f_1(v(a, \lambda)) > -(\mu_4/2). \qquad (10.5.17)$$

From the definition of w, we deduce, in view of (10.5.10) and (10.5.17), that

$$\begin{aligned} f_1(w(a, \lambda)) &= f_1(v(a, \lambda) - \delta) \\ &> f_1(v(a, \lambda)) - (\mu_4/2) \\ &> v_x(a, \lambda) \\ &= w_x(a, \lambda). \end{aligned} \qquad (10.5.18)$$

Similarly, from (10.5.10), (10.5.12), (10.5.14), and (10.5.15), we derive

$$f_2(w(b, \lambda)) = f_2(v(b, \lambda) - \delta) < f_2(v(b, \lambda)) + (\mu_4/2)$$
$$< v_x(b, \lambda) = w_x(b, \lambda). \qquad (10.5.19)$$

Clearly, (10.5.18) and (10.5.19) hold for all $\lambda \in [\lambda^*, \lambda_2]$. On the basis of Theorem 10.2.4, it follows that

$$m(t, x) > u(t, x, \psi), \qquad x \in [a, b], \qquad t \in [0, T_1],$$

since $m(0, x) > \psi(x)$, $x \in [a, b]$. From the definition of m and T_1, we get

$$v(x, \lambda^0) > u(T_1, x, \psi), \qquad x \in [a, b].$$

Thus, by Corollary 10.5.1,

$$v(x, \lambda^0) > u(t, x, \psi), \qquad x \in [a, b], \qquad t > T_1,$$

which, together with (10.5.9), gives us (10.5.8).

The next step of the proof is to show that there exists a $T_2 > 0$ such that

$$\max_{\substack{x \in [a,b] \\ t > T_2}} [u(t, x, \phi) - u(t, x, \psi)] < \epsilon. \qquad (10.5.20)$$

The proof of this consists of showing that there is a lower bound for $u(t, x, \psi)$ which can be increased with time until it is within ϵ of $u(t, x, \phi)$ at some time T_2. The proof of this fact is similar to the first part and differs only in minor details.

Let $T = \max[T_1, T_2]$. Then, from (10.5.8) and (10.5.20), we obtain

$$\max_{\substack{x \in [a,b] \\ t > T}} |u(t, x, \phi) - u(t, x, \psi)| < \epsilon.$$

This completes the proof of the theorem.

COROLLARY 10.5.2. The conclusion of Theorem 10.5.2 remains valid if (10.5.6) is replaced by either

(i) $f_v = 0$ and $f_{v_x}(x, v(x, \lambda), v_x(x, \lambda), v_{xx}(x, \lambda)) \neq 0$; or

(ii) $f_v = 0$, $f_{v_x} = 0$, and $f_{v_{xx}}(x, v(x, \lambda), v_x(x, \lambda), v_{xx}(x, \lambda)) \neq 0$ for $x \in [a, b]$, $\lambda \in [\lambda_1, \lambda_2]$.

We now give an example to illustrate Theorems 10.5.1 and 10.5.2. Consider the partial differential equation

$$u_. = (1 + u^2) u_{xx} - u u_x^2, \qquad x \in [1, 2],$$

subject to the boundary conditions

$$u_x(t, 1) = f_1(u(t, 1)) \quad \text{and} \quad u_x(t, 2) = f_2(u(t, 2)).$$

The equation

$$(1 + u^2) u_{xx} - u u_x^2 = 0$$

has a one-parameter family of solutions $v(x, \lambda)$ given by $v(x, \lambda) = \sinh \lambda x$. Notice that

$$v_x(x, \lambda) = x^{-1}[1 + v^2(x, \lambda)]^{1/2} \sinh^{-1} v(x, \lambda).$$

Suppose that

$$f_1(u) > (1 + u^2)^{1/2} \sinh^{-1} u, \quad u > 0, \quad f_1(0) = 0,$$
$$f_1(u) < (1 + u^2)^{1/2} \sinh^{-1} u, \quad u < 0,$$
$$f_2(u) < \tfrac{1}{2}(1 + u^2)^{1/2} \sinh^{-1} u, \quad u > 0, \quad f_2(0) = 0,$$
$$f_2(u) > \tfrac{1}{2}(1 + u^2)^{1/2} \sinh^{-1} u, \quad u < 0.$$

Then, by Theorem 10.5.1, $u(t, x, \phi) \equiv 0$ is stable. To apply Theorem 10.5.2, we observe that

$$f_v(x, v, v_x, v_{xx}) = 2vv_{xx} - v_x^2$$
$$= 2\lambda^2 \sinh^2 \lambda x - \lambda^2 \cosh^2 \lambda x$$
$$= \lambda^2(\sinh^2 \lambda x - 1).$$

Thus, $f_v < 0$ if $\sinh^2 \lambda x < 1$ or $u < 1$. Hence, $u(t, x, \phi) = 0$ is asymptotically stable, by Theorem 10.5.2.

10.6. Systems of parabolic differential inequalities in bounded domains

Let us consider now a partial differential system of the type

$$u_t^i = f^i(t, x, u, u_x^i, u_{xx}^i), \quad i = 1, 2, \ldots, N,$$

where

$$u_x^i = (u_{x_1}^i, u_{x_2}^i, \ldots, u_{x_n}^i),$$

$$u_{xx}^i = (u_{x_1 x_1}^i, u_{x_1 x_2}^i, \ldots, u_{x_n x_n}^i).$$

For convenience, we shall write the preceding system in the form

$$u_t = f(t, x, u, u_x^i, u_{xx}^i), \qquad (10.6.1)$$

where $f \in C[\bar{H} \times R^N \times R^n \times R^{n^2}, R^N]$ and each function f^i is elliptic, so that the system is parabolic.

We shall first state the following lemmas, which are extensions of Lemmas 10.1.1 and 10.1.2.

LEMMA 10.6.1. Suppose that

(i) $m \in C[\bar{H}, R^N]$, $m(t, x)$ possesses continuous partial derivatives m_t, m_x, m_{xx} in H;

(ii) $m(t, x) < 0$ on $P_{t_0} \cup \partial H$;

(iii) for any $(t_1, x_1) \in P_{t_1}$ and an index j, $1 \leqslant j \leqslant N$, if $m^j(t_1, x_1) = 0$, $m^i(t_1, x_1) \leqslant 0$, $i \neq j$, $m_x^j(t_1, x_1) = 0$, and the quadratic form

$$\sum_{i,k=1}^n \frac{\partial^2 m^j(t_1, x_1)}{\partial x_i \, \partial x_k} \lambda_i \lambda_k \leqslant 0,$$

λ being an arbitrary vector, then $m^j(t_1, x_1) < 0$.

Under these assumptions, we have

$$m(t, x) < 0 \quad \text{on} \quad \bar{H}. \tag{10.6.2}$$

Proof. Assume, if possible, that the set

$$Z = \bigcup_{i=1}^N [(t, x) \in \bar{H} : m^i(t, x) \geqslant 0]$$

is nonempty. Let Z_t be the projection of Z on t axis and $t_1 = \inf Z$. It follows from condition (ii) that $t_1 > t_0$. Since the set Z is closed and condition (ii) holds, we conclude that

$$m^i(t_1, x) \leqslant 0 \quad \text{on} \quad \bar{H} \cap [t_0, t_1],$$

for $i = 1, 2, ..., N$, and there is an index j, $1 \leqslant j \leqslant N$, and an $x_1 \in \text{int } P_{t_1}$ such that

$$m^j(t_1, x_1) = 0.$$

We therfore have

$$m_t^j(t_1, x_1) \geqslant 0. \tag{10.6.3}$$

On the other hand, since $m^j(t, x)$ attains its maximum at $(t_1, x_1) \in \text{int } P_{t_1}$, we get

$$m_x^j(t_1, x_1) = 0,$$

and, for some nonzero vector λ,

$$\sum_{i,k=1}^{n} \frac{\partial^2 m^j(t_1, x_1)}{\partial x_i \, \partial x_k} \lambda_i \lambda_k \leqslant 0.$$

By condition (iii), it follows that

$$m_t^j(t_1, x_1) < 0,$$

contradicting (10.6.3). Hence, the inequality (10.6.2) is true.

Let $\alpha \in C[\partial H, R^N]$, and denote by ∂H_α^i that part of the boundary ∂H on which $\alpha^i(t, x) > 0$. Corresponding to this change, we may define, as before, for each $i = 1, 2, ..., N$, and $(t, x) \in \partial H_\alpha^i$, $\partial u^i/\partial \tau^i$, $\partial u^i/\partial \tau_0^i$, τ^i, τ_0^i being the directions. Then, we have an extension of Lemma 10.1.2 which may be proved by combining the proofs of Lemmas 10.1.2 and 10.6.1.

LEMMA 10.6.2. Let condition (ii) of Lemma 10.6.1 be replaced by

(iia) $m^i(t, x) < 0$ on P_{t_0} and $\partial H - \partial H_\alpha^i$;

(iib) for any index j, $1 \leqslant j \leqslant N$ and $(t_1, x_1) \in \partial H_\alpha^j$, if $m^j(t_1, x_1) = 0$, $m^i(t_1, x_1) = 0$, $i \neq j$, then

$$\frac{\partial m^j(t_1, x_1)}{\partial \tau^j} < 0,$$

where, for each $(t, x) \in \partial H_\alpha^i$, $\partial m^i/\partial \tau$ is assumed to exist.

Then, the assertion (10.6.2) is true if the other assumptions remain the same.

We are now in a position to extend the previous results to systems of parabolic inequalities. We shall present only some typical theorems.

THEOREM 10.6.1. Assume that

(i) $u, v \in C[\bar{H}, R^N]$, the partial derivatives u_t, u_x, u_{xx}, v_t, v_x, v_{xx} exist and are continuous in H;

(ii) $f \in C[\bar{H} \times R^N \times R^n \times R^{n^2}, R^N]$, the differential operator T is parabolic, the function $f(t, x, u, P, R)$ is quasi-monotone nondecreasing in u for each (t, x, P, R), and

$$T[u] < T[v] \quad \text{on} \quad H;$$

(iii) $u(t, x) < v(t, x)$ on $P_{t_0} \cup \partial H$.

Then, we have

$$u(t, x) < v(t, x) \quad \text{on} \quad \bar{H}. \qquad (10.6.4)$$

Proof. We shall show that the vector function
$$m(t, x) = u(t, x) - v(t, x)$$
satisfies the hypotheses of Lemma 10.6.1. In view of conditions (i) and (ii) of Theorem 10.6.1, it is sufficient to check assumption (iii) of Lemma 10.6.1. Suppose that $(t_1, x_1) \in P_{t_1}$, $m^j(t_1, x_1) = 0$, $1 \leqslant j \leqslant N$, $m^i(t_1, x_1) \leqslant 0$, $i \neq j$, $m_x^j(t_1, x_1) = 0$, and
$$\sum_{i,k=1}^n \frac{\partial^2 m^j(t_1, x_1)}{\partial x_i \, \partial x_k} \lambda_i \lambda_k \leqslant 0,$$
for an arbitrary vector λ. This means that
$$u^j(t_1, x_1) = v^j(t_1, x_1),$$
$$u^i(t_1, x_1) \leqslant v^i(t_1, x_1), \qquad i \neq j, \tag{10.6.5}$$
$$u_x^j(t_1, x_1) = v_x^j(t_1, x_1),$$
and
$$\sum_{i,k=1}^n \frac{\partial^2 [u^j(t_1, x_1) - v^j(t_1, x_1)]}{\partial x_i \, \partial x_k} \lambda_i \lambda_k \leqslant 0.$$

Since f^j is elliptic, we get
$$f^j(t_1, x_1, u(t_1, x_1), u_x^j(t_1, x_1), u_{xx}^j(t_1, x_1))$$
$$\leqslant f^j(t_1, x_1, u(t_1, x_1), u_x^j(t_1, x_1), v_{xx}^j(t_1, x_1)).$$

This, together with (10.6.5) and quasi-monotonicity of f^j, implies that
$$f^j(t_1, x_1, u(t_1, x_1), u_x^j(t_1, x_1), u_{xx}^j(t_1, x_1))$$
$$\leqslant f^j(t_1, x_1, v(t_1, x_1), v_x^j(t_1, x_1), v_{xx}^j(t_1, x_1)). \tag{10.6.6}$$

The inequality $T[u] < T[v]$ now yields
$$m_t^j(t_1, x_1) < f^j(t_1, x_1, u(t_1, x_1), u_x^j(t_1, x_1), u_{xx}^j(t_1, x_1))$$
$$- f^j(t_1, x_1, v(t_1, x_1), v_x^j(t_1, x_1), v_{xx}^j(t_1, x_1))$$
$$\leqslant 0,$$
in view of (10.6.6). Hence, the validity of (10.6.4) follows by Lemma 10.6.1.

10.6. SYSTEMS OF PARABOLIC INEQUALITIES

THEOREM 10.6.2. Let assumptions (i) and (ii) of Theorem 10.6.1 hold. Suppose further that

(iii*) $u^i(t, x) < v^i(t, x)$ on P_{t_0} and $\partial H - \partial H_\alpha^i$, $i = 1, 2, ..., N$;
(iv) for each $(t, x) \in \partial H_\alpha^i$, $\partial u^i/\partial \tau^i$, $\partial v^i/\partial \tau^i$ exist, and

$$\alpha^i(t, x) \frac{\partial u^i(t, x)}{\partial \tau^i} + Q^i(t, x, u(t, x))$$
$$< \alpha^i(t, x) \frac{\partial v^i(t, x)}{\partial \tau^i} + Q^i(t, x, v(t, x)) \quad \text{on} \quad \partial H_\alpha^i,$$

where $Q \in C[\partial H \times R^N, R^N]$, $Q(t, x, u)$ is quasimonotone nondecreasing in u for each (t, x).

Then, the inequality (10.6.4) remains valid.

Proof. The proof proceeds as in Theorem 10.6.1. To verify the assumptions of Lemma 10.6.2, we have only to check condition (iib).

Let $1 \leqslant j \leqslant N$ and $(t_1, x_1) \in \partial H_\alpha^j$. Suppose that $m^j(t_1, x_1) = 0$ and $m^i(t_1, x_1) \leqslant 0$, $i \neq j$. This implies that

$$u^j(t_1, x_1) = v^j(t_1, x_1),$$
$$u^i(t_1, x_1) \leqslant v^i(t_1, x_1), \quad i \neq j.$$

Hence, by the quasi-monotonicity character of Q, we have

$$Q^j(t_1, x_1, u(t_1, x_1)) \leqslant Q^j(t_1, x_1, v(t_1, x_1)).$$

Thus, there results the inequality

$$\alpha^j(t_1, x_1) \frac{\partial m^j(t_1, x_1)}{\partial \tau^j} < Q^j(t_1, x_1, v(t_1, x_1)) - Q^j(t_1, x_1, v(t_1, x_1))$$
$$= 0,$$

which assures that $\partial m^j/\partial \tau^j(t_1, x_1) < 0$, since $\alpha^j(t_1, x_1) > 0$ on ∂H_α^j. This completes the proof.

It is now easy to formulate and prove comparison theorems, componentwise bounds, and error estimates for systems of parabolic differential equations. We shall not include such results. However, the following comparison theorem will be needed later, and, hence, we shall merely state it.

THEOREM 10.6.3. Suppose that

(i) $m \in C[\bar{H}, R_+^N]$, $m(t, x)$ possesses continuous partial derivatives m_t, m_x, m_{xx} in H;

(ii) $f \in C[\bar{H} \times R^N \times R^n \times R^{n^2}, R^N]$, the differential operator T is parabolic, and $T[m] \leq 0$ on H;

(iii) $g \in C[J \times R_+^N, R^N]$, $g(t, y)$ is quasi-monotone nondecreasing in y for each $t \in J$, $r(t, t_0, y_0) \geq 0$ is the maximal solution of the differential system

$$y' = g(t, y), \quad y(t_0) = y_0 \geq 0$$

existing for $t \geq t_0$, and

$$f(t, x, z, 0, 0) \leq g(t, z), \quad z \geq 0;$$

(iv) $m(t, x) \leq r(t, t_0, y_0)$ on $P_{t_0} \cup \partial H$.

Then,

$$m(t, x) \leq r(t, t_0, y_0) \quad \text{on} \quad \bar{H}.$$

10.7. Lyapunov-like functions

Let us continue to consider the partial differential system (10.6.1). We shall restrict ourselves to the first initial-boundary-value problem of such systems. In what follows, a solution will always mean according to Definition 10.3.1, unless specified otherwise.

Let $V \in C[\bar{H} \times R^N, R_+]$ and $V = V(t, x, u)$ possess continuous partial derivatives with respect to t and the components of x, u. Let

$$V_x = \left[\frac{\partial V}{\partial x^\nu} + \frac{\partial u_\mu}{\partial x^\nu} \frac{\partial V}{\partial u_\mu} ; \begin{matrix} \mu = 1, 2, \ldots, N, \\ \nu = 1, 2, \ldots, n \end{matrix} \right],$$

$$V_{xx} = \left[\frac{\partial^2 V}{\partial x^\nu \partial x^{\nu_1}} + \frac{\partial^2 u_\mu}{\partial x^\nu \partial x^{\nu_1}} \frac{\partial V}{\partial u_\mu} + \frac{\partial u_\mu}{\partial x^\nu} \frac{\partial^2 V}{\partial u_\mu \partial u_{\mu_1}} \frac{\partial u_{\mu_1}}{\partial x^{\nu_1}} ; \right.$$

$$\left. \begin{matrix} \mu, \mu_1 = 1, 2, \ldots, N, \\ \nu, \nu_1 = 1, 2, \ldots, n \end{matrix} \right].$$

The following theorem offers an estimate of solutions in terms of Lyapunov-like functions and is useful subsequently.

THEOREM 10.7.1. Suppose that

(i) $f \in C[\bar{H} \times R^N \times R^n \times R^{n^2}, R^N]$, $G \in C[\bar{H} \times R_+ \times R^n \times R^{n^2}, R]$, G is elliptic, and

$$\frac{\partial V}{\partial t} + \frac{\partial V}{\partial u} f(t, x, u, u_x^i, u_{xx}^i) \leq G(t, x, V(t, x, u), V_x, V_{xx}), \quad (10.7.1)$$

10.7. LYAPUNOV-LIKE FUNCTIONS

where $V \in C[\bar{H} \times R^N, R_+]$ and V_x, V_{xx} are given in the foregoing;

(ii) $g \in C[J \times R_+, R]$ and

$$G(t, x, z, 0, 0) \leqslant g(t, z); \qquad (10.7.2)$$

(iii) the maximal solution $r(t, t_0, y_0)$ of the differential equation (10.2.6) exists for $t \geqslant t_0$;

(iv) $u(t, x)$ is any solution of the partial differential system (10.6.1) such that

$$V(t, x, \phi(t, x)) \leqslant r(t, t_0, y_0) \quad \text{on} \quad P_{t_0} \cup \partial H. \qquad (10.7.3)$$

These assumptions imply

$$V(t, x, u(t, x)) \leqslant r(t, t_0, y_0) \quad \text{on} \quad \bar{H}. \qquad (10.7.4)$$

Proof. Let $u(t, x)$ be any solution of (10.6.1) satisfying (10.7.3). Define the function

$$m(t, x) = V(t, x, u(t, x)).$$

Then,

$$m(t, x) \leqslant r(t, t_0, y_0) \quad \text{on} \quad P_{t_0} \cup \partial H.$$

Moreover, because of (10.7.1), we obtain

$$\frac{\partial m(t, x)}{\partial t} \leqslant G(t, x, m(t, x), m_x(t, x), m_{xx}(t, x)).$$

Now, a straightforward application of Theorem 10.2.1 yields the estimate (10.7.4).

THEOREM 10.7.2. *Let the assumptions of Theorem 10.7.1 remain the same except that the relations (10.7.1) and (10.7.3) are replaced by*

$$A(t) \left[\frac{\partial V}{\partial t} + \frac{\partial V}{\partial u} \cdot f(t, x, u, u_x^i, u_{xx}^i) \right] + V(t, x, u) A'(t)$$
$$\leqslant G(t, x, A(t) V(t, x, u), A(t) V_x, A(t) V_{xx}), \qquad (10.7.5)$$

where $A(t) > 0$ is continuously differentiable on J and

$$A(t) V(t, x, \phi(t, x)) \leqslant r(t, t_0, y_0) \quad \text{on} \quad P_{t_0} \cup \partial H.$$

Then, the inequality (10.7.4) takes the form

$$A(t) V(t, x, u(t, x)) \leqslant r(t, t_0, y_0) \quad \text{on} \quad \bar{H}. \qquad (10.7.6)$$

Proof. Defining
$$V_1(t, x, u) = A(t) \, V(t, x, u),$$
it can be easily checked that $V_1(t, x, u)$ preserves the properties of $V(t, x, u)$ in Theorem 10.7.1, and hence (10.7.6) follows from Theorem 10.7.1.

REMARK 10.7.1. Taking $A(t) \equiv 1$, we see that Theorem 10.7.2 reduces to Theorem 10.7.1. Since Theorem 10.7.1 is an important result in itself in the study of various problems of partial differential equations, we have listed it separately. We note that $g(t, z)$ in (10.7.2) need not be nonnegative, and, hence, this has an advantage in obtaining sharper bounds. For example, taking $V = \|u\|^2$ and $g(t, u) = L(t)u$, where $L(t)$ is continuous on J, one can get an upper bound, from Theorem 10.7.1, as follows:

$$\|u(t, x)\| \leqslant y_0 \exp\left[\frac{1}{2} \int_{t_0}^{t} L(s) \, ds\right] \quad \text{on} \quad \bar{H},$$

provided

$$\|\phi(t, x)\| \leqslant y_0 \exp\left[\frac{1}{2} \int_{t_0}^{t} L(s) \, ds\right] \quad \text{on} \quad P_{t_0} \cup \partial H,$$

where
$$\max_{x \in P_{t_0}} \|\phi(t_0, x)\| \leqslant y_0.$$

If we assume that $V(t, x, u) \equiv 0$ if and only if $u = 0$, Theorem 10.7.1 may be used to prove a uniqueness result.

THEOREM 10.7.3. Assume that

(i) $f \in C[\bar{H} \times R^N \times R^n \times R^{n^2}, R^N]$, $G \in C[\bar{H} \times R_+ \times R^n \times R^{n^2}, R]$, G is elliptic, and

$$\frac{\partial V(t, u-v)}{\partial t} + \frac{\partial V(t, u-v)}{\partial u} \cdot [f(t, x, u, u_x^i, u_{xx}^i) - f(t, x, v, v_x^i, v_{xx}^i)]$$
$$\leqslant G(t, x, V(t, x, u-v), V_x(t, x, u-v), V_{xx}(t, x, u-v));$$

(ii) $g \in C[J \times R_+, R]$, $g(t, 0) \equiv 0$, and
$$G(t, x, z, 0, 0) \leqslant g(t, z), \quad z > 0;$$

(iii) the maximal solution $r(t, t_0, 0)$ of (10.2.9) is identically zero.

Then, the parabolic differential system (10.6.1) admits a unique solution.

10.7. LYAPUNOV-LIKE FUNCTIONS

Let $V \in C[\bar{H} \times R^N, R_+^N]$, and suppose that $V(t, x, u)$ possesses continuous partial derivatives V_t, V_x, V_{xx}, V_u. Then, the following extension of Theorem 10.7.1 may be proved.

THEOREM 10.7.4. Let the following assumptions hold:

(i) $f \in C[\bar{H} \times R^N \times R^n \times R^{n^2}, R^N]$, $G \in C[\bar{H} \times R_+^N \times R^n \times R^{n^2}, R^N]$, G is elliptic, and

$$\frac{\partial V}{\partial t} + \frac{\partial V}{\partial u} \cdot f(t, x, u, u_x{}^i, u_{xx}^i) \leqslant G(t, x, V(t, x, u), V_x{}^i, V_{xx}^i); \quad (10.7.7)$$

(ii) $g \in C[J \times R_+^N, R^N]$, $g(t, y)$ is quasi-monotone in y for each $t \in J$, and

$$G(t, x, z, 0, 0) \leqslant g(t, z), \qquad z \geqslant 0; \quad (10.7.8)$$

(iii) the maximal solution $r(t, t_0, y_0)$ of the ordinary differential system

$$y' = g(t, y), \qquad y(t_0) = y_0 \geqslant 0$$

exists for $t \geqslant t_0$;

(iv) $u(t, x)$ is any solution of the system (10.7.1) such that

$$V(t, x, \phi(t, x)) \leqslant r(t, t_0, y_0) \quad \text{on} \quad P_{t_0} \cup \partial H. \quad (10.7.9)$$

Under these assumptions, we have

$$V(t, x, u(t, x)) \leqslant r(t, t_0, y_0) \quad \text{on} \quad \bar{H}.$$

Proof. Consider the vector function

$$m(t, x) = V(t, x, u(t, x)),$$

where $u(t, x)$ is any solution of (10.6.1) such that (10.7.9) holds. By the assumption (10.7.7), we deduce the inequality

$$m(t, x) \leqslant G(t, x, m(t, x), m_x{}^i(t, x), m_{xx}^i(t, x)).$$

Since $m(t, x)$ satisfies the assumptions of Theorem 10.6.3, it follows that

$$m(t, x) \leqslant r(t, t_0, y_0) \quad \text{on} \quad \bar{H},$$

and the proof is complete.

10.8. Stability and boundedness

Let $u(t, x)$ be any solution of the partial differential system (10.6.1). We shall assume that the system (10.6.1) possesses the identically zero solution. Denote

$$\| u(t, \cdot) \|_{P_t} = \max_{x \in P_t} \| u(t, x) \|.$$

DEFINITION 10.8.1. The trivial solution of the partial differential system (10.6.1) is said to be *equistable* if, for each $\epsilon > 0$ and $t_0 \in J$, there exists a positive function $\delta(t_0, \epsilon)$, which is continuous in t_0 for each ϵ, such that

(i) $\quad \| \phi(t_0, \cdot) \|_{P_{t_0}} \leq \delta(t_0, \epsilon);$

(ii) $\quad \| \phi(t, \cdot) \|_{\partial H} < \epsilon, \quad t \geq t_0,$

imply

$$\| u(t, \cdot) \|_{P_t} < \epsilon, \quad t \geq t_0.$$

On the basis of this definition, it is easy to formulate the various definitions of stability and boundedness analogous to those in the earlier chapters.

We shall now give sufficient conditions for stability and boundedness of solutions of the parabolic differential system (10.6.1).

THEOREM 10.8.1. Assume that there exist functions $V(t, x, u)$ and $g(t, y)$ satisfying the following conditions:

(i) $g \in C[J \times R_+, R]$, $g(t, 0) \equiv 0$;

(ii) $V \in C[\bar{H} \times R^N, R_+]$, $V(t, x, u)$ possesses continuous partial derivatives V_t, V_x, V_{xx}, V_u in H, and

$$b(\| u \|) \leq V(t, x, u) \leq a(t, \| u \|), \tag{10.8.1}$$

where $b \in \mathcal{K}$, $a \in C[J \times R_+, R_+]$, $a \in \mathcal{K}$ for each $t \in J$;

(iii) $f \in C[\bar{H} \times R^N \times R^n \times R^{n^2}, R^N]$, $G \in C[\bar{H} \times R_+ \times R^n \times R^{n^2}, R]$ G is elliptic, and

$$\frac{\partial V}{\partial t} + \frac{\partial V}{\partial u} \cdot f(t, x, u, u_x^i, u_{xx}^i) \leq G(t, x, V(t, x, u), V_x, V_{xx});$$

(iv) $G(t, x, z, 0, 0) \leq g(t, z), z \geq 0.$

10.8. STABILITY AND BOUNDEDNESS

Then, the equistability of the trivial solution of the ordinary scalar differential equation

$$y' = g(t, y), \quad y(t_0) = y_0 \geq 0 \qquad (10.8.2)$$

implies the equistability of the trivial solution of the partial differential system (10.6.1).

Proof. Let $\epsilon > 0$ and $t_0 \in J$ be given. Assume that the trivial solution of (10.8.2) is equistable. Then, given $b(\epsilon) > 0$ and $t_0 \in J$, there exists a $\delta = \delta(t_0, \epsilon) > 0$ that is continuous in t_0 for each ϵ such that

$$y(t, t_0, y_0) < b(\epsilon), \quad t \geq t_0, \qquad (10.8.3)$$

provided $y_0 \leq \delta$, where $y(t, t_0, y_0)$ is any solution of (10.8.2). Let $u(t, x)$ be any solution of (10.6.1) such that

$$\begin{aligned} V(t_0, x, \phi(t_0, x)) &\leq y_0, & x \in P_{t_0}, \\ V(t, x, \phi(t, x)) &\leq r(t, t_0, y_0), & (t, x) \in \partial H. \end{aligned} \qquad (10.8.4)$$

Then, by Theorem 10.7.1, we obtain

$$V(t, x, u(t, x)) \leq r(t, t_0, y_0) \quad \text{on} \quad \bar{H}, \qquad (10.8.5)$$

where $r(t, t_0, y_0)$ is the maximal solution of (10.8.2). Choose y_0 such that $a(t_0, \| \phi(t_0, \cdot)\|_{P_{t_0}}) = y_0$. Then, there exists a $\delta_1 = \delta_1(t_0, \epsilon)$ such that

$$\| \phi(t_0, \cdot)\|_{P_{t_0}} \leq \delta_1 \quad \text{and} \quad a(t_0, \| \phi(t_0, \cdot)\|_{P_{t_0}}) \leq \delta \qquad (10.8.6)$$

hold simultaneously. It is clear that δ_1 is continuous in t_0 for each ϵ. We claim that, if

(a) $\qquad \| \phi(t_0, \cdot)\|_{P_{t_0}} \leq \delta;$

(b) $\qquad \| \phi(t, \cdot)\|_{\partial H} < \epsilon, \quad t \geq t_0,$

then

$$\| u(t, \cdot)\|_{P_t} < \epsilon, \quad t \geq t_0.$$

If this is not true, suppose that, for some $t_1 > t_0$, we have

$$\| u(t_1, \cdot)\|_{P_{t_1}} \geq \epsilon.$$

Then, by condition (b), there exists an $x_0 \in \text{int } P_{t_1}$ such that

$$\| u(t_1, x_0)\| = \epsilon.$$

It now follows from relations (10.8.1), (10.8.3), and (10.8.5) that

$$b(\epsilon) \leqslant V(t_1, x_0, u(t_1, x_0)) \leqslant r(t_1, t_0, y_0) < b(\epsilon),$$

which is a contradiction. This proves the equistability of the trivial solution of (10.6.1), and the proof is complete.

THEOREM 10.8.2. *Let the assumptions of Theorem 10.8.1 hold except that the function $a(t, y)$ occurring in (10.8.1) is independent of t, that is, $a(t, y) \equiv a(y) \in \mathcal{K}$. Then, the uniform stability of the trivial solution of (10.8.2) implies the uniform stability of the trivial solution of (10.6.1).*

Proof. In this case, it is enough to choose $\delta_1 = a^{-1}(\delta)$. Since δ is independent of t_0, it is clear that δ_1 is independent of t_0. The rest of the proof is very much the same as that of Theorem 10.8.1.

THEOREM 10.8.3. *Under the assumptions of Theorem 10.8.1, the equi-asymptotic stability of the trivial solution of (10.8.2) implies the equi-asymptotic stability of the null solution of (10.6.1).*

Proof. Assume that the solution $y = 0$ of (10.8.2) is equi-asymptotically stable. This implies that (S_1^*) and (S_3^*) hold. Hence, we need to prove only the quasi-equiasymptotic stability of the trivial solution of (10.6.1), as, by Theorem 10.8.1, the equistability is guaranteed.

Let $\epsilon > 0$ and $t_0 \in J$ be given. It then follows, on account of (S_3^*), that, given $b(\epsilon) > 0$ and $t_0 \in J$, there exist positive numbers $\delta_0 = \delta_0(t_0)$ and $T = T(t_0, \epsilon)$ such that

$$y(t, t_0, y_0) < b(\epsilon), \qquad t \geqslant t_0 + T, \qquad (10.8.7)$$

provided $y_0 \leqslant \delta_0$. Choosing $y_0 = a(t_0, \|\phi(t_0, \cdot)\|_{P_{t_0}})$, as before, we can show the existence of a positive number $\delta_0 = \delta_0(t_0)$ such that

$$\|\phi(t_0, \cdot)\|_{P_{t_0}} \leqslant \delta_0 \quad \text{and} \quad a(t_0, \|\phi(t_0, \cdot)\|_{P_{t_0}}) \leqslant \delta_0$$

hold at the same time Suppose now that

(a) $\|\phi(t_0, \cdot)\|_{P_{t_0}} \leqslant \delta_0$;
(b) $\|\phi(t, \cdot)\|_{\partial H} < \epsilon$, $t \geqslant t_0 + T$.

Let there exist a sequence $\{t_k\}$, $t_k \geqslant t_0 + T$ and $t_k \to \infty$ as $k \to \infty$, such that $\|u(t_k, \cdot)\|_{P_{t_k}} \geqslant \epsilon$ for some solution $u(t, x)$. Then, there exist $x_k \in \text{int } P_{t_k}$ satisfying $\|u(t_k, x_k)\| = \epsilon$ in view of condition (b). Thus, using relations (10.8.1), (10.8.5), and (10.8.7), we arrive at the contradiction

$$b(\epsilon) \leqslant V(t_k, x_k, u(t_k, x_k)) \leqslant r(t_k, t_0, y_0) < b(\epsilon).$$

Thus, the quasi-equi-asymptotic stability holds, and, as a result, the trivial solution of (10.6.1) is equi-asymptotically stable.

THEOREM 10.8.4. Under the assumptions of Theorem 10.8.2, the uniform asymptotic stability of the trivial solution of (10.8.2) implies the uniform asymptotic stability of the trivial solution of (10.6.1).

Proof. Assume that the trivial solution of (10.8.2) is uniformly asymptotically stable. Then, we have (S_2^*) and (S_4^*). By Theorem 10.8.2, the uniform stability of the trivial solution of (10.6.1) follows. To prove the quasi-uniform asymptotic stability of the trivial solution, we proceed as in Theorem 10.8.3 and choose $\delta_0 = a^{-1}(\delta_0)$, observing that δ_0 and T are both independent of t_0. The proof is complete.

THEOREM 10.8.5. Assume that there exist functions $V(t, x, u)$, $g(t, y)$, and $A(t)$ satisfying the following conditions:

(i) $A(t) > 0$ is continuously differentiable for $t \in J$, and $A(t) \to \infty$ as $t \to \infty$;

(ii) $g \in C[J \times R_+, R]$, and $g(t, 0) \equiv 0$;

(iii) $V \in C[\bar{H} \times R^N, R_+]$, $V(t, x, u)$ possesses continuous partial derivatives V_t, V_x, V_{xx}, V_u in H, and (10.8.1) holds;

(iv) $f \in C[\bar{H} \times R^N \times R^n \times R^{n^2}, R^N]$, $G \in C[\bar{H} \times R_+ \times R^n \times R^{n^2}, R]$, G is elliptic, and

$$A(t)\left[\frac{\partial V}{\partial t} + \frac{\partial V}{\partial u} \cdot f(t, x, u, u_x^i, u_{xx}^i)\right] + V(t, x, u) A'(t)$$
$$\leqslant G(t, x, A(t) V(t, x, u), A(t) V_x, A(t) V_{xx});$$

(v) $G(t, x, z, 0, 0) \leqslant g(t, z)$, $z \geqslant 0$.

Then, the equistability of the trivial solution of (10.8.2) guarantees the equi-asymptotic stability of the trivial solution of (10.6.1).

Proof. Let $\epsilon > 0$ and $t_0 \in J$ be given, and let $\sigma = \min_{t \in J} A(t)$. By assumption (i), $\sigma > 0$. Set $\eta = \sigma b(\epsilon)$. Assume that (S_1^*) holds. Then, given $\eta > 0$, $t_0 \in J$, there exists a $\delta = \delta(t_0, \epsilon)$ such that

$$y(t, t_0, y_0) < \eta, \quad t \geqslant t_0, \qquad (10.8.8)$$

whenever $y_0 \leqslant \delta$. Let $u(t, x)$ be any solution of (10.6.1) such that

$$A(t_0) V(t_0, x, \phi(t_0, x)) \leqslant y_0, \quad x \in P_{t_0},$$

and

$$A(t) V(t, x, \phi(t, x)) \leqslant r(t, t_0, y_0) \quad \text{on} \quad \partial H.$$

Then, by Theorem 10.7.2, it follows that

$$A(t) V(t, x, u(t, x)) \leqslant r(t, t_0, y_0) \quad \text{on} \quad \bar{H}, \tag{10.8.9}$$

where $r(t, t_0, y_0)$ is the maximal solution of (10.8.2). We choose y_0 so that $a(t_0, \|\phi(t_0, \cdot)\|_{P_{t_0}}) A(t_0) = y_0$. Then, we can assert the existence of a $\delta = \delta(t_0, \epsilon)$ such that the inequalities

$$\|\phi(t_0, \cdot)\|_{P_{t_0}} \leqslant \delta \quad \text{and} \quad A(t_0) a(t_0, \|\phi(t_0, \cdot)\|_{P_{t_0}}) \leqslant \delta$$

hold together. With this δ, the equistability of the trivial solution of (10.6.1) is verified. For, otherwise, we arrive at the contradiction, proceeding as in the proof of Theorem 10.8.1,

$$\eta = \sigma b(\epsilon) \leqslant A(t_1) V(t_1, x_0, u(t_1, x_0)) \leqslant r(t_1, t_0, y_0) < \eta.$$

For a fixed $\epsilon = \epsilon_1$, we designate by $\delta_0 = \delta_0(t_0)$ the number $\delta(t_0, \epsilon_1)$. Let $0 < \epsilon < \epsilon_1$ and $t_0 \in J$ be given. Suppose that $\|\phi(t_0, \cdot)\|_{P_{t_0}} \leqslant \delta_0$. Since $A(t) \to \infty$ as $t \to \infty$, there is a number $T = T(t_0, \epsilon)$ satisfying

$$b(\epsilon) A(t) > \eta, \quad t \geqslant t_0 + T. \tag{10.8.10}$$

It is easy to show that, with this δ_0 and T, the quasi-equi-asymptotic stability holds. Suppose that this is not true. Then, there exists a sequence $\{t_k\}$, $t_k \geqslant t_0 + T$, and $t_k \to \infty$ as $k \to \infty$ such that $\|u(t_k, \cdot)\|_{P_{t_k}} \geqslant \epsilon$ for some solution $u(t, x)$ satisfying $\|\phi(t_0, \cdot)\|_{P_{t_0}} \leqslant \delta_0$ and $\|\phi(t, \cdot)\|_{\partial H} < \epsilon$, $t \geqslant t_0 + T$. Also, there exist $x_k \in \text{int } P_{t_k}$ such that $\|u(t_k, x_k)\| = \epsilon$. The relations (10.8.8) and (10.8.9) yield

$$A(t_k) b(\epsilon) < \eta,$$

which is an absurdity in view of (10.8.10). It therefore follows that the trivial solution of (10.6.1) is equi-asymptotically stable, and the proof is complete.

Let us consider the following example. Let $L(u)$ denote the differential form

$$L(u) = \sum_{j,k=1}^{n} a_{jk}(t, x) u_{x_j x_k} + \sum_{k=1}^{n} b_k(t, x) u_{x_k}, \tag{10.8.11}$$

where $a_{j,k}(t, x)$ and $b_k(t, x)$ are continuous functions on $\bar{H}$, and the quadratic form

$$\sum_{j,k=1}^{n} a_{j,k}(t, x) \lambda_j \lambda_k \geqslant 0, \tag{10.8.12}$$

on $\bar{H}$, λ being an arbitrary vector. Let $F \in C[\bar{H} \times R^N, R^N]$. Consider the system
$$u_t = L(u) + F(t, x, u). \qquad (10.8.13)$$
Assume that $F(t, x, 0) \equiv 0$ and
$$\sum_{i=1}^{n} u^i F^i(t, x, u) \leqslant \lambda(t) \sum_{i=1}^{n} (u^i)^2,$$
where $\lambda \in C[J, R]$. Taking $V(t, x, u) = \| u \|^2$ and making use of the inequality (10.8.12), we obtain
$$\partial V/\partial t \leqslant L(V) + 2\lambda(t)V = G(t, x, V, V_x, V_{xx}).$$
Since
$$G(t, x, V, 0, 0) \leqslant 2\lambda(t)V = g(t, V),$$
it follows, from Theorem 10.7.1, that
$$\| u(t, x) \|^2 \leqslant y_0 \exp\left[2 \int_{t_0}^{t} \lambda(s) \, ds \right], \qquad (t, x) \in \bar{H},$$
provided
$$\| \phi(t, x) \|^2 \leqslant y_0 \exp\left[2 \int_{t_0}^{t} \lambda(s) \, ds \right], \qquad (t, x) \in P_{t_0} \cup \partial H,$$
where
$$\max_{x \in P_t} \| \phi(t_0, x) \| \leqslant y_0.$$
If, in addition,
$$\int_{t_0}^{t} \lambda(s) \, ds < \infty,$$
the application of Theorem 10.8.1 yields equistability of the trivial solution of (10.8.13). On the other hand, the assumption
$$\int_{t_0}^{\infty} \lambda(s) \, ds = -\infty$$
implies the equi-asymptotic stability of the trivial solution of (10.8.13), by Theorem 10.8.3.

Regarding the boundedness of solutions of the partial differential system (10.6.1), we have the following result.

196   CHAPTER 10

THEOREM 10.8.6. Assume that there exist functions $V(t, x, u)$ and $g(t, y)$ fulfilling the following hypotheses:

(i) $V \in C[\bar{H} \times R^N, R_+]$, $V(t, x, u)$ possesses continuous partial derivatives V_t, V_x, V_{xx}, V_u in H, and

$$b(\|u\|) \leq V(t, x, u) \leq a(t, \|u\|), \qquad (10.8.14)$$

where $b \in \mathscr{K}$, $b(y) \to \infty$ as $y \to \infty$, $a \in C[J \times R_+, R_+]$, and $a \in \mathscr{K}$ for each $t \in J$;

(ii) $f \in C[\bar{H} \times R^N \times R^n \times R^{n^2}, R^N]$, $G \in C[\bar{H} \times R_+ \times R^n \times R^{n^2}, R]$, G is elliptic, and

$$\frac{\partial V}{\partial t} + \frac{\partial V}{\partial u} \cdot f(t, x, u, u_x{}^i, u_{xx}^i) \leq G(t, x, V(t, x, u), V_x, V_{xx});$$

(iii) $G(t, x, z, 0, 0) \leq g(t, z)$, $z \geq 0$, where $g \in C[J \times R_+, R]$.

Then, the equiboundedness of Eq. (10.8.2) implies the equiboundedness of the system (10.6.1).

Proof. Let $\alpha > 0$ and $t_0 \in J$ be given. Suppose that $\|\phi(t_0, \cdot)\|_{P_{t_0}} \leq \alpha$. Define $\alpha_1 = a(t_0, \alpha)$. Let Eq. (10.8.2) be equibounded. Then, given $\alpha_1 > 0$ and $t_0 \in J$, there exists a $\beta_1 = \beta_1(t_0, \alpha)$ such that

$$y(t, t_0, y_0) < \beta_1, \qquad t \geq t_0, \qquad (10.8.15)$$

if $y_0 \leq \alpha_1$. Since $b(y) \to \infty$ as $y \to \infty$, we can find a $\beta = \beta(t_0, \alpha)$ such that

$$b(\beta) \geq \beta_1. \qquad (10.8.16)$$

Let $u(t, x)$ be any solution of (10.6.1). Then, if (10.8.4) holds, we have (10.8.5). Choose $y_0 = a(t_0, \|\phi(t_0, \cdot)\|_{P_{t_0}})$. Suppose that there exists a solution $u(t, x)$ of (10.6.1) such that

(a) $\|\phi(t_0, \cdot)\|_{P_{t_0}} \leq \alpha$,
(b) $\|\phi(t, \cdot)\|_{\partial H} < \beta$, $t \geq t_0$,

and $\|u(t_1, \cdot)\|_{P_{t_1}} \geq \beta$ for some $t_1 > t_0$. This implies the existence of an $x_0 \in \text{int } P_{t_1}$ satisfying $\|u(t_1, x_0)\| = \beta$. We are now lead to an absurdity, in view of the relations (10.8.14), (10.8.5), (10.8.15), and (10.8.16):

$$b(\beta) \leq V(t_1, x_0, u(t_1, x_0)) \leq r(t_1, t_0, y_0) < \beta_1 \leq b(\beta).$$

This proves the equiboundedness of the differential system (10.6.1), and the theorem is proved.

10.8. STABILITY AND BOUNDEDNESS

On the strength of Theorem 10.8.6 and the parallel boundedness and Lagrange stability results in ordinary differential equations, the following theorem on various boundedness results and Lagrange stability may be proved. We only state the theorem, leaving the construction of the proof as an exercise.

THEOREM 10.8.7. Let assumptions (i), (ii), and (iii) of Theorem 10.8.6 hold, except that the relation (10.8.14) is strengthened to

$$b(\| u \|) \leqslant V(t, x, u) \leqslant a(\| u \|),$$

where $a, b \in \mathcal{K}$ and $b(y) \to \infty$ as $y \to \infty$. Then, if the ordinary differential equation (10.8.2) satisfies one of the notions (B_1^*) to (B_8^*), the partial differential system verifies the corresponding one of the concepts (B_1) to (B_8).

Corresponding to the system (10.6.1), we may consider the perturbed system

$$u_t = f(t, x, u, u_x^i, u_{xx}^i) + F(t, x, u), \tag{10.8.17}$$

where $F(t, x, u)$ is a perturbation term. Then, we have the following result.

THEOREM 10.8.8. Assume that

(i) $V \in C[\bar{H} \times R^N, R_+]$, $V(t, x, u)$ possesses continuous partial derivatives V_t, V_x, V_{xx}, V_u in H, $\| \partial V(t, x, u)/\partial u \| \leqslant K$, and

$$b(\| u \|) \leqslant V(t, x, u) \leqslant a(\| u \|),$$

where $a, b \in \mathcal{K}$;

(ii) $f \in C[\bar{H} \times R^N \times R^n \times R^{n^2}, R^N]$, $F \in C[\bar{H} \times R^N, R^N]$, $G \in C[\bar{H} \times R_+ \times R^n \times R^{n^2}, R]$, G is elliptic, and

$$\frac{\partial V}{\partial t} + \frac{\partial V}{\partial u} \cdot f(t, x, u, u_x^i, u_{xx}^i) \leqslant -\alpha V(t, x, u) + G(t, x, V, V_x, V_{xx}),$$

for some $\alpha > 0$;

(iii) $G(t, x, z, 0, 0) \leqslant g(t, z)$, $z \geqslant 0$, where $g \in C[J \times R_+, R]$, $g(t, 0) \equiv 0$;

(iv) $\| F(t, x, u) \| \leqslant \eta V(t, x, u)$, and $\alpha \geqslant K\eta$.

Then, one of the stability notions of the trivial solution of (10.8.2) implies the corresponding one of the stability results of the trivial solution of the perturbed system (10.8.17).

Proof. Using the respective assumptions in (i), (ii), and (iv), we find

$$\frac{\partial V}{\partial t} + \frac{\partial V}{\partial u} \cdot [f(t, x, u, u_x^{\,i}, u_{xx}^{\,i}) + F(t, x, u)]$$

$$\leqslant \frac{\partial V}{\partial t} + \frac{\partial V}{\partial u} \cdot f(t, x, u, u_x^{\,i}, u_{xx}^{\,i}) + \left\|\frac{\partial V}{\partial u}\right\| \|F(t, x, u)\|$$

$$\leqslant -\alpha V(t, x, u) + G(t, x, V, V_x, V_{xx}) + K\eta V(t, x, u)$$

$$\leqslant G(t, x, V, V_x, V_{xx}).$$

It is evident, from this inequality, that we can directly apply Theorems 10.8.1–10.8.4 to obtain the desired result. The proof is complete.

Although we can prove a number of results by the techniques just used, that is, by reducing the study of partial differential system to the study of ordinary differential equations, in certain situations, this method does not yield all the information about the given system. For instance, consider again the example (10.8.11). Suppose we now assume that the quadratic form $\sum_{j,k=1}^{n} a_{j,k}(t, x) \lambda_j \lambda_k$ is positive definite instead of positive semidefinite, as demanded in (10.8.12). This stronger hypothesis has no effect. In other words, we do not get more information because of this assumption. To be more specific, suppose $F \equiv 0$ so that $g \equiv 0$. Then, we can conclude by Theorem 10.8.1 that the trivial solution of (10.8.11) is stable. This conclusion remains the same even when the preceding quadratic form is assumed to be positive definite. In such situations, the following theorem is more fruitful.

THEOREM 10.8.9. Assume that

(i) $V \in C[\bar{H} \times R^N, R_+]$, $V(t, x, u)$ possesses continuous partial derivatives V_t, V_x, V_{xx}, V_u in H, and

$$b(\|u\|) \leqslant V(t, x, u) \leqslant a(t, \|u\|),$$

where $b \in \mathcal{K}$, $a \in C[J \times R_+, R_+]$, and $a \in \mathcal{K}$ for each $t \in J$;

(ii) $f \in C[\bar{H} \times R^N \times R^n \times R^{n^2}, R^N]$, $G \in C[\bar{H} \times R_+ \times R^n \times R^{n^2}, R]$, G is elliptic, $G(t, x, 0, 0, 0) \equiv 0$, and

$$\frac{\partial V}{\partial t} + \frac{\partial V}{\partial u} \cdot f(t, x, u, u_x^{\,i}, u_{xx}^{\,i}) \leqslant G(t, x, V, V_x, V_{xx});$$

(iii) $G(t, x, z_1, P, R) - G(t, x, z_2, P, R) \leqslant g(t, z_1 - z_2)$, $z_1 \geqslant z_2$,

10.8. STABILITY AND BOUNDEDNESS

where $g \in C[J \times R_+, R]$, $g(t, 0) \equiv 0$, and the maximal solution $r(t, t_0, 0)$ of
$$y' = g(t, y), \quad y(t_0) = 0$$
is identically zero.

Then, the equistability of the trivial solution of
$$z_t = G(t, x, z, z_x, z_{xx}) \tag{10.8.18}$$
implies the equistability of the trivial solution of (10.6.1).

Proof. Let $u(t, x)$ be any solution of (10.6.1) such that
$$V(t_0, x, \phi(t_0, x)) \leqslant z(t_0, x) \quad \text{on} \quad P_{t_0}$$
and
$$V(t, x, \phi(t, x)) \leqslant z(t, x) \quad \text{on} \quad \partial H,$$
where $z(t, x) \geqslant 0$ is the solution of (10.8.18). Define
$$m(t, x) = V(t, x, u(t, x)).$$
Then, we get
$$\frac{\partial m(t, x)}{\partial t} = \frac{\partial V}{\partial t} + \frac{\partial V}{\partial u} \cdot f(t, x, u, u_x{}^i, u_{xx}^i)$$
$$\leqslant G(t, x, m(t, x), m_x(t, x), m_{xx}(t, x)).$$

If we write $T[v] = v_t - G(t, x, v, v_x, v_{xx}]$, then it is clear that $T[m] \leqslant T[z]$. Furthermore, $m(t, x) \leqslant z(t, x)$ on $P_{t_0} \cup \partial H$. All the assumptions of Theorem 10.2.2 being verified, we deduce that
$$V(t, x, u(t, x)) \leqslant z(t, x) \quad \text{on} \quad \overline{H}.$$

Let $\epsilon > 0$ and $t_0 \in J$ be given. Assume that the trivial solution (10.8.18) is equistable. Then, given $b(\epsilon) > 0$ and $t_0 \in J$, there exists a $\delta = \delta(t_0, \epsilon)$ such that

(i) $\max_{x \in P_{t_0}} z(t_0, x) \leqslant \delta$,
(ii) $\max_{x \in \partial H} z(t, x) < \epsilon, t \geqslant t_0$,

implies
$$\max_{x \in P_t} z(t, x) < \epsilon, \quad t \geqslant t_0.$$

Let $\max_{x \in P_{t_0}} z(t_0, x) = a(t_0, \|\phi(t_0, \cdot)\|_{P_{t_0}})$, and let $\delta_1 = \delta_1(t_0, \epsilon)$ be the same number chosen according to the inequalities (10.8.6) in the proof of Theorem 10.8.1. Suppose that $\|\phi(t_0, \cdot)\|_{P_{t_0}} \leqslant \delta_1$ and $\|\phi(t, \cdot)\|_{\partial H} < \epsilon$,

$t \geq t_0$. Assume that there exists a solution $u(t, x)$ of (10.6.1) such that, for some $t_1 > t_0$, $\| u(t_1, \cdot) \| \geq \epsilon$. It then follows that there is an $x_0 \in \text{int } P_{t_1}$ satisfying $\| u(t_1, x_0) \| = \epsilon$. From this, we deduce the inequality

$$b(\epsilon) \leq V(t_1, x_0, u(t_1, x_0)) \leq z(t_1, x_0) < b(\epsilon).$$

This contradiction proves the equistability of the trivial solution of (10.6.1).

On the basis of this theorem, we can formulate other stability results in this setup. We notice, however, that we now have the problem of knowing the stability behavior of partial differential equation (10.8.18). In the cases where the function $G(t, x, z, z_x, z_{xx})$ is simple enough to know the behavior of its solutions by other methods, this technique is useful.

10.9. Conditional stability and boundedness

In this section, we shall consider the partial differential system of the type

$$u_t = f(t, x, u, u_x, u_{xx}), \tag{10.9.1}$$

where

$$u_x = \left(\frac{\partial u_1}{\partial x_1}, \ldots, \frac{\partial u_1}{\partial x_n}, \frac{\partial u_2}{\partial x_1}, \ldots, \frac{\partial u_2}{\partial x_n}, \ldots, \frac{\partial u_N}{\partial x_1}, \ldots, \frac{\partial u_N}{\partial x_n} \right)$$

and

$$u_{xx} = \left(\frac{\partial^2 u_1}{\partial x_1^2}, \frac{\partial^2 u_1}{\partial x_1 \partial x_2}, \ldots, \frac{\partial^2 u_1}{\partial x_n^2}, \ldots, \frac{\partial^2 u_N}{\partial x_1^2}, \ldots, \frac{\partial^2 u_N}{\partial x_n^2} \right).$$

It will be assumed that the first initial-boundary-value problem with respect to (10.9.1) admits the trivial solution and that all solutions exist on $\bar{H}$. In the sequel, a solution of (10.9.1) will always mean a solution of the first initial-boundary-value problem.

Let $k < N$ and $M_{(N-k)}$ denote a manifold of $(N - k)$ dimensions containing the origin. Let $S(\alpha)$ and $\bar{S}(\alpha)$ represent the sets, as before,

$$S(\alpha) = [u : \| u \| < \alpha] \quad \text{and} \quad \bar{S}(\alpha) = [u : \| u \| \leq \alpha],$$

respectively. Parallel to the conditional stability and conditional boundedness definitions (C_1) to (C_{16}) of Sect. 4.4, we can formulate the definitions of conditional stability and boundedness of the trivial solution of (10.9.1.). Corresponding to (C_1), we have

10.9. CONDITIONAL STABILITY AND BOUNDEDNESS

DEFINITION 10.9.1. The trivial solution of the partial differential system (10.9.1) is said to be *conditionally equistable* if, for each $\epsilon > 0$ and $t_0 \in J$, there exists a positive function $\delta(t_0, \epsilon)$, which is continuous in t_0 for each ϵ, such that, if

(i) $\phi(t_0, x) \subset \bar{S}(\delta) \cup M_{(N-k)}$, $x \in P_{t_0}$,
(ii) $\phi(t, x) \subset S(\epsilon)$, $(t, x) \in \partial H$,

then
$$u(t, x) \subset S(\epsilon), \quad (t, x) \in \bar{H}.$$

Sufficient conditions for the conditional stability of the trivial solution of (10.9.1) are given by the following result.

THEOREM 10.9.1. Assume that

(i) $g \in C[J \times R_+^N, R^N]$, $g(t, 0) \equiv 0$, and $g(t, y)$ is quasi-monotone nondecreasing in y for each $t \in J$;

(ii) $V \in C[\bar{H} \times R^N, R_+^N]$, $V(t, x, u)$ possesses continuous partial derivatives V_t, V_x, V_{xx}, V_u in H, and

$$b(\|u\|) \leqslant \sum_{i=1}^{N} V_i(t, x, u) \leqslant a(t, \|u\|),$$

where $b \in \mathscr{K}$, $a \in C[J \times R_+, R_+]$, and $a \in \mathscr{K}$ for each $t \in J$;

(iii)
$$f \in C[\bar{H} \times R^N \times R^{Nn} \times R^{Nn^2}, R^N], G \in C[\bar{H} \times R_+^N \times R^N \times R^{N^2}, R^N],$$
$$G(t, x, z, z_x^i, z_{xx}^i)$$

is elliptic, and
$$\frac{\partial V}{\partial t} + \frac{\partial V}{\partial u} \cdot f(t, x, u, u_x, u_{xx}) \leqslant G(t, x, V(t, x, u), V_x^i, V_{xx}^i);$$

(iv) $G(t, x, z, 0, 0) \leqslant g(t, z)$, $z \geqslant 0$;
(v) $V_i(t, x, u) \equiv 0$ ($i = 1, 2, ..., k$), $k < N$, if $u \in M_{(N-k)}$, where $M_{(N-k)}$ is an $(N-k)$-dimensional manifold containing the origin.

Then, if the trivial solution of the ordinary differential system

$$y' = g(t, y), \quad y(t_0) = y_0 \geqslant 0, \quad t_0 \geqslant 0, \quad (10.9.2)$$

is conditionally equistable (in the sense of Definition 4.4.2), the trivial solution of the partial differential system (10.9.1) is conditionally equistable.

Proof. For any $\epsilon > 0$, if $\|u\| = \epsilon$, we have from assumption (ii) that

$$b(\epsilon) \leqslant \sum_{i=1}^{N} V_i(t, x, u). \tag{10.9.3}$$

Suppose that the trivial solution of (10.9.2) is conditionally equistable. Then, given $b(\epsilon) > 0$ and $t_0 \in J$, there exists a $\delta = \delta(t_0, \epsilon) > 0$ such that

$$\sum_{i=1}^{N} y_i(t, t_0, y_0) < b(\epsilon), \qquad t \geqslant t_0, \tag{10.9.4}$$

provided

$$\sum_{i=1}^{N} y_{i0} \leqslant \delta \quad \text{and} \quad u_{i0} = 0 \quad (i = 1, 2, ..., k), \tag{10.9.5}$$

where $y(t, t_0, y_0)$ is any solution of (10.9.2). Suppose that $u(t, x)$ is any solution of (10.9.1). It follows by Theorem 10.6.4 that

$$\sum_{i=1}^{N} V_i(t, x, u(t, x)) \leqslant \sum_{i=1}^{N} r_i(t, t_0, y_0), \qquad (t, x) \in \bar{H}, \tag{10.9.6}$$

whenever

$$\sum_{i=1}^{N} V_i(t_0, x, \phi(t_0, x)) \leqslant \sum_{i=1}^{N} y_{i0}, \qquad x \in P_{t_0},$$

and

$$\sum_{i=1}^{N} V_i(t, x, \phi(t, x)) \leqslant \sum_{i=1}^{N} r_i(t, t_0, y_0), \qquad (t, x) \in \partial H,$$

where $r(t, t_0, y_0)$ is the maximal solution of (10.9.2). Choose y_0 such that $\sum_{i=1}^{N} y_{i0} = a(t_0, \|\phi(t_0, \cdot)\|_{P_{t_0}})$ and $\phi(t_0, x) \subset M_{(N-k)}$, $x \in P_{t_0}$, so that $y_{i0} = V_i(t_0, x, \phi(t_0, x)) \equiv 0$ $(i = 1, 2, ..., k)$, by assumption (v). In view of the properties of $a(t, r)$, there exists a $\delta_1 = \delta_1(t_0, \epsilon) > 0$ such that

$$\|\phi(t_0, \cdot)\|_{P_{t_0}} \leqslant \delta_1 \quad \text{and} \quad a(t_0, \|\phi(t_0, \cdot)\|_{P_{t_0}}) \leqslant \delta$$

at the same time. From the choice of y_{i0} and condition (ii), it follows that, whenever $\phi(t_0, x) \subset \bar{S}(\delta_1) \cap M_{(N-k)}$, we have the inequality (10.9.6).

Suppose now that there exists a solution $u(t, x)$ of (10.9.1) which satisfies

(a) $\phi(t_0, x) \subset \bar{S}(\delta_1) \cap M_{(N-k)}$, $x \in P_{t_0}$,
(b) $\phi(t, x) \subset S(\epsilon)$, $(t, x) \in \partial H$,

and has the property that $u(t, x) \notin S(\epsilon)$ for some $t_1 > t_0$ and $x \in P_{t_1}$. Because of relation (b), there exists an $x_0 \in \text{int } P_{t_1}$ such that $\| u(t_1, x_0)\| = \epsilon$. Hence, by (10.9.3), (10.9.4), and (10.9.6), we are led to the following absurdity:

$$b(\epsilon) \leqslant \sum_{i=1}^{N} V_i(t_1, x_0, u(t_1, x_0)) \leqslant \sum_{i=1}^{N} r_i(t_1, t_0, y_0) < b(\epsilon).$$

Consequently, the trivial solution of (10.9.1) is conditionally equistable, and the proof is complete.

On the strength of Theorem 10.9.1 and the parallel theorems on conditional stability and boundedness (in Sect. 4.4), we have the following

THEOREM 10.9.2. Assume that the hypotheses of Theorem 10.9.1 hold, and suppose that $a(t, r) \equiv a(r)$, $a \in \mathcal{K}$. Then, one of the notions (C_1^*) to (C_{16}^*) relative to the ordinary differential system (10.9.2) implies the corresponding one of the conditional stability concepts (C_1) to (C_{16}).

The following example, in addition to demonstrating the conditional stability, serves to show that the system (10.9.1) need not be parabolic. Consider the system

$$\frac{\partial u_1}{\partial t} = \sum_{i,j=1}^{N} (a_{ij} + b_{ij}) \frac{\partial^2 u_1}{\partial x_i \partial x_j} + \sum_{i,j=1}^{n} (a_{ij} - b_{ij}) \frac{\partial^2 u_2}{\partial x_i \partial x_j}$$

$$+ \sum_{i,j=1}^{n} (b_{ij} - a_{ij}) \frac{\partial^2 u_3}{\partial x_i \partial x_j} + F_1;$$

$$\frac{\partial u_2}{\partial t} = \sum_{i,j=1}^{n} (a_{ij} - c_{ij}) \frac{\partial^2 u_1}{\partial x_i \partial x_j} + \sum_{i,j=1}^{n} (a_{ij} + c_{ij}) \frac{\partial^2 u_2}{\partial x_i \partial x_j}$$

$$+ \sum_{i,j=1}^{n} (c_{ij} - a_{ij}) \frac{\partial^2 u_3}{\partial x_i \partial x_j} + F_2;$$

$$\frac{\partial u_3}{\partial t} = \sum_{i,j=1}^{n} (b_{ij} - c_{ij}) \frac{\partial^2 u_1}{\partial x_i \partial x_j} + \sum_{i,j=1}^{n} (c_{ij} - b_{ij}) \frac{\partial^2 u_2}{\partial x_i \partial x_j}$$

$$+ \sum_{i,j=1}^{n} (c_{ij} + b_{ij}) \frac{\partial^2 u_3}{\partial x_i \partial x_j} + F_3,$$

where

$$F_1 = (1 + \cos t)u_1 + (1 - \cos t)u_2 + (\cos t - 1)u_3;$$
$$F_2 = (1 - e^{-t})u_1 + (1 + e^{-t})u_2 + (e^{-t} - 1)u_3;$$
$$F_3 = (\cos t - e^{-t})u_1 + (e^{-t} - \cos t)u_2 + (e^{-t} + \cos t)u_3.$$

Assume that the quadratic forms $\sum_{i,j=1}^{n} a_{ij}\lambda_i\lambda_j$, $\sum_{i,j=1}^{n} b_{ij}\lambda_i\lambda_j$, and $\sum_{i,j=1}^{n} c_{ij}\lambda_i\lambda_j$ are all nonnegative for arbitrary vector λ.

Choosing the vector Lyapunov function $V = (V_1, V_2, V_3)$ such that

$$V_1 = (u_1 + u_2 - u_3)^2,$$
$$V_2 = (u_1 - u_2 + u_3)^2,$$
$$V_3 = (-u_1 + u_2 + u_3)^2,$$

and noting that

$$\sum_{i=1}^{3} V_i(t, u) = u_1^2 + u_2^2 + u_3^2 + (u_1 - u_2)^2 + (u_2 - u_3)^2 + (u_3 - u_1)^2,$$

we observe that the functions $b(r)$ and $a(t, r)$ reduce to

$$b(r) = [(u_1^2 + u_2^2 + u_3^2)^{1/2}]^2 \quad \text{and} \quad a(t, r) = 5[(u_1^2 + u_2^2 + u_3^2)^{1/2}]^2.$$

Furthermore, the function $G = (G_1, G_2, G_3)$ takes the form

$$G_1 = \sum_{i,j=1}^{n} a_{ij} \frac{\partial^2 V_1}{\partial x_i \, \partial x_j} + 4V_1,$$

$$G_2 = \sum_{i,j=1}^{n} b_{ij} \frac{\partial^2 V_2}{\partial x_i \, \partial x_j} + 4\cos t V_2,$$

$$G_3 = \sum_{i,j=1}^{n} c_{ij} \frac{\partial^2 V_3}{\partial x_i \, \partial x_j} + 4e^{-t} V_3.$$

The differential system (10.9.2) can be reduced to

$$y_1' = 4y_1 = g_1(t, y),$$
$$y_2' = 4\cos t y_2 = g_2(t, y),$$
$$y_3' = 4e^{-t} y_3 = g_3(t, y).$$

We find that $g = (g_1, g_2, g_3)$ fulfills the monotonic requirements. Choose $k = 1$. Then the solution $r(t, t_0, y_0)$ of (10.9.2) is given by

$$r_1(t, t_0, y_0) = y_{10} \exp[4(t - t_0)],$$
$$r_2(t, t_0, y_0) = y_{20} \exp[\sin t - \sin t_0],$$
$$r_3(t, t_0, y_0) = y_{30} \exp[4e^{-t_0} - 4e^{-t}].$$

Here we have $M_{(N-k)} = M_2$, the set of points u such that $(u_1 + u_2 - u_3)^2 = 0$. It is clear that the condition (C_1^*) holds, which,

10.10. Parabolic differential inequalities in unbounded domains

Let D be a region in R^{n+1} of (t, x) space, satisfying the following conditions:

(i) D is open and contained in the zone $t_0 < t < \infty$;

(ii) for any $t_1 \in [t_0, \infty]$, the intersection S_{t_1} of $\bar{D}$ with the plane $t = t_1$ is nonempty and unbounded;

(iii) for any t_1, S_{t_1} is identical with the intersection of the plane $t = t_1$ with the closure of that part of D which is contained in the zone $t_0 \leqslant t \leqslant t_1$.

We denote by ∂D that part of the boundary of D which is contained in the zone $t_0 < t < \infty$.

DEFINITION 10.10.1. For any $\psi \in C[\bar{D}, R]$, if the inequality

$$\psi(t, x) \leqslant M \exp(K \| x \|^2)$$

holds, where M, K are positive constants, we shall say that $\psi \in \bar{E}_2(M, K)$. Similarly, $\psi \in \hat{E}_2(M, K)$ implies that

$$\psi(t, x) \geqslant -M \exp(K \| x \|^2).$$

If $\psi \in \bar{E}_2(M, K)$ and $\psi \in \hat{E}_2(M, K)$ simultaneously, we shall say that $\psi \in E_2(M, K)$.

DEFINITION 10.10.2. A function $f \in C[D \times R \times R^n \times R^{n^2}, R]$ is said to satisfy an $\mathscr{L}$ condition if there exist positive constants L_i ($i = 0, 1, 2, 3, 4$) such that

$$|f(t, x, u, P, R) - f(t, x, \bar{u}, \bar{P}, \bar{R})| \leqslant L_0 \sum_{i,k=1}^{n} |R_{ik} - \bar{R}_{ik}|$$

$$+ (L_1 \| x \| + L_2) \sum_{i=1}^{n} |P_i - \bar{P}_i| + (L_3 \| x \|^2 + L_4) |u - \bar{u}|.$$

THEOREM 10.10.1. Assume that

(i) $u, v \in C[\bar{D}, R]$, the partial derivatives $u_t, u_x, u_{xx}, v_t, v_x, v_{xx}$ exist and are continuous in D;

(ii) $f \in C[D \times R \times R^n \times R^{n^2}, R]$, f satisfies an $\mathscr{L}$ condition, the differential operator T is parabolic, and

$$T[u] \leqslant T[v] \quad \text{on} \quad D;$$

(iii) $u \in \bar{E}_2(M, K_0)$ and $v \in \hat{E}_2(M, K_0)$;
(iv) $u(t, x) \leqslant v(t, x)$ on $\bar{S}_{t_0} \cup \partial D$.

Then, everywhere in D, we have

$$u(t, x) \leqslant v(t, x). \tag{10.10.1}$$

Proof. Let D_R denote an open bounded domain separated from D by the cylindrical surface Γ_R, with the equation $\|x\| = R$. We shall denote by $S_R^{t_0}$, ∂D_R the parts of the surfaces S_{t_0}, ∂D, respectively, lying inside and on Γ_R. Let C_R represent the part of Γ_R contained in D. Furthermore, let us designate by D^h, $D_R{}^h$, ∂D^h, $\partial D_R{}^h$, $C_R{}^h$ the parts of the sets D, D_R, ∂D, ∂D_R, C_R, respectively, contained in the strip $t_0 \leqslant t < t_0 + h$. We shall prove the theorem for a domain D^h, where h will be chosen conveniently.

Consider the function

$$m = m(t, x) = u(t, x) - v(t, x).$$

Define $u = \bar{u}H$, $v = \bar{v}H$, where

$$H = H(t, x; K) = \exp\left[\frac{K\|x\|^2}{1 - \mu(t - t_0)} + \gamma t\right]$$

is the growth damping factor. Here $\mu = \mu(K)$, $\gamma = \gamma(K)$ are positive constants, which will be chosen later, for $K > K_0$. Setting

$$z = z(t, x) = \bar{u}(t, x) - \bar{v}(t, x),$$

we have $m = zH$ defined in D^h, where $h < \mu^{-1}$. It then follows that

$$\frac{\partial z}{\partial t}H + z\frac{\partial H}{\partial t} = \frac{\partial u}{\partial t} - \frac{\partial v}{\partial t}$$

$$\leqslant f(t, x, u, \bar{u}_x H + \bar{u}H_x, R^{(1)})$$
$$-f(t, x, u, \bar{u}_x H + \bar{u}H_x, R^{(2)})$$
$$+f(t, x, u, \bar{u}_x H + \bar{u}H_x, R^{(2)})$$
$$-f(t, x, v, \bar{v}_x H + \bar{v}H_x, R^{(3)}), \tag{10.10.2}$$

where

$$R^{(1)} = [\bar{u}_{x_i x_j} H + \bar{u}_{x_i} H_{x_j} + \bar{u}_{x_j} H_{x_i} + \bar{u} H_{x_i x_j}],$$

$$R^{(2)} = [\bar{v}_{x_i x_j} H + \bar{u}_{x_i} H_{x_j} + \bar{u}_{x_j} H_{x_i} + \bar{u} H_{x_i x_j}],$$

$$R^{(3)} = [\bar{v}_{x_i x_j} H + \bar{v}_{x_i} H_{x_j} + \bar{v}_{x_j} H_{x_i} + \bar{v} H_{x_i x_j}].$$

Let $\{R_\alpha\}$ be an increasing sequence such that $R_\alpha \to \infty$ as $\alpha \to \infty$. For a fixed α, consider the domain $D^h_{R_\alpha}$. Denote

$$A = \sup_{(t,x) \in D^h_{R_\alpha}} z(t, x).$$

Then, there exists a $(t_1, x_1) \in \overline{D^h_{R_\alpha}} = D^h_{R_\alpha} + S^h_{R_\alpha} + \partial D^h_{R_\alpha} + C^h_{R_\alpha}$ such that $A = z(t_1, x_1)$. We shall show that the constants μ, ν, and h may be chosen such that $(t_1, x_1) \in D^h_{R_\alpha} + S^h_{R_\alpha}$ implies the inequality $z(t_1, x_1) \leqslant 0$. Indeed, supposing the contrary, we have, at (t_1, x_1),

$$\partial z/\partial t\, (t_1, x_1) \geqslant 0, \qquad z_x(t_1, x_1) = 0,$$

and

$$\sum_{i,k=1}^{n} \frac{\partial^2 z(t_1, x_1)}{\partial x_i \, \partial x_k} \lambda_i \lambda_k \leqslant 0,$$

for an arbitrary vector λ. This means that

$$\bar{u}_x(t_1, x_1) = \bar{v}_x(t_1, x_1),$$

$$\sum_{i,k=1}^{n} \frac{\partial^2 [\bar{u}(t_1, x_1) - \bar{v}(t_1, x_1)]}{\partial x_i \, \partial x_k} \lambda_i \lambda_k \leqslant 0.$$

Since f is elliptic, it follows that

$$f(t_1, x_1, u(t_1, x_1), \bar{u}_x H + \bar{u} H_x, R^{(1)})$$
$$\leqslant f(t_1, x_1, u(t_1, x_1), \bar{u}_x H + \bar{u} H_x, R^{(2)}).$$

Hence, using the $\mathscr{L}$ condition and the relation (10.10.2), we get

$$0 \leqslant \frac{\partial z}{\partial t} H \leqslant z \left[\left(L_0 \sum_{i,k=1}^{n} |H_{x_i x_k}| \right) + \left((L_1 \|x\| + L_2) \sum_{i=1}^{n} |H_{x_i}| \right) \right.$$
$$\left. + (L_3 \|x\|^2 + L_4) H - \frac{\partial H}{\partial t} \right] = zFH \text{ (say)}.$$

From the definition of H, we deduce that

$$FH \leqslant H\left[\frac{4K^2L_0}{\{1-\mu(t-t_0)\}^2}\sum_{i,k=1}^{n}|x_ix_k| + \frac{2KL_0n}{1-\mu(t-t_0)}\right.$$
$$+ \frac{2K}{1-\mu(t-t_0)}(L_1\|x\|+L_2)\sum_{i=1}^{n}|H_{x_i}| + (L_3\|x\|^2+L_4)$$
$$\left. - \frac{\mu K\|x\|^2}{\{1-\mu(t-t_0)\}^2} - \gamma\right].$$

Setting $L = \max[L_i, L_2]$, and, taking into consideration the inequalities

$$|x_i| \leqslant \|x\|, \qquad \|x\| \leqslant \|x\|^2 + 1, \qquad 1 - \mu(t-t_0) \leqslant 1,$$

we have

$$FH \leqslant \frac{H}{\{1-\mu(t-t_0)\}^2}[4K^2L_0n^2\|x\|^2 + 2KL_0n + 4KLn\|x\|^2$$
$$+ 2KLn + L_3\|x\|^2 + L_4 - \mu K\|x\|^2 - \gamma\{1-\mu(t-t_0)\}^2].$$

Let us now choose

$$K\mu = 4K^2L_0n^2 + 4KLn + L_3 + \lambda, \qquad \lambda > 0,$$

and

$$h = (1-\beta)/\mu, \qquad 0 < \beta < 1.$$

This choice leads to the inequality

$$\beta \leqslant 1 - \mu(t-t_0) \leqslant 1 \qquad \text{in} \quad D^h,$$

and, hence, letting

$$\gamma = \frac{2KL_0n + 2KLn + L_4 + N}{\beta^2}, \qquad N > 0,$$

we obtain

$$FH \leqslant -\lambda H\|x\|^2 - NH \leqslant -NH < 0.$$

Thus, we have a contradiction to the fact that $\partial z(t_1, x_1)/\partial t \geqslant 0$. Hence, if $(t_1, x_1) \in D_{R_\alpha}^h \cup S_{R_\alpha}^h$, it follows that $z(t_1, x_1) \leqslant 0$.

Now, if $(t_1, x_1) \in \partial D_{R_\alpha}^h$, it turns out, by assumption (iv), that $z(t_1, x_1) \leqslant 0$. Finally, suppose that $(t_1, x_1) \in C_{R_\alpha}^h$. Then,

$$z(t_1, x_1) \leqslant \frac{M \exp K_0 R_\alpha^2}{\exp\{[KR_\alpha^2/1 - \mu(t-t_0)] + \gamma t_1\}}. \tag{10.10.3}$$

Let (t, x) be an arbitrary point in $\overline{D^h}$. Given $\epsilon > 0$, there exists an $\alpha_0 = \alpha_0(\epsilon)$ such that, for $\alpha > \alpha_0$, the point $(t, x) \in D^h_{R_x}$ and the right-hand side of (10.10.3) is less than ϵ. Since $z(t, x) \leqslant z(t_1, x_1)$, it follows that $z(t, x) \leqslant \epsilon$, if $\alpha > \alpha_0(\epsilon)$. Consequently, $z(t, x) \leqslant 0$ in $\overline{D^h}$, and, hence,

$$m(t, x) \leqslant 0 \quad \text{in} \quad \overline{D^h}.$$

This implies that

$$u(t, x) \leqslant v(t, x) \quad \text{on} \quad \overline{D^h}.$$

In particular, we have the inequality (10.10.1) in the intersection of the closure of D^h with the plane $t = t_0 + h$. Since this intersection, by proposition (iii) in the definition of the domain D, is identical with S_{t_0+h}, we have (10.10.1) for $(t, x) \in S_{t_0+h}$. We may therefore repeat the arguments starting from the plane $t = t_0 + h$ instead of $t = t_0$ and obtain the inequality (10.10.1) in the intersection of D with the zone $t_0 + h \leqslant t \leqslant t_0 + 2h$.

In this way, the validity of (10.10.1) at any point of D follows after a finite number of steps. The proof is therefore complete.

THEOREM 10.10.2. Assume that

(i) $u, v \in C[\overline{D}, R]$, the partial derivatives u_t, u_x, u_{xx}, v_t, v_x, v_{xx} exist and are continuous in D;

(ii) $f_1, f_2 \in C[D \times R \times R^n \times R^{n^2}, R]$,

$$f_1(t, x, u, P, R) \leqslant f_2(t, x, u, P, R),$$

and one of the functions f_1, f_2 is elliptic and satisfies an $\mathscr{L}$ condition;

(iii) $u \in \overline{E}_2(M, K_0)$, $v \in \hat{E}_2(M, K_0)$, and, in D,

$$u_t \leqslant f_1(t, x, u, u_x, u_{xx}),$$
$$v_t \geqslant f_2(t, x, v, v_x, v_{xx});$$

(iv) $u(t, x) \leqslant v(t, x)$ on $\overline{S}_{t_0} \cup \partial D$.

These assumptions imply

$$u(t, x) \leqslant v(t, x) \quad \text{everywhere in} \quad D.$$

By repeating the proof of Theorem 10.10.1 with obvious modifications, this theorem can be proved. We leave the details to the reader.

COROLLARY 10.10.1. Assume that

(i) $v \in C[\bar{D}, R]$, $v(t, x)$ possesses continuous partial derivatives v_t, v_x, v_{xx} in D, and $v \in \hat{E}_2(M, K_0)$;
(ii) $f \in C[D \times R \times R^n \times R^{n^2}, R]$, f satisfies an $\mathscr{L}$ condition, the differential operator T is parabolic, and $T[v] \geqslant 0$ in D;
(iii) $v(t, x) \geqslant 0$ on $\bar{S}_{t_0} \cup \partial D$.

These hypotheses imply that

$$v(t, x) \geqslant 0 \quad \text{everywhere in} \quad D.$$

Consider, as an example, the operator

$$T[u] = \frac{\partial u}{\partial t} - \left[\sum_{i,k=1}^{n} a_{ik}(t, x) u_{x_i x_k} + \sum_{i=1}^{n} b_i(t, x) u_{x_i} + c(t, x) u\right],$$

in $\Omega_0 = (0, \infty) \times R^n$. Let T be parabolic, and, for $i, k = 1, 2, ..., n$, $(t, x) \in \Omega_0$, suppose that

$$|a_{ik}(t, x)| \leqslant L_0, \quad |b_i(t, x)| \leqslant L_1 \|x\| + L_2,$$

$$c(t, x) \leqslant L_3 \|x\|^2 + L_4.$$

Let us assume that $T[u] \geqslant 0$ in Ω_0 and that $u \in \hat{E}_2(M, K_0)$ in Ω, where $\Omega = [0, \infty) \times R^n$. If $u(0, x) \geqslant 0$ in R^n, then it follows from Corollary 10.10.1 that $u(t, x) \geqslant 0$ in Ω.

On the basis of the foregoing results, one can state and prove theorems on comparison principle, bounds and error estimates for equations in unbounded domains, and corresponding parallel results for systems. We leave them as exercises.

10.11. Uniqueness

We shall consider the parabolic differential system

$$u_t = f(t, x, u, u_x^i, u_{xx}^i), \qquad (10.11.1)$$

where $f \in C[D \times R^N \times R^n \times R^{n^2}, R^N]$.

DEFINITION 10.11.1. Given a function $\phi \in C[\bar{S}_{t_0} \cup \partial D, R^N]$, any function $u \in C[\bar{D}, R^N]$ that has continuous partial derivatives u_t, u_x, u_{xx} in D and that satisfies (10.11.1) in D such that $u(t, x) = \phi(t, x)$ on $\bar{S}_{t_0} \cup \partial D$ is called a *solution of the first boundary-value problem of Fourier*.

10.11. UNIQUENESS

DEFINITION 10.11.2. We shall say that a vector function f satisfies an $\mathscr{L}$ condition if, for each $i = 1, 2, ..., N$,

$$|f^i(t, x, u, P^i, R^i) - f^i(t, x, \bar{u}, \overline{P^i}, \overline{R^i})|$$
$$\leq L_0 \sum_{j,k=1}^{n} |R^i_{jk} - \overline{R^i_{jk}}| + (L_1 \|x\| + L_2) \sum_{j=1}^{n} |P^i_j - \overline{P^i_j}|$$
$$+ (L_3 \|x\|^2 + L_4) \sum_{j=1}^{N} |u_j - \bar{u}_j|.$$

THEOREM 10.11.1. *If $f \in C[D \times R^N \times R^n \times R^{n^2}, R^N]$ and satisfies an $\mathscr{L}$ condition, then the first boundary-value problem of Fourier possesses at most one solution $u(t, x)$ such that each component of $u(t, x)$ belongs to $E_2(M, K_0)$.*

Proof. Suppose that there exist two such solutions, $u(t, x)$ and $v(t, x)$. Then the difference

$$m(t, x) = u(t, x) - v(t, x)$$

obeys the equation

$$\frac{\partial m(t, x)}{\partial t} = f(t, x, u(t, x), u_x^i(t, x), u_{xx}^i(t, x))$$
$$-f(t, x, v(t, x), v_x^i(t, x), v_{xx}^i(t, x)).$$

Define $u = \bar{u}H$, $v = \bar{v}H$, where

$$H = H(t, x, K) = \exp\left[\frac{K \|x\|^2}{1 - \mu(t - t_0)} + \gamma t\right]$$

is the growth damping factor and $K > K_0$. We shall retain the meaning of the symbols D_R, Γ_R, $S_R^{t_0}$, ∂D_R, C_R and D^h, D_R^h, ∂D^h, ∂D_R^h, C_R^h as defined in the proof of Theorem 10.10.1. Set

$$z = z(t, x) = \bar{u}(t, x) - \bar{v}(t, x).$$

Then, we have $m = zH$ defined in D^h, where $h < \mu^{-1}$. Moreover, we observe that

$$\frac{\partial z}{\partial t} H + z \frac{\partial H}{\partial t} = u_t - v_t$$
$$= f(t, x, \bar{u}H, \overline{u_x^i}H + \overline{u^i}H_x, R^{(1)})$$
$$- f(t, x, \bar{u}H, \overline{u_x^i}H + \overline{u^i}H_x, R^{(2)})$$
$$+ f(t, x, \bar{u}H, \overline{u_x^i}H + \overline{u^i}H_x, R^{(2)})$$
$$- f(t, x, \bar{v}H, \overline{v_x^i}H + \overline{v^i}H_x, R^{(3)}), \quad (10.11.2)$$

where

$$R^{(1)} = [\overline{u^i_{x_j x_k}}H + \overline{u^i_{x_j}}H_{x_k} + \overline{u^i_{x_k}}H_{x_j} + \overline{u^i}H_{x_j x_k}],$$

$$R^{(2)} = [\overline{v^i_{x_j x_k}}H + \overline{u^i_{x_j}}H_{x_k} + \overline{u^i_{x_k}}H_{x_j} + \overline{u^i}H_{x_j x_k}],$$

$$R^{(3)} = [\overline{v^i_{x_j x_k}}H + \overline{v^i_{x_j}}H_{x_k} + \overline{v^i_{x_k}}H_{x_j} + \overline{v^i}H_{x_j x_k}].$$

Let $\{R_\alpha\}$ be an increasing sequence, $R_\alpha \to \infty$ as $\alpha \to \infty$. For a fixed α, we consider the domain $D^h_{R_\alpha}$. Let us denote

$$A_\alpha = \max_i [\sup_{(t,x) \in D^h_{R_\alpha}} |z^i(t,x)|].$$

Then, there exists an index i_α and a point $(t_\alpha, x_\alpha) \in \overline{D^h_{R_\alpha}}$ such that $A_\alpha = |z^{i_\alpha}(t_\alpha, x_\alpha)|$. We shall show that μ, γ, and h may be conveniently chosen such that $(t_\alpha, x_\alpha) \in D^h_{R_\alpha} \cup S^h_{R_\alpha}$ implies $|z^{i_\alpha}(t_\alpha, x_\alpha)| = 0$. Let us suppose the contrary. Then, there are two cases to be considered:

(a) $z^{i_\alpha}(t_\alpha, x_\alpha) > 0$, and
(b) $z^{i_\alpha}(t_\alpha, x_\alpha) < 0$.

In case (a) holds, we have, for an arbitrary vector λ,

$$\frac{\partial z^{i_\alpha}(t_\alpha, x_\alpha)}{\partial t} \geq 0, \qquad \frac{\partial z^{i_\alpha}(t_\alpha, x_\alpha)}{\partial x} = 0,$$

and

$$\sum_{j,k=1}^n \frac{\partial^2 z^{i_\alpha}(t_\alpha, x_\alpha)}{\partial x_j \partial x_k} \lambda_j \lambda_k \leq 0.$$

Since f^{i_α} is elliptic, it follows that

$$f^{i_\alpha}(t_\alpha, x_\alpha, u(t_\alpha, x_\alpha), \overline{u^i_\alpha}H + \overline{u^i_\alpha}H_x, R^{(1)})$$
$$\leq f^{i_\alpha}(t_\alpha, x_\alpha, u(t_\alpha, x_\alpha), \overline{u^i_\alpha}H + \overline{u^i_\alpha}H_x, R^{(2)}).$$

Hence, we get, from (10.11.2),

$$0 \leq \frac{\partial z^{i_\alpha}(t_\alpha, x_\alpha)}{\partial t} H \leq z^{i_\alpha} \left[L_0 \sum_{j,k=1}^n \left| \frac{\partial^2 H}{\partial x_j \partial x_k} \right| \right.$$
$$\left. + (L_1 \|x\| + L_2) \sum_{j=1}^n \left| \frac{\partial H}{\partial x_j} \right| + (L_3 \|x\|^2 + L_4) NH - \frac{\partial H}{\partial t} \right],$$

because of the $\mathscr{L}$ condition and the fact $\sum_{j=1}^{N} |z^j(t,x)| \leq N\bar{z}^{i_\alpha}(t_\alpha, x_\alpha)$. We now proceed, as in the proof of Theorem 10.10.1, to show that

$$\frac{\partial z^{i_\alpha}(t_\alpha, x_\alpha)}{\partial t} < 0.$$

The choice μ, h, and γ would be as follows:

$$\mu K = 4K^2 L_0 n^2 + 4KLn + L_3 N + \lambda, \qquad \lambda > 0,$$
$$h = (1-\beta)/\mu, \qquad 0 < \beta < 1,$$

and

$$\gamma = \frac{2KL_0 n + 2KLn + L_4 N + N_1}{\beta^2}, \qquad N_1 > 0.$$

Thus, we have a contradiction to $\partial z^{i_\alpha}(t_\alpha, x_\alpha)/\partial t \geq 0$.

If $z^{i_\alpha}(t_\alpha, x_\alpha) < 0$, we repeat the foregoing reasoning with $\overline{z^{i_\alpha}}(t,x) = -z^{i_\alpha}(t,x)$, and, again, we get a contradiction to $\partial z^{i_\alpha}(t_\alpha, x_\alpha)/\partial t \geq 0$.

If $(t_\alpha, x_\alpha) \in \partial D_{R_\alpha}^h$, $z^{i_\alpha}(t_\alpha, x_\alpha) = 0$ by assumption. Finally, if $(t_\alpha, x_\alpha) \in C_{R_\alpha}^h$, then

$$|z^{i_\alpha}(t_\alpha, x_\alpha)| \leq \frac{M \exp(K_0 \|x_\alpha\|^2)}{\exp\{[K\|x_\alpha\|^2/1 - \mu(t_\alpha - t_0)] + \gamma t_\alpha\}}$$
$$= \frac{M \exp(K_0 R_\alpha^2)}{\exp\{[KR_\alpha^2/1 - \mu(t_\alpha - t_0)] + \gamma t_\alpha\}}.$$

Since $|z^i(t,x)| \leq |z^{i_\alpha}(t_\alpha, x_\alpha)|$ in $\overline{D_{R_\alpha}^h}$, again arguing as in Theorem 10.10.1, we can show that $z^i(t,x) = 0$ in $\overline{D^h}$. Consequently, it follows that $m^i(t,x) \equiv 0$ ($i = 1, 2, ..., N$) in D, proving the uniqueness of solutions in $\overline{D^h}$.

The validity of the uniqueness of solutions at any point of D may now be proved after a finite number of steps as in Theorem 10.10.1. This proves the theorem.

10.12. Exterior boundary-value problem and uniqueness

Let Δ be a bounded and closed domain R^n, and let S be its complementary domain. Assume that the boundary $\partial \Delta$ of Δ is represented by the equation

$$\Gamma(x) = 0, \qquad (10.12.1)$$

where $\Gamma(x)$ is a continuously differentiable function in $\bar{S}$, having bounded second derivatives in S and satisfying the condition

$$|\text{grad } \Gamma(x)| \geq \Gamma_0 > 0. \qquad (10.12.2)$$

We define, for $t_0 \geqslant 0$,

$$D = (t_0, \infty) \times S \quad \text{and} \quad \sigma = (t_0, \infty) \times \partial \Delta. \tag{10.12.3}$$

For every $(t, x) \in \sigma$ and each $i = 1, 2,..., N$, let τ_i be a straight half-line entering the interior of D and parallel to the plane $t = t_0$. It is assumed that there exists a positive constant γ_0 such that

$$\cos(\tau_i, n_0) \geqslant \gamma_0 \quad (i = 1, 2,..., N), \quad (t, x) \in \sigma,$$

where n_0 is the normal to σ directed to the interior of D.

We consider the partial differential system

$$u_t = f(t, x, u, u_x^i, u_{xx}^i), \tag{10.12.4}$$

where $f \in C[D \times R^N \times R^n \times R^{n^2}, R^N]$.

DEFINITION 10.12.1. Given an initial function $\phi \in C[S, R^N]$, a *solution* of (10.12.4) is any function $u \in C[\bar{D}, R^N]$ satisfying the following conditions:

(i) $u(t_0, x) = \phi(x)$ on S;
(ii) $du(t, x)/d\tau$ exists in σ, and

$$[du(t, x)/d\tau] + G(t, x, u(t, x)) = 0 \quad \text{on} \quad \sigma,$$

where $G \in C[\sigma \times R^N, R^N]$;

(iii) $u(t, x)$ possesses continuous partial derivatives u_t, u_x, u_{xx} in D and satisfies (10.12.4) in D.

The problem of finding a solution of (10.12.4) is called an *exterior boundary-value problem*. The following theorem concerns the uniqueness of solutions of such a problem.

THEOREM 10.12.1. Assume that

(i) $f \in C[D \times R^N \times R^n \times R^{n^2}, R^N]$, f is elliptic and satisfies an $\mathscr{L}$ condition;
(ii) $G \in C[\sigma \times R^N, R^N]$, and, for each $i = 1, 2,..., N$,

$$G_i(t, x, y) - G_i(t, x, \bar{y}) \leqslant L \sum_{i=1}^{N} |y_i - \bar{y}_i|, \quad L > 0.$$

Then, the exterior boundary-value problem has no more than one solution belonging to class $E_2(M_1, K_0)$ in D.

10.12. EXTERIOR BOUNDARY-VALUE PROBLEM

Proof. Suppose, if possible, there are two solutions $u(t, x)$ and $v(t, x)$ of (10.12.4) belonging to $E_2(M_1, K_0)$ in D. Then,

$$m(t, x) = u(t, x) - v(t, x)$$

belongs to $E_2(M, K_0)$, where $M = 2M_1$.

We may assume that

$$\Gamma(x) = \|x\| \quad \text{for} \quad \|x\| > R_0,$$

where R_0 is the radius of the sphere $\|x\| = R_0$ situated in R^n and containing the boundary $\partial \Delta$ in its interior. It then follows that there exist two constants $A, B > 0$ such that

$$\left| \sum_{j=1}^{n} \Gamma_{x_j}(x) \right| \leqslant A \quad \text{and} \quad \left| \sum_{j,k=1}^{n} \Gamma_{x_j x_k}(x) \right| \leqslant B \quad \text{in} \quad S. \quad (10.12.5)$$

We shall use the function

$$H(t, x; K) = \exp\left(\frac{K[\Gamma(x) - p]^2}{1 - \mu(t - t_0)} + vt\right), \quad K > K_0, \quad (10.12.6)$$

where

$$p = \frac{NL + 1}{2K\Gamma_0\gamma_0}, \quad (10.12.7)$$

$$\mu = 4KL_0 A^2 + 2L_1 A + \frac{NL_3 + 1}{K}, \quad (10.12.8)$$

and

$$v = \max\left\{ \frac{(KL_0 B + KL_1 Ap + L_3 pN)^2 + 2KL_0 A^2 + L_3 p^2 N + L_4 N + 1}{\gamma^2}, \right.$$

$$\left. \frac{[KL_0 B + KA(L_1 R_0 + L_2)]^2 + 2KL_0 A^2 + N(L_3 R_0^2 + L_4) + 1}{\gamma^2} \right\}, \quad (10.12.9)$$

γ being arbitrarily chosen so that $0 < \gamma < 1$.

First of all, we shall prove the theorem for the part

$$D^{h_0} = (t_0, t_0 + h_0] \times S \quad \text{of} \quad D,$$

where

$$h_0 = (1 - \gamma)/\mu. \quad (10.12.10)$$

We shall write

$$FH = FH(t, x, K) = L_0 \sum_{j,k=1}^{n} \left| \frac{\partial^2 H}{\partial x_j \partial x_k} \right| + (L_1 \|x\| + L_2)$$

$$\times \sum_{j=1}^{n} \left| \frac{\partial H}{\partial x_j} \right| + (L_3 \|x\|^2 + L_4) NH - \frac{\partial H}{\partial t}. \quad (10.12.11)$$

It will be shown that the function $H(t, x; K)$ verifies the inequality

$$FH \leqslant -H(t, x, K) \quad \text{for} \quad (t, x) \in D^{h_0}. \tag{10.12.12}$$

Indeed, (10.12.10) yields

$$0 < \gamma \leqslant 1 - \mu(t - t_0) \leqslant 1. \tag{10.12.13}$$

In view of (10.12.5), (10.12.6), and (10.12.13), we get

$$FH \leqslant \frac{H}{[1 - \mu(t - t_0)]^2} \{4K^2 L_0 A(\Gamma - p)^2 + 2KL_0 A^2 + 2KBL_0 \mid \Gamma - p \mid$$
$$+ (L_1 \parallel x \parallel + L_2) 2KA \mid \Gamma - p \mid + (L_3 \parallel x \parallel^2 + L_4) N - \mu K(\Gamma - p)^2 - \nu\gamma^2\}. \tag{10.12.14}$$

If $\parallel x \parallel > R_0$, the assumption that $\Gamma(x) = \parallel x \parallel$, the inequality $\parallel x \parallel \leqslant \mid \parallel x \parallel - p \mid + p$, and (10.12.8) yield

$$FH \leqslant \frac{H}{[1 - \mu(t - t_0)]^2} \{-[\mid \parallel x \parallel - p \mid - (KL_0 B + pKL_1 A + NpL_3)]^2$$
$$+ (KL_0 B + pKL_1 A + NpL_3)^2 + 2KL_0 A^2 + Np^2 L_3 + NL_4 - \nu\gamma^2\}.$$

Hence, by (10.12.9), there results (10.12.12).

If $\parallel x \parallel \leqslant R_0$, then, from (10.12.8) and (10.12.14), it follows that

$$FH \leqslant \frac{H}{[1 - \mu(t - t_0)]^2} \{-[\mid \Gamma - p \mid - (KL_0 B + KA(L_1 R_0 + L_2))]^2$$
$$+ (KL_0 B + KA(L_1 R_0 + L_2))^2 + 2KL_0 A^2 + N(L_3 R_0^2 + L_4) - \nu\gamma^2\}.$$

By virtue of (10.12.9), we see that the inequality (10.12.12) is again verified.

Let us introduce the functions $\bar{u}$ and $\bar{v}$ such that

$$\begin{aligned} u &= u(t, x) = \bar{u} H(t, x; K), \\ v &= v(t, x) = \bar{v} H(t, x; K). \end{aligned} \tag{10.12.15}$$

Set $z = z(t, x) = \bar{u} - \bar{v}$, so that

$$m = m(t, x) = z(t, x) H(t, x; K). \tag{10.12.16}$$

Let us choose an increasing sequence $\{R_\alpha\}$, $R_\alpha > R_0$, and $R_\alpha \to \infty$ as $\alpha \to \infty$. Denote by S^{t_0} the part of the boundary of D lying on the plane $t = t_0$. Furthermore, let $D_\alpha^{h_0}$ and $S_\alpha^{t_0}$ denote the parts of the domains D^{h_0} and S^{t_0} contained inside the cylindrical surface C_α with the equation $\|x\| = R_\alpha$. Put $C_\alpha^{h_0} = D^{h_0} \cap C_\alpha$. Define

$$A_\alpha = \max_i [\max_{(t,x) \in \overline{D_\alpha^{h_0}}} |z^i(t, x)|].$$

Observe that the sequence $\{A_\alpha\}$ is nondecreasing and $A_\alpha \geqslant 0$ for every α. Hence, it is enough to show that $A_\alpha \to 0$ as $\alpha \to \infty$. For this purpose, notice that, for any α, there exists an index i_α and a point $(t_\alpha, x_\alpha) \in \overline{D_\alpha^{h_0}}$ such that $A_\alpha = |z^{i_\alpha}(t_\alpha, x_\alpha)|$. The following cases are now possible:

(i) $(t_\alpha, x_\alpha) \in S_\alpha^{t_0}$,
(ii) $(t_\alpha, x_\alpha) \in D_\alpha^{h_0}$,
(iii) $(t_\alpha, x_\alpha) \in \sigma^{h_0}$,
(iv) $(t_\alpha, x_\alpha) \in C_\alpha^{h_0}$.

Evidently, in case (i), $z^{i_\alpha}(t_\alpha, x_\alpha) = 0$, and therefore $A_\alpha = 0$. In case (ii) holds, we have either

(iia) $z^{i_\alpha}(t_\alpha, x_\alpha) = 0$

or

(iib) $z^{i_\alpha}(t_\alpha, x_\alpha) > 0$

or

(iic) $z^{i_\alpha}(t_\alpha, x_\alpha) < 0$.

Clearly, case (iia) implies $A_\alpha = 0$. If case (iib) is true, then, using a similar argument as in the proof of Theorem 10.10.1, we arrive at the inequality

$$0 \leqslant \frac{\partial z^{i_\alpha}(t_\alpha, x_\alpha)}{\partial t} H(t_\alpha, x_\alpha; K) \leqslant z^{i_\alpha}(t_\alpha, x_\alpha) FH(t_\alpha, x_\alpha; K), \quad (10.12.17)$$

and thus, by (10.12.12), the impossibility of case (iib) follows. The case (iic) may be reduced to case (iib) by the substitution $\hat{z}^{i_\alpha} = -z^{i_\alpha}$.

When case (iii) holds, we have again three possibilities:

(iiia) $z^{i_\alpha}(t_\alpha, x_\alpha) = 0$,
(iiib) $z^{i_\alpha}(t_\alpha, x_\alpha) > 0$,
(iiic) $z^{i_\alpha}(t_\alpha, x_\alpha) < 0$.

If case (iiib) holds, then

$$\frac{dz^{i_\alpha}(t_\alpha, x_\alpha)}{d\tau_{i_\alpha}} \leqslant 0, \quad (10.12.18)$$

since, otherwise, there would exist a point $(t_1, x_1) \in \mathrm{int}\ D^{h_0}$ such that $z^{i_\alpha}(t_1, x_1) > z^{i_\alpha}(t_\alpha, x_\alpha)$, and this contradicts the definition of $z^{i_\alpha}(t_\alpha, x_\alpha)$.

On the other hand, we obtain, by the relations (10.12.15), (10.12.16), and the condition (ii) in the Definition 10.12.1,

$$\frac{dz^{i_\alpha}}{d\tau_{i_\alpha}} H + z^{i_\alpha} \frac{dH}{d\tau_{i_\alpha}} + G_{i_\alpha}(t_\alpha, x_\alpha, \bar{u}H) - G_{i_\alpha}(t_\alpha, x_\alpha, \bar{v}H) = 0,$$

whence, according to assumption (ii) of the theorem, we get

$$-\frac{dz^{i_\alpha}}{d\tau_{i_\alpha}} H \leqslant z^{i_\alpha} \frac{dH}{d\tau_{i_\alpha}} + L \sum_{i=1}^{N} |z^i| H$$

$$\leqslant z^{i_\alpha} \left(\frac{dH}{d\tau_{i_\alpha}} + LHN \right). \qquad (10.12.19)$$

The sign of $\Gamma(x)$ in (10.12.1) may be so chosen that

$$\frac{\partial \Gamma(x)}{\partial x_i} = |\mathrm{grad}\ \Gamma(x)| \cos(x_i, n_0). \qquad (10.12.20)$$

Hence, taking into account the relations (10.12.1), (10.12.2), (10.12.7), (10.12.19), and the fact $\cos(\tau_i, n_0) \geqslant \gamma_0$, we derive the inequality

$$-\frac{dz^{i_\alpha}}{d\tau_{i_\alpha}} H \leqslant z^{i_\alpha} H \left[\frac{-2Kp}{1 - \mu(t - t_0)} |\mathrm{grad}\ \Gamma(x)| \cos(\tau_{i_\alpha}, n_0) + NL \right]$$

$$\leqslant z^{i_\alpha} H(-2Kp\Gamma_0 \gamma_0 + NL)$$

$$= -z^{i_\alpha} H < 0,$$

which contradicts (10.12.18). We can repeat the same reasoning for the function $\hat{z}^{i_\alpha} = -z^{i_\alpha}$ to show that case (iiic) is also impossible. We have now proved that, in each of the cases (i), (ii), and (iii), $A_\alpha = 0$.

When case (iv) holds, we obtain, by (10.12.16) and the fact that $m \in E_2(M, K_0)$, the estimate

$$|z^{i_\alpha}(t_\alpha, x_\alpha)| \leqslant \frac{M \exp(K_0 \|x_\alpha\|^2)}{\exp\{[K(\Gamma - p)^2 / 1 - \mu(t_\alpha - t_0)] + \nu t_\alpha\}}$$

$$= \frac{M \exp(K_0 R_\alpha^2)}{\exp\{[K(R_\alpha - p)^2 / 1 - \mu(t_\alpha - t_0)] + \nu t_\alpha\}}$$

$$\to 0 \quad \text{as} \quad \alpha \to \infty.$$

Consequently, in all cases, $A_\alpha \to 0$ as $\alpha \to \infty$. The validity of the uniqueness of solutions at any point of D may now be proved after a finite number of steps as in Theorem 10.11.1. The theorem is therefore proved.

10.13. Notes

Lemmas 10.1.1 and 10.1.2 are new. For theorems of the type 10.1.1 and 10.1.2, see Friedman [22], Mlak [3], Nagumo and Simoda [1], Szarski [8], Walter [8], and Westphal [1]. For a systematic use of inequalities in the theory of partial differential equation, see Picone [1]. Theorem 10.2.1 is taken from Lakshmikantham [5]. For Theorems 10.2.2 and 10.2.3, see Brzychczy [1], Mlak [6], Szarski [8], and Walter [8]. Theorem 10.2.4 is due to Maple and Peterson [1].

The results of Sects. 10.3 and 10.4 are adapted from similar results in Szarski [8] and Walter [8]. Section 10.5 consists of the work of Maple and Peterson [1]. For the results analogous to the results of Sect. 10.6, see Szarski [8] and Walter [8].

The results of Sects. 10.9 and 10.10 are due to Lakshmikantham [5]. See also Bellman [2], Mlak [2], and Narasimhan [1]. Section 10.11 contains the work of Lakshmikantham and Leela [1]. The rest of the chapter contains the results based on the work of Besala [1–4], Krzyzanski [2, 8], and Szarski [4, 5, 7].

For related results, see Aronson [1–3], Aronson and Besala [1], Barrar [1], Brzychczy [2], Cameron [1], Ciliberto [2], Eidelman [1–5], Foias *et al.*, [2], Friedman [1–24], Ilin *et al.* [1], Ito [1], Kaplan [1], Komatsu [1], Krzyzanski [1–10], Ladyzhenskaja [1], Lax and Milgram [1], Lees and Protter [1], Lions and Malgrange [1], McNabb [1], Milicer-Gruzewska [1, 2], Mizohata [1–3], Mlak [5, 6, 8, 10, 16], Murakami [1, 2], Nickel [1–3], Nirenberg [1, 2], Pini [1–3], Pogorzelski [1, 2], Prodi [1–3], Protter [2], Serrin [1], Slobodetski [1], Smirnova [1], Vishik [2], and Zeragia [1, 2].

Chapter 11

11.0. Introduction

This chapter is concerned with hyperbolic differential equations. Main results associated with hyperbolic differential inequalities are discussed, including under and over functions. Certain uniqueness criteria, growth, and error estimates are treated utilizing the comparison principle.

11.1. Hyperbolic differential inequalities

Let $f \in C[R_0 \times R^3, R]$, where R_0 is the rectangle defined by $R_0 = [0 \leqslant x \leqslant a, 0 \leqslant y \leqslant b]$. We shall denote by $P\phi$ the expression $\phi_{xy} - f(x, y, \phi, \phi_x, \phi_y)$ in what follows.

A fundamental result in hyperbolic differential inequalities is the following.

THEOREM 11.1.1. Assume that

(i) $u, v \in C[R_0, R]$, the partial derivatives u_x, u_y, u_{xy}, v_x, v_y, v_{xy} exist and are continuous in R_0;

(ii) $f \in C[R_0 \times R^3, R]$, $f(x, y, u, p, q)$ is monotonic nondecreasing in u, p, q, and, in R_0,
$$pu < pv;$$

(iii) $u(0, 0) < v(0, 0)$, $u_x(x, 0) < v_x(x, 0)$, and $u_y(0, y) < v_y(0, y)$ for $0 \leqslant x \leqslant a$, $0 \leqslant y \leqslant b$.

Then we have, on R_0,

$$u(x, y) < v(x, y), \quad u_x(x, y) < v_x(x, y), \quad u_y(x, y) < v_y(x, y). \qquad (11.1.1)$$

Proof. It is enough to show that, in R_0,

$$u_x(x,y) < v_x(x,y) \quad \text{and} \quad u_y(x,y) < v_y(x,y).$$

Assume that this is false, and let t_0 be the greatest lower bound of numbers $t > x + y$ such that the two inequalities are satisfied for $x + y < t_0$. Then there exists a point (x_0, y_0) with $x_0 + y_0 = t_0$ for which one of the inequalities is not true. Let us suppose that

$$u_x(x_0, y_0) = v_x(x_0, y_0). \tag{11.1.2}$$

Clearly, by assumption (iii), $y_0 > 0$. We have

$$u_x(x_0, y_0 - h) < v_x(x_0, y_0 - h), \quad h > 0, \tag{11.1.3}$$

and

$$u(x_0, y_0) \leqslant v(x_0, y_0), \quad u_y(x_0, y_0) \leqslant v_y(x_0, y_0). \tag{11.1.4}$$

From (11.1.2) and (11.1.3), we obtain

$$u_{xy}(x_0, y_0) \geqslant v_{xy}(x_0, y_0). \tag{11.1.5}$$

Furthermore, since f is monotonic nondecreasing in u, p, q, it follows, from (11.1.4), that

$$f(x_0, y_0, u(x_0, y_0), u_x(x_0, y_0), u_y(x_0, y_0))$$
$$\leqslant f(x_0, y_0, v(x_0, y_0), u_x(x_0, y_0), v_y(x_0, y_0)).$$

This, together with (11.1.5) and the assumption $Pu < Pv$, yields

$$f(x_0, y_0, v(x_0, y_0), u_x(x_0, y_0), v_y(x_0, y_0))$$
$$> f(x_0, y_0, v(x_0, y_0), v_x(x_0, y_0), v_y(x_0, y_0)),$$

which is an absurdity, in view of (11.1.2).

It is clear that, if we suppose that $u_y(x_0, y_0) = v_y(x_0, y_0)$ instead of (11.1.2), an argument similar to the foregoing leads to a contradiction. Hence, the desired result (11.1.1) is true.

DEFINITION 11.1.1. A function $u \in C[R_0, R]$, possessing continuous partial derivatives u_x, u_y, u_{xy} in R_0 and satisfying the hyperbolic differential inequality

$$u_{xy} < f(x, y, u, u_x, u_y)$$

in R_0, is said to be an *under function* with respect to the hyperbolic differential equation

$$u_{xy} = f(x, y, u, u_x, u_y). \tag{11.1.6}$$

If, on the other hand, u satisfies the reversed inequality, it is said to be an *over function*.

The following result can be proved by a repeated application of Theorem 11.1.1.

THEOREM 11.1.2. Let u, w be under and over functions with respect to the hyperbolic differential equation (11.1.6), respectively. Suppose that v is a solution of (11.1.6) such that

$$u(0, 0) < v(0, 0) < w(0, 0),$$
$$u_x(x, 0) < v_x(x, 0) < w_x(x, 0),$$
$$u_y(0, y) < v_y(0, y) < w_y(0, y).$$

Then we have, on R_0,

$$u(x, y) < v(x, y) < w(x, y),$$
$$u_x(x, y) < v_x(x, y) < w_x(x, y),$$
$$u_y(x, y) < v_y(x, y) < w_y(x, y).$$

THEOREM 11.1.3. Assume that

(i) $u, v \in C[R_0, R]$, the partial derivatives u_x, u_y, u_{xy}, v_x, v_y, v_{xy} exist and are continuous in R_0;

(ii) $f \in C[R_0 \times R^2, R]$, $f(x, y, u, p)$ is monotonic nondecreasing in u, and, in R_0, $Pu < Pv$, where $P\phi = \phi_{xy} - f(x, y, \phi, \phi_x)$;

(iii) $u(0, 0) < v(0, 0)$ and $u_x(x, 0) < v_x(x, 0)$ for $0 \leqslant x \leqslant a$.

Then we have, on R_0,

$$u(x, y) < v(x, y) \quad \text{and} \quad u_x(x, y) < v_x(x, y).$$

This theorem is a special case of Theorem 11.1.1. However, it is to be noted that the function f needs to be monotonic in u alone, instead of u and p, as one might expect. The reason is obvious, if we follow the proof of Theorem 11.1.1.

11.2. Uniqueness criteria

We wish to consider a general uniqueness theorem of Perron type for the hyperbolic differential equation

$$u_{xy} = f(x, y, u, u_x, u_y), \qquad (11.2.1)$$

where $f \in C[R_0 \times R^3, R]$, subject to the conditions

$$u(x, 0) = \sigma(x), \quad u(0, y) = \tau(y), \quad \sigma(0) = \tau(0) = u_0,$$

the functions σ and τ being Lipschitz continuous on $0 \leqslant x \leqslant a$, $0 \leqslant y \leqslant b$, respectively.

THEOREM 11.2.1. Assume that

(i) $f \in C[R_0 \times R^3, R]$ and

$$|f(x, y, u, p, q) - f(x, y, \bar{u}, \bar{p}, \bar{q})|$$
$$\leqslant g(x, y, |u - \bar{u}|, |p - \bar{p}|, |q - \bar{q}|),$$

where $g \in C[R_0 \times R_+^3, R_+]$, $g(x, y, 0, 0, 0) \equiv 0$, g is monotonic nondecreasing in z, p, q and bounded;

(ii) $z(x, y) \equiv 0$ is the only solution of the hyperbolic equation

$$z_{xy} = g(x, y, z, z_x, z_y), \tag{11.2.2}$$

such that

$$z(0, 0) = 0, \quad z(x, 0) \equiv 0, \quad z(0, y) \equiv 0. \tag{11.2.3}$$

Then, there is at most one solution for Eq. (11.2.1).

Proof. Suppose that there exist two solutions $u(x, y)$ and $v(x, y)$ for Eq. (11.2.1) on R_0. We define

$$A(x, y) = |u(x, y) - v(x, y)|,$$
$$B(x, y) = |u_x(x, y) - v_x(x, y)|,$$
$$C(x, y) = |u_y(x, y) - v_y(x, y)|.$$

Since we have

$$u(x, 0) = v(x, 0) = \sigma(x), \quad u_x(x, 0) = v_x(x, 0) = \sigma'(x), \quad 0 \leqslant x \leqslant a,$$
$$u(0, y) = v(0, y) = \tau(y), \quad u_y(0, y) = v_y(0, y) = \tau'(y), \quad 0 \leqslant y \leqslant b,$$

it follows that

$$A(0, 0) = 0, \quad B(x, 0) \equiv 0, \quad C(0, y) \equiv 0.$$

Furthermore, by condition (i), we obtain

$$A(x, y) \leqslant \int_0^x \int_0^y g(s, t, A(s, t), B(s, t), C(s, t))\, ds\, dt,$$

$$B(x, y) \leqslant \int_0^y g(x, t, A(x, t), B(x, t), C(x, t))\, dt,$$

$$C(x, y) \leqslant \int_0^x g(s, y, A(s, y), B(s, y), C(s, y))\, ds.$$

We note that $A(x, y)$ is continuous, $B(x, y)$ and $C(x, y)$ are uniformly Lipschitz continuous in y and x, respectively. Let us define the sequence of successive approximations to the solution of (11.2.2) and (11.2.3) as follows:

$$\alpha_0 = \alpha_0(x, y) = A(x, y), \quad \beta_0 = \beta_0(x, y) = B(x, y), \quad \gamma_0 = \gamma_0(x, y) = C(x, y),$$

and, for $n \geq 1$,

$$\alpha_{n+1}(x, y) = \int_0^x \int_0^y g(s, t, \alpha_n(s, t), \beta_n(s, t), \gamma_n(s, t))\, ds\, dt,$$

$$\beta_{n+1}(x, y) = \int_0^y g(x, t, \alpha_n(x, t), \beta_n(x, t), \gamma_n(x, t))\, dt,$$

$$\gamma_{n+1}(x, y) = \int_0^x g(s, y, \alpha_n(s, y), \beta_n(s, y), \gamma_n(s, y))\, ds.$$

Since $\alpha_0 \leq \alpha_1$, $\beta_0 \leq \beta_1$, $\gamma_0 \leq \gamma_1$, and g is nondecreasing in z, p, q, it follows by induction that

$$\alpha_n \leq \alpha_{n+1}, \quad \beta_n \leq \beta_{n+1}, \quad \gamma_n \leq \gamma_{n+1}.$$

Also, the functions α_n, β_n, γ_n are uniformly bounded in view of the fact that g is assumed to be bounded. Hence, we get $\lim_{n \to \infty} \alpha_n(x, y) = \alpha(x, y)$, $\lim_{n \to \infty} \beta_n(x, y) = \beta(x, y)$, and $\lim_{n \to \infty} \gamma_n(x, y) = \gamma(x, y)$, uniformly on R_0. It is easy to see that $\alpha(x, y)$ is a solution of (11.2.2) and (11.2.3). Consequently, on R_0,

$$A(x, y) \leq \alpha(x, y),$$
$$B(x, y) \leq \beta(x, y),$$
$$C(x, y) \leq \gamma(x, y).$$

By assumption, identically zero is the only solution for the problem (11.2.2) and (11.2.3). This proves $A(x, y) \equiv 0$, $B(x, y) \equiv 0$, and $C(x, y) \equiv 0$. The theorem is therefore proved.

The next uniqueness theorem is under Nagumo's condition, and the interest is rather in the elementary method employed in the proof.

THEOREM 11.2.2. Assume that

(i) $f(x, y, u, p, q)$ is defined for $(x, y) \in R_0$ and $u, p, q \in R$, and

$$xy\, |f(x, y, u_1, p_1, q_1) - f(x, y, u_2, p_2, q_2)|$$
$$\leq \alpha(x, y)\, |u_1 - u_2| + \beta(x, y)\, |p_1 - p_2| + \gamma(x, y)\, y\, |q_1 - q_2|,$$

where α, β, $\gamma \geqslant 0$ are continuous functions on R_0 such that $\alpha + \beta + \gamma = 1$;

(ii)
$$x \mid f(x, 0, u, p, q_1) - f(x, 0, u, p, q_2)\mid \, \leqslant \mid q_1 - q_2 \mid,$$
and
$$y \mid f(0, y, u, p_1, q) - f(0, y, u, p_2, q)\mid \, \leqslant \mid p_1 - p_2 \mid.$$

Then, there is at most one solution on R_0 for Eq. (11.2.1).

Proof. Suppose that $u(x, y)$ and $v(x, y)$ are two solutions of (11.2.1) existing on R_0. Then,
$$u(x, 0) = v(x, 0) = \sigma(x), \quad u_x(x, 0) = v_x(x, 0) = \sigma'(x), \quad 0 \leqslant x \leqslant a.$$

It can be shown that
$$u_y(x, 0) = v_y(x, 0) \quad \text{for} \quad 0 \leqslant x \leqslant a.$$

For, putting $y = 0$ in the partial differential equations satisfied by $u(x, y)$ and $v(x, y)$, we get the ordinary differential equation
$$w' = f(x, 0, \sigma(x), \sigma'(x), w). \tag{11.2.4}$$

We observe that $u_y(x, 0)$ and $v_y(x, 0)$ are solutions of (11.2.4) satisfying the initial condition $u_y(0, 0) = v_y(0, 0) = \tau'(0)$. Furthermore, the function f of (11.2.4) verifies the following condition:
$$x \mid f(x, 0, \sigma(x), \sigma'(x), w) - f(x, 0, \sigma(x), \sigma'(x), w_1)\mid \, \leqslant \mid w - w_1 \mid.$$

This being exactly Nagumo's uniqueness condition, it follows that
$$u_y(x, 0) = v_y(x, 0) \quad \text{for} \quad 0 \leqslant x \leqslant a,$$
as desired. Similarly, we have
$$u(0, y) = v(0, y) = \tau(y),$$
$$u_y(0, y) = v_y(0, y) = \tau'(y),$$
and
$$u_x(0, y) = v_x(0, y) \quad \text{for} \quad 0 \leqslant y \leqslant b.$$

Consequently, we obtain
$$u_{xy}(x, 0) = v_{xy}(x, 0), \quad 0 \leqslant x \leqslant a,$$
and
$$u_{xy}(0, y) = v_{xy}(0, y), \quad 0 \leqslant y \leqslant b.$$

We next consider the function defined by

$$F(x, y) = \alpha(x, y) \frac{|u(x, y) - v(x, y)|}{xy} + \beta(x, y) \frac{|u_x(x, y) - v_x(x, y)|}{y}$$
$$+ \gamma(x, y) \frac{|u_y(x, y) - v_y(x, y)|}{x},$$

for $0 < x \leqslant a$, $0 < y \leqslant b$. We shall show that $F(x, y)$ is continuous on R_0. Clearly, it is continuous for $0 < x \leqslant a$, $0 < y \leqslant b$. Hence, it remains only to verify that

$$\lim_{(x,y)\to(\bar{x},0)} F(x, y) = 0, \qquad 0 \leqslant \bar{x} \leqslant a,$$

and that

$$\lim_{(x,y)\to(0,\bar{y})} F(x, y) = 0, \qquad 0 \leqslant \bar{y} \leqslant b.$$

Since α, β, γ are continuous on R_0, it suffices to prove that the three functions

$$\frac{u(x, y) - v(x, y)}{xy}, \quad \frac{u_x(x, y) - v_x(x, y)}{y}, \quad \frac{u_y(x, y) - v_y(x, y)}{x}, \quad (11.2.5)$$

where $0 < x \leqslant a$, $0 < y \leqslant b$, approach zero as $(x, y) \to (\bar{x}, 0)$, with $0 \leqslant \bar{x} \leqslant a$, or as $(x, y) \to (0, \bar{y})$, $0 \leqslant \bar{y} \leqslant b$.

We have, by mean value theorem,

$$\frac{u(x, y) - v(x, y)}{xy} = \frac{u_y(x, \eta) - v_y(x, \eta)}{x}, \qquad 0 < \eta < y.$$

Similarly, it results that

$$\frac{u(x, y) - v(x, y)}{xy} = u_{yx}(\xi, \eta) - v_{yx}(\xi, \eta),$$

where $0 < \xi < x$, $0 < \eta < y$. Proceeding in a similar way, we are lead to the equalities

$$\frac{u_x(x, y) - v_x(x, y)}{y} = u_{xy}(x, \eta^*) - v_{xy}(x, \eta^*), \qquad 0 < \eta^* < y,$$

and

$$\frac{u_y(x, y) - v_y(x, y)}{x} = u_{yx}(\xi^*, y) - v_{yx}(\xi^*, y), \qquad 0 < \xi^* < x.$$

But the function $u_{xy} - v_{xy}$ is continuous on R_0 and vanishes when $x = 0$

and $y = 0$. Thus, three functions given in (11.2.5) tend to zero as asserted, and this implies the continuity of the function $F(x, y)$ as desired.

Suppose now, contrary to what we want to prove, that $u(x, y) - v(x, y)$ is not identically zero on R_0. Then, $F(x, y)$ must have a positive maximum at a point (x_0, y_0) such that $0 < x_0 \leqslant a$, $0 < y_0 \leqslant b$, and, furthermore, $F(x, y) < F(x_0, y_0)$, whenever $0 \leqslant x + y < x_0 + y_0$ with $0 \leqslant x \leqslant a$ and $0 \leqslant y \leqslant b$.

At this stage, we get a contradiction by applying the mean value theorem twice to the α term of $F(x_0, y_0)$, once to each of the other two terms of $F(x_0, y_0)$, and then using the Nagumo condition that is satisfied by f. We obtain, for example,

$$\begin{aligned}
\frac{|u(x_0, y_0) - v(x_0, y_0)|}{x_0 y_0} &= |u_{xy}(\xi, \eta) - v_{xy}(\xi, \eta)| \\
&= |f(\xi, \eta, u(\xi, \eta), u_x(\xi, \eta), u_y(\xi, \eta)) \\
&\quad - f(\xi, \eta, v(\xi, \eta), v_x(\xi, \eta), v_y(\xi, \eta))| \\
&\leqslant \alpha(\xi, \eta) \frac{|u(\xi, \eta) - v(\xi, \eta)|}{\xi \eta} \\
&\quad + \beta(\xi, \eta) \frac{|u_x(\xi, \eta) - v_x(\xi, \eta)|}{\eta} \\
&\quad + \gamma(\xi, \eta) \frac{|u_y(\xi, \eta) - v_y(\xi, \eta)|}{\xi} = F(\xi, \eta),
\end{aligned}$$

where $0 < \xi < x_0$, $0 < \eta < y_0$. Similarly,

$$\frac{|u_x(x_0, y_0) - v_x(x_0, y_0)|}{y_0} \leqslant F(x_0, \eta^*), \qquad 0 < \eta^* < y_0,$$

and

$$\frac{|u_y(x_0, y_0) - v_y(x_0, y_0)|}{x_0} \leqslant F(\xi^*, y_0), \qquad 0 < \xi^* < x_0.$$

Putting together the last three inequalities and using the definition of F, we obtain

$$F(x_0, y_0) \leqslant \alpha(x_0, y_0) F(\xi, \eta) + \beta(x_0, y_0) F(x_0, \eta^*) + \gamma(x_0, y_0) F(\xi^*, y_0),$$

where $\xi + \eta$, $x_0 + \eta^*$, $\xi^* + y_0$ are less than $x_0 + y_0$. This contradicts the choice of (x_0, y_0), since, by definition of (x_0, y_0), we must have

$$\left.\begin{aligned} F(\xi, \eta) \\ F(x_0, \eta^*) \\ F(\xi^*, y_0) \end{aligned}\right\} < F(x_0, y_0),$$

while, at the same time, $\alpha(x_0, y_0) \geqslant 0$, $\beta(x_0, y_0) \geqslant 0$, $\gamma(x_0, y_0) \geqslant 0$, and $\alpha(x_0, y_0) + \beta(x_0, y_0) + \gamma(x_0, y_0) = 1$.
The proof is complete.

11.3. Upper bounds and error estimates

Let us first prove some results that give *a priori* estimates of the solutions of hyperbolic equation (11.2.1).

THEOREM 11.3.1.   Assume that

(i) $f \in C[R_0 \times R^3, R]$, $f(x, y, u, p, q)$ is monotonic nondecreasing in u, p, q, and

$$|f(x, y, u, p, q)| \leqslant g(x, y, |u|, |p|, |q|), \qquad (11.3.1)$$

where $g \in C[R_0 \times R_+^3, R_+]$;

(ii) $z \in C[R_0, R_+]$, $z(x, y)$ possesses continuous partial derivatives z_x, z_y, $z_{xy} > 0$, and

$$z_{xy}(x, y) > g(x, y, z(x, y), z_x(x, y), z_y(x, y));$$

(iii) $u(x, y)$ is any solution of (11.2.1) existing on R_0 such that

$$|u_0| < z(0, 0), \qquad |\sigma'(x)| < z_x(x, 0), \qquad |\tau'(y)| < z_y(0, y),$$

for $0 \leqslant x \leqslant a$, $0 \leqslant y \leqslant b$.

Then we have, on R_0,

$$|u(x, y)| < z(x, y), \qquad |u_x(x, y)| < z_x(x, y), \qquad |u_y(x, y)| < z_y(x, y).$$

Proof. The proof proceeds as in the first part of Theorem 11.1.1. It is enough to prove that, on R_0,

$$|u_x(x, y)| < z_x(x, y) \qquad \text{and} \qquad |u_y(x, y)| < z_y(x, y).$$

Assume that this is false. Let t_0 be the greatest lower bound of numbers $t > x + y$ such that the inequalities are satisfied for $x + y < t_0$. Then, there exists a point (x_0, y_0) with $x_0 + y_0 = t_0$ for which one of the two inequalities is not true. Suppose that

$$|u_x(x_0, y_0)| = z_x(x_0, y_0). \qquad (11.3.2)$$

It then follows that

$$|u(x, y)| \leq z(x, y), \quad |u_x(x, y)| \leq z_x(x, y),$$
$$|u_y(x, y)| \leq z_y(x, y) \qquad (11.3.3)$$

for $x + y \leq t_0$. In view of the condition (11.3.1), we have

$$|u_x(x_0, y_0)| \leq |\sigma'(x_0)| + \int_0^{y_0} |f(x_0, t, u(x_0, t), u_x(x_0, t), u_y(x_0, t))| \, dt$$

$$\leq |\sigma'(x_0)| + \int_0^{y_0} g(x_0, t, |u(x_0, t)|, |u_x(x_0, t)|, |u_y(x_0, t)|) \, dt$$

$$\leq z_x(x_0, 0) + \int_0^{y_0} g(x_0, t, z(x_0, t), z_x(x_0, t), z_y(x_0, t)) \, dt$$

because of condition (iii), the relations (11.3.3), and the monotonic character of g in u, p, q. Assumption (ii) now leads to the inequality

$$|u_x(x_0, y_0)| < z_x(x_0, y_0),$$

which contradicts (11.3.2).

On the other hand, if we suppose that $|u_y(x_0, y_0)| = z_y(x_0, y_0)$, proceeding similarly, we get the inequality $|u_y(x_0, y_0)| < z_y(x_0, y_0)$, which is again a contradiction. Hence, the stated estimates are true, and the theorem is proved.

Using a similar argument with obvious modifications, we can prove the next theorem, which offers an estimate for the difference of any two solutions of (11.2.1).

THEOREM 11.3.2. *Assume that hypotheses* (i) *and* (ii) *of Theorem* 11.3.1 *hold except that the condition* (11.3.1) *is replaced by*

$$|f(x, y, u_1, p_1, q_1) - f(x, y, u_2, p_2, q_2)|$$
$$\leq g(x, y, |u_1 - u_2|, |p_1 - p_2|, |q_1 - q_2|).$$

Let $u(x, y)$ *and* $v(x, y)$ *be any two solutions of* (11.2.1) *satisfying*

$$u(0, 0) = u_0, \quad v(0, 0) = v_0, \quad u(x, 0) = \sigma_1(x),$$
$$v(x, 0) = \sigma_2(x), \quad u(0, y) = \tau_1(y), \quad v(0, y) = \tau_2(y),$$

such that

$$|u_0 - v_0| < z(0, 0), \quad |\sigma_1'(x) - \sigma_2'(x)| < z_x(x, 0),$$
$$|\tau_1'(y) - \tau_2'(y)| < z_y(0, y)$$

11.3. UPPER BOUNDS AND ERROR ESTIMATES

for $0 \leqslant x \leqslant a$, $0 \leqslant y \leqslant b$. Under these assumptions, we have, on R_0,

$$|u(x, y) - v(x, y)| < z(x, y),$$

$$|u_x(x, y) - v_x(x, y)| < z_x(x, y),$$

$$|u_y(x, y) - v_y(x, y)| < z_y(x, y).$$

Next we shall consider the hyperbolic differential inequality

$$|u_{xy} - f(x, y, u, u_x, u_y)| \leqslant \delta(x, y), \tag{11.3.4}$$

where $f \in C[R_0 \times R^3, R]$ and $\delta \in C[R_0, R_+]$.

DEFINITION 11.3.1. Any function $u \in C[R_0, R]$, possessing continuous partial derivatives u_x, u_y, u_{xy} in R_0 and satisfying (11.3.4) in R_0, is said to be a δ-*approximate* solution of (11.2.1), if it verifies the boundary conditions specified in Sect. 11.2.

The following theorem estimates the error between a solution and a δ-approximate solution (11.2.1).

THEOREM 11.3.3. Assume that

(i) $u(x, y)$ is a solution and $v(x, y)$ is a δ-approximate solution of (11.2.1);

(ii) $z \in C[R_0, R_+]$, $z(x, y)$ possesses continuous partial derivatives z_x, z_y, $z_{xy} > 0$ in R_0, and

$$z_{xy}(x, y) > g(x, y, z(x, y), z_x(x, y), z_y(x, y)) + \delta(x, y);$$

(iii) $f \in C[R_0 \times R^3, R]$, $\delta \in C[R_0, R_+]$, $f(x, y, u, p, q)$ is monotonic nondecreasing in u, p, q, and

$$|f(x, y, u_1, p_1, q_1) - f(x, y, u_2, p_2, q_2)|$$
$$\leqslant g(x, y, |u_1 - u_2|, |p_1 - p_2|, |q_1 - q_2|),$$

where $g \in C[R_0 \times R_+^3, R_+]$.

Then, the inequalities

$$|v_0 - u_0| < z(0, 0), \qquad |v_x(x, 0) - \sigma'(x)| < z_x(x, 0),$$

$$|v_y(0, y) - \tau'(y)| < z_y(0, y) \quad \text{for} \quad 0 \leqslant x \leqslant a, \quad 0 \leqslant y \leqslant b,$$

imply
$$|v(x,y) - u(x,y)| < z(x,y),$$
$$|v_x(x,y) - u_x(x,y)| < z_x(x,y),$$
$$|v_y(x,y) - u_y(x,y)| < z_y(x,y),$$
on R_0.

Proof. We proceed as in Theorem 11.3.1 to show that there exists a $t_0 = x_0 + y_0$ such that either
$$|v_x(x_0, y_0) - u_x(x_0, y_0)| = z_x(x_0, y_0)$$
or
$$|v_y(x_0, y_0) - u_y(x_0, y_0)| = z_y(x_0, y_0),$$
and, in either case, we get
$$|v(x,y) - u(x,y)| \leqslant z(x,y),$$
$$|v_x(x,y) - u_x(x,y)| \leqslant z_x(x,y),$$
$$|v_y(x,y) - u_y(x,y)| \leqslant z_y(x,y)$$
for $x + y < t_0$. Then we have, successively,

$|v_x(x_0, y_0) - u_x(x_0, y_0)|$

$\leqslant \left| \int_0^{y_0} v_{xy}(x_0, t) \, dt + v_x(x_0, 0) - \sigma'(x_0) \right.$

$\left. - \int_0^{y_0} f(x_0, t, u(x_0, t), u_x(x_0, t), u_y(x_0, t)) \, dt \right|$

$\leqslant \int_0^{y_0} |v_{xy}(x_0, t) - f(x_0, t, v(x_0, t), v_x(x_0, t), v_y(x_0, t))| \, dt$

$+ |v_x(x, 0) - \sigma'(x_0)| + \int_0^{y_0} |[f(x_0, t, v(x_0, t), v_x(x_0, t), v_y(x_0, t))$

$- f(x_0, t, u(x_0, t), u_x(x_0, t), u_y(x_0, t))]| \, dt$

$\leqslant z_x(x_0, 0) + \int_0^{y_0} [g(x_0, t, |v(x_0, t) - u(x_0, t)|,$

$|v_x(x_0, t) - u_x(x_0, t)|, |v_y(x_0, t) - u_y(x_0, t)|) + \delta(x_0, t)] \, dt$

$\leqslant z_x(x_0, 0) + \int_0^{y_0} \{g(x_0, t, z(x_0, t), z_x(x_0, t), z_y(x_0, t)) + \delta(x_0, t)\} \, dt$

$< z_x(x_0, y_0),$

contradicting the fact that $|v_x(x_0, y_0) - u_x(x_0, y_0)| = z_x(x_0, y_0)$. A similar contradiction occurs in the other case, too. Consequently, the desired result follows.

11.4. Notes

See Walter [8] for the hyperbolic differential inequalities given in Sect. 11.1. Theorem 11.2.1 is taken from Shanahan [1], whereas Theorem 11.2.2 is due to Diaz and Walter [8]. See also Lakshmikantham [2]. For the estimates of the type given in Sect. 11.3, see Walter [8].

For the global existence theorems using Tychonoff's fixed point theorem, see Aziz and Maloney [1]. See also Bielecki [1]. Concerning periodic solutions of hyperbolic differential equations, refer to Aziz [2], Cesari [1, 2], and Hale [11]. For the application of contraction mapping theorem in generalized metric spaces to existence and uniqueness theorems of a particular type of hyperbolic differential equation, see Wong [1, 2].

For related results, see Alexiewicz and Orlicz [1], Aziz [1], Aziz and Diaz [1], Chu [1], Ciliberto [1, 3–5], Conlan [1, 2], Conti [1], Diaz [1, 2], Guglielmino [1–4], Kisynski [1, 3], Lakshmikantham [2], Palczewski [1], Palczewski and Pawelski [1], Pelczar [1, 2, 4, 5], Phillips [1], Protter [1], Santoro [1], Szarski [8], Szmydt [1–5], Volkov [1], and Walter [1–6, 8].

DIFFERENTIAL EQUATIONS
IN ABSTRACT SPACES

Chapter 12

12.0. Introduction

A study of differential equations in abstract spaces is the content of this chapter. A variety of results on existence, uniqueness, continuous dependence, and method of averaging are given. A major part of the chapter is devoted to nonlinear evolution equation. A number of results are obtained regarding existence, estimates on approximate solutions, Chapligin's method, asymptotic behavior, and stability and boundedness of solutions.

12.1. Existence

Let E be a Banach space, and, for any $u \in E$, let $|u|$ denote the norm of u. Suppose that $f(t, u)$ is a mapping from $[t_0, t_0 + a] \times E$ to E. We consider the differential equation

$$u' = f(t, u). \qquad (12.1.1)$$

DEFINITION 12.1.1. Any function $u(t)$ is said to be a solution of (12.1.1) if the following conditions are verified:

(i) $u \in C[[t_0, t_0 + a], E]$;

(ii) $u(t_0) = u_0$;

(iii) $u(t)$ is strongly differentiable in t for $t_0 < t \leqslant t_0 + a$ and satisfies Eq. (12.1.1) for $t_0 < t \leqslant t_0 + a$.

We shall now prove a local existence theorem.

THEOREM 12.1.1. Assume that

(i) $f \in C[[t_0, t_0 + a] \times S_b, E]$, where

$$S_b = [u \in E : |u - u_0| \leqslant b],$$

and $f(t, u)$ maps bounded sets into bounded sets so that, for any $b > 0$, there exists a positive number M such that

$$|f(t, u)| \leqslant M, \qquad (t, u) \in [t_0, t_0 + a] \times S_b;$$

(ii) $V \in C[[t_0, t_0 + a] \times S_b \times S_b, R_+]$, $V(t, u, v) > 0$ if $u \neq v$, $V(t, u, v) \equiv 0$ if $u = v$, $v(t, u_n, v_n) \to 0$ implies $u_n - v_n \to 0$ for each t;

(iii) $V(t, u, v)$ has continuous (bounded additive Fréchet) derivatives and

$$\frac{\partial V}{\partial t} + \frac{\partial V}{\partial u} f(t, u) + \frac{\partial V}{\partial v} f(t, v) \leqslant 0;$$

(iv) for any positive number M, $\partial V/\partial t$, $(\partial V/\partial u)x$, $(\partial V/\partial v)x$ are continuous in (u, v) uniformly for $(t, u, v) \in [t_0, t_0 + a] \times S_h \times S_b$ and $|x| \leqslant M$.

Then, Eq. (12.1.1) possesses a solution on $[t_0, t_0 + \alpha]$, $\alpha > 0$ being the number satisfying $\alpha M \leqslant b$.

Proof. Let Δ be a subdivision of $[t_0, t_0 + \alpha]$, namely,

$$t_0 < t_1 < \cdots < t_n = t_0 + \alpha.$$

For $t_{k-1} \leqslant t \leqslant t_k$, we define

$$\phi_\Delta(t) = \phi_\Delta(t_{k-1}) + \int_{t_{k-1}}^{t} f(s, \phi_\Delta(t_{k-1})) \, ds,$$

$$\phi_\Delta(t_0) = u_0.$$

Then,

$$\phi_\Delta(t) = \phi_\Delta(t_0) + \int_{t_0}^{t_1} f(s, \phi_\Delta(t_0)) \, ds + \cdots$$

$$+ \int_{t_{k-2}}^{t_{k-1}} f(s, \phi_\Delta(t_{k-2})) \, ds + \int_{t_{k-1}}^{t} f(s, \phi_\Delta(t_{k-1})) \, ds$$

$$= u_0 + \int_{t_0}^{t} f_\Delta(s) \, ds,$$

and

$$\phi_\Delta'(t) = f(t, \phi_\Delta(t_{k-1})) = f_\Delta(t) \qquad \text{for} \qquad t_{k-1} < t < t_k.$$

Let Δ and Δ_0 be two subdivisions of $[t_0, t_0 + \alpha]$, and consider $\phi_\Delta(t)$

12.1. EXISTENCE

and $\phi_{\Delta_0}(t)$ as before. If t is not a subdivision point of either Δ or Δ_0, $t_{k-1} < t < t_k$ and $t_{j-1}^0 < t < t_j^0$, say, then

$$\frac{d}{dt}[V(t, \phi_\Delta(t), \phi_{\Delta_0}(t))]$$

$$= \frac{\partial V}{\partial t} + \frac{\partial V}{\partial u}\phi_\Delta'(t) + \frac{\partial V}{\partial v}\phi_{\Delta_0}'(t)$$

$$= \frac{\partial V}{\partial t} + \frac{\partial V}{\partial u}f(t, \phi_\Delta(t_{k-1})) + \frac{\partial V}{\partial v}f(t, \phi_{\Delta_0}(t_{j-1}^0))$$

$$= \left(\frac{\partial V}{\partial t} - \frac{\partial \tilde{V}}{\partial t}\right) + \frac{\partial \tilde{V}}{\partial t} + \left[\left(\frac{\partial V}{\partial u} - \frac{\partial \tilde{V}}{\partial u}\right) + \frac{\partial \tilde{V}}{\partial u}\right]$$

$$\times f(t, \phi_\Delta(t_{k-1})) + \left[\left(\frac{\partial V}{\partial v} - \frac{\partial \tilde{V}}{\partial v}\right) + \frac{\partial \tilde{V}}{\partial v}\right]f(t, \phi_{\Delta_0}(t_{j-1}^0)),$$

where

$$\frac{\partial V}{\partial t} = \frac{\partial V(t, \phi_\Delta(t), \phi_{\Delta_0}(t))}{\partial t},$$

$$\frac{\partial \tilde{V}}{\partial t} = \frac{\partial V(t, \phi_\Delta(t_{k-1}), \phi_{\Delta_0}(t_{j-1}^0))}{\partial t},$$

with similar expressions for $\partial V/\partial u$, $\partial V/\partial v$.

By assumption (iv), for any $\epsilon > 0$, there exists a $\delta > 0$ such that, if we take $|\Delta| = \max(t_k - t_{k-1}) < \delta$ and $|\Delta_0| = \max(t_j^0 - t_{j-1}^0) < \delta$, then we have

$$\left|\frac{\partial V}{\partial t} - \frac{\partial \tilde{V}}{\partial t}\right| < \frac{\epsilon}{3},$$

$$\left|\left(\frac{\partial V}{\partial u} - \frac{\partial \tilde{V}}{\partial u}\right)f(t, \phi_\Delta(t_{k-1}))\right| < \frac{\epsilon}{3},$$

and

$$\left|\left(\frac{\partial V}{\partial v} - \frac{\partial \tilde{V}}{\partial v}\right)f(t, \phi_{\Delta_0}(t_{j-1}^0))\right| < \frac{\epsilon}{3}.$$

Thus, we deduce that

$$\frac{dV(t, \phi_\Delta(t), \phi_{\Delta_0}(t))}{dt} < \epsilon.$$

Hence, it follows that

$$V(t, \phi_\Delta(t), \phi_{\Delta_0}(t)) = V(t, \phi_\Delta(t), \phi_{\Delta_0}(t)) - V(t_0, \phi_\Delta(t_0), \phi_{\Delta_0}(t_0))$$
$$\leqslant \epsilon\alpha,$$

and this proves that there exists a $\phi(t)$ such that

$$\phi_\Delta(t) \to \phi(t) \quad \text{for each} \quad t.$$

Let now $t \in [t_0, t_0 + \alpha]$ be fixed. Then, for each subdivision Δ, there exists a k such that $t_{k-1} \leqslant t \leqslant t_k$. Furthermore, we obtain

$$|\phi_\Delta(t_{k-1}) - \phi_\Delta(t)| \leqslant M(t - t_{k-1}) \leqslant M|\Delta|,$$

which shows that

$$\phi_\Delta(t_{k-1}) \to \phi(t).$$

Consequently,

$$f(t, \phi_\Delta(t_{k-1})) = f_\Delta(t) \to f(t, \phi(t)),$$

and, by dominated convergence,

$$\phi(t) = u_0 + \int_{t_0}^{t} f(s, \phi(s))\, ds.$$

The proof is therefore complete.

In Theorem 12.1.1, we have assumed that $f(t, u)$ is continuous. It can be shown that the conclusion of Theorem 12.1.1 is true even under weaker hypotheses on f. For this purpose, we need the following

DEFINITION 12.1.2. Let $f(t, u)$ be a mapping from $[t_0, t_0 + a] \times E$ to E. We say that $f(t, u)$ is *demicontinuous* if it is continuous from $[t_0, t_0 + a] \times E$ with the strong topology to E with the weak topology.

THEOREM 12.1.2. Suppose that E is a reflexive Banach space. Then, the conclusion of Theorem 12.1.1 remains valid if we replace the continuity of f by demicontinuity, other assumptions being the same. (In this case, the differentiation is in the sense of weak topology.)

Proof. Notice that $\partial V/\partial u$ and $\partial V/\partial v$ are bounded additive functionals. Therefore, by an argument similar to that used in the proof of Theorem 12.1.1, we can prove that

$$f_\Delta(t) \rightharpoonup f(t, \phi(t)),$$

so that we have

$$\phi'(t) = f(t, \phi(t)),$$

where $\rightharpoonup$ means weak convergence.

REMARK 12.1.1. Suppose now that E is a Hilbert space and that $f(t, u)$ satisfies the monotonicity condition, that is,

$$\operatorname{Re}(f(t, u) - f(t, v), u - v) \leqslant M|u - v|^2, \qquad (12.1.2)$$

where we denote the scalar product by (x, y) and the norm by $|x| = (x, x)^{1/2} \geqslant 0$. Then $V(t, u, v) = e^{-2Mt}|u - v|^2$ satisfies all the hypotheses of Theorem 12.1.1, and, consequently, the conclusion of Theorem 12.1.1 is true.

We shall now give an example to show that, even when $f(t, u)$ does not satisfy the monotonicity condition, there does exist a $V(t, u, v)$ satisfying the hypotheses of Theorem 12.1.1. Consider the example

$$\frac{du}{dt} = f(t, u) = \begin{cases} 1 + u^{1/2}, & u \geqslant 0, \\ 1, & u < 0, \end{cases}$$

assuming that $E = R$. Clearly (12.1.2) is not fulfilled. However, there exists a $V(t, u, v)$ defined as follows:

$$V(t, u, v) = \begin{cases} (\sqrt{u} - \sqrt{v} - \log(1 + \sqrt{u}) + \log(1 + \sqrt{v})^2, & u \geqslant 0, \quad v \geqslant 0, \\ (\sqrt{u} - \log(1 + \sqrt{u}) - \tfrac{1}{2}v)^2, & u \geqslant 0, \quad v < 0, \\ (\tfrac{1}{2}u - \sqrt{v} + \log(1 + \sqrt{v}))^2, & u < 0, \quad v \geqslant 0, \\ \tfrac{1}{4}(u - v)^2, & u < 0, \quad v < 0. \end{cases}$$

12.2. Nonlocal existence

We shall use the functional method of Leray–Schauder to prove the nonlocal existence of solutions of the differential equation (12.1.1).

DEFINITION 12.2.1. Let A and B be completely continuous operators defined for $u \in S_\rho$, where $S_\rho = [u \in E: |u| \leqslant \rho]$, with values in E and $Au \in E$, $Bu \in E$ for $u \in S_\rho$. Then we say that A and B are *homotopic* if there exists an operator $T(u, \lambda)$, which is completely continuous on $E \times [0, 1]$ such that $T(u, 0) = Au$ and $T(u, 1) = Bu$ for $u \in S_\rho$ and $T(u, \lambda) \neq u$ for $|u| = \rho$.

We need the following lemma of Leray–Schauder.

LEMMA 12.2.1. Let the completely continuous operator A be homotopic to the operator identically equal to zero. Then, there exists at least one solution u of the equation $Au = u$ such that $|u| < \rho$.

Let $V(u)$ be a functional that possesses Fréchet differential L. Then, we have

$$V(u + h) - V(u) = L(u, h) + \omega(u, h),$$

where $\omega(u, h)/|h| \to 0$ as $|h| \to 0$.

As an application of Lemma 12.2.1, we prove the following existence result.

THEOREM 12.2.1. Assume that

(i) there exist functionals $V_i(u)$ possessing Fréchet differentials L_i such that
$$L_i(u, f(t, u)) \leqslant g_i(t, V_1(u),..., V_n(u)) \qquad (i = 1, 2,..., n), \qquad (12.2.1)$$
where $f(t, u)$ is completely continuous for $t \in [t_0, t_0 + a]$, $u \in E$;

(ii) $g \in C[t_0, t_0 + a] \times R_+^n, R^n]$, $g(t, y)$ is quasi-monotone nondecreasing in y for each t, and the maximal solution $r(t, t_0, y_0)$ of
$$y' = g(t, y), \qquad y(t_0) = y_0$$
exists on $[t_0, t_0 + a]$;

(iii) $\Phi(u) = \max_i V_i(u)$ and $\Phi(u) \to \infty$ as $|u| \to \infty$.

Then, for every $u_0 \in E$, there exists at least one solution $u(t)$ of the differential equation (12.1.1) defined on $[t_0, t_0 + a]$ such that $u(t_0) = u_0$.

Proof. Let us consider the space B of all continuous functions $u(t)$ with values lying in E, continuous on the interval $[t_0, t_0 + a]$. Define $|u| = \max_{t_0 \leqslant t \leqslant t_0 + a} |u(t)|$. Also, consider the operator
$$u(t) \to \lambda u_0 + \lambda \int_{t_0}^{t} f(s, u(s))\, ds, \qquad 0 \leqslant \lambda \leqslant 1,$$
which maps B into itself. Denote this operator by $T(u, \lambda)$. Since $f(t, u)$ is completely continuous, it follows that $T(u, \lambda)$ is completely continuous on $B \times [0, 1]$. Suppose now that $u(t, \lambda)$ is a solution of the equation
$$u(t) = \lambda u_0 + \lambda \int_{t_0}^{t} f(s, u(s))\, ds.$$
Hence, $u'(t, \lambda) = \lambda f(t, u(t, \lambda))$ and $u(t_0, \lambda) = \lambda u_0$. Because of (12.2.1), we get
$$L_i(u(t, \lambda), u'(t, \lambda)) \leqslant \lambda g_i(t, V_1(u(t, \lambda)),..., V_n(u(t, \lambda))).$$
The last inequality implies that
$$D^+ m_i(t) \leqslant g_i(t, m_1(t),..., m_n(t)),$$
where $m_i(t) = V_i(u(t, \lambda))$. By Corollary 1.7.1, we obtain
$$V_i(u(t, \lambda)) \leqslant r_i(t, t_0, V_1(\lambda u_0),..., V_n(\lambda u_0)),$$

choosing $y_0 = V(\lambda u_0)$. Write

$$K = \max_{i,\lambda,t} r_i(t, t_0, V_1(\lambda u_0), ..., V_n(\lambda u_0)).$$

Then, it follows that

$$V_i(u(t, \lambda)) \leqslant K, \quad t \in [t_0, t_0 + a], \quad 0 \leqslant \lambda \leqslant 1. \quad (12.2.2)$$

There is such an M that $|u(t, \lambda)| \leqslant M$ for $0 \leqslant \lambda \leqslant 1$ and $t_0 \leqslant t \leqslant t_0 + a$. Indeed, if it were not so, then there would exist a sequence $\lambda_n \in [0, 1]$ and a sequence $t_n \in [t_0, t_0 + a]$ such that $|u(t_n, \lambda_n)| \to \infty$. Hence, $\Phi(u(t_n, \lambda_n)) \to \infty$. This contradicts (12.2.2). Thus, $|u(t, \lambda)| \leqslant M$, $0 \leqslant \lambda \leqslant 1$, $t \in [t_0, t_0 + a]$.

We conclude that $T(u, \lambda) \neq u$ for $0 \leqslant \lambda \leqslant 1$ and $|u| = M + \epsilon$, $\epsilon > 0$. It is easily seen that $T(u, 1)$ is homotopic to zero in the region $|u| \leqslant M + \epsilon$. By Lemma 12.2.1, we find that, in the space B, there exists at least one solution of the equation

$$u(t) = u_0 + \int_{t_0}^{t} f(s, u(s))\, ds.$$

This completes the proof.

12.3. Uniqueness

Let E be a normed space, $D(t)$, $t \in [t_0, t_0 + a]$ a subset of E, and $D = [(t, u) : t \in [t_0, t_0 + a], u \in D(t)]$. We consider Eq. (12.1.1), where $f(t, u)$ is defined on D. We note that $f(t, u)$ need not be continuous.

THEOREM 12.3.1. Assume that

(i) $u_0 \in \overline{D(0)}$, $f(t, u)$ is defined on the set D;
(ii) $V \in C[Q, R_+]$, where

$$Q = [(t, u, v) : (t, u) \in D, (t, v) \in D],$$

$V(t, u, v) > 0$ if $u \neq v$, $V(t, u, v) = 0$ if $u = v$;

(iii) $V(t, u, v)$ has continuous (bounded additive Fréchet) derivatives and

$$\frac{\partial V}{\partial t} + \frac{\partial V f(t, u)}{\partial u} + \frac{\partial V f(t, v)}{\partial v} \leqslant 0.$$

Then, there is at most one solution of (12.1.1). Furthermore, if $V(t, u, v)$ satisfies the condition

$$V(t, u_n, v_n) \to 0 \quad \text{implies} \quad u_n - v_n \to 0 \quad \text{for each} \quad t, \quad (12.3.1)$$

then the solution $u(t)$ depends continuously on u_0 for each t.

Proof. Suppose that there are two solutions $u(t)$, $v(t)$ of (12.1.1) such that $u(t_0) = v(t_0) = u_0$. Then, we derive that

$$\frac{dV(t, u(t), v(t))}{dt} = \left[\frac{\partial V}{\partial t} + \frac{\partial V f(t, u(t))}{\partial u} + \frac{\partial V f(t, v(t))}{\partial v} \right]$$

$$\leqslant 0 \quad \text{for} \quad t > t_0.$$

Hence, it follows that

$$V(t, u(t), v(t)) \leqslant V(t_0, u(t_0), v(t_0)). \quad (12.3.2)$$

Since $V(t_0, u(t_0), v(t_0)) = 0$, we see that $V(t, u(t), v(t)) \equiv 0$, and, consequently, $u(t) \equiv v(t)$. This proves uniqueness.

Since $V(t, u, v)$ is continuous, $|u(t_0) - v(t_0)| \to 0$ implies

$$V(t_0, u(t_0), v(t_0)) \to 0.$$

This implies that $V(t, u(t), v(t)) \to 0$ by (12.3.2). If $V(t, u, v)$ satisfies (12.3.1), we have

$$|u(t) - v(t)| \to 0$$

when $|u(t_0) - v(t_0)| \to 0$.

The proof of the theorem is complete.

Consider the following simple parabolic equation,

$$u_t = u_{xx} + F(t, x, u), \quad (12.3.3)$$

in a region bounded by $t = t_0$, $t = t_0 + a$, $x = \lambda_1(t)$, and $x = \lambda_2(t)$, where $\lambda_1(t) < \lambda_2(t)$ for $t_0 < t \leqslant t_0 + a$. The initial and boundary conditions are as follows: $u = g(x)$ on $t = t_0$, $u = h_1(t)$, $u = h_2(t)$ on $x = \lambda_1(t)$, $x = \lambda_2(t)$, respectively, where g, h_1, and h_2 are continuous and $g(\lambda_1(t_0)) = h_1(t_0)$, $g_2(\lambda_2(t_0)) = h_2(t_0)$. Suppose that

$$F(t, x, u_1) - F(t, x, u_2) \leqslant K(u_1 - u_2) \quad \text{for} \quad u_1 > u_2.$$

Then, we can use Theorem 12.3.1 to prove uniqueness of solutions of (12.3.3). Let $D(t)$ be a set of functions x that are continuous on

12.3. UNIQUENESS

$\lambda_1(t) \leqslant x \leqslant \lambda_2(t)$, twice continuously differentiable on $\lambda_1(t) < x < \lambda_2(t)$, and take values $h_1(t), h_2(t)$ at $x = \lambda_1(t), x = \lambda_2(t)$, respectively. Define

$$V(t, u, v) = e^{-2Kt} \int_{\lambda_1(t)}^{\lambda_2(t)} |u(x) - v(x)|^2 \, dx$$

and

$$f(t, u) = u_{xx}(x) + F(t, x, u(x))$$

for $u \in D(t)$. Then, we have

$$\frac{\partial V}{\partial t} + \frac{\partial V f(t, u)}{\partial u} + \frac{\partial V f(t, v)}{\partial v}$$

$$= \int_{\lambda_1(t)}^{\lambda_2(t)} [u_{xx}(x) + F(t, x, u(x)) - v_{xx}(x) - F(t, x, v(x))](u(x) - v(x)) \, dx$$

$$- K \int_{\lambda_1(t)}^{\lambda_2(t)} |u(x) - v(x)|^2 \, dx$$

$$\leqslant - \int_{\lambda_1(t)}^{\lambda_2(t)} \left(\frac{d(u(x) - v(x))}{dx} \right)^2 dx$$

$$\leqslant 0,$$

so that the solution is unique by Theorem 12.3.1.

THEOREM 12.3.2. *Let E be a Banach space and $f(t, u)$ be defined and demicontinuous on a set $D \subset [(t, u): t_0 < t \leqslant t_0 + a, u \in E]$ for which we assume that $D(t) = [u: (t, u) \in D]$ is closed in E for each $t \in (t_0, t_0 + a]$. Suppose further that*

$$|f(t, u)| \leqslant M(t) \quad \text{on} \quad D,$$

where $M(t)$ is a summable function. Assume that there exists a real valued function $V(t, u, v)$ satisfying conditions (ii), (iii), and (12.3.1) of Theorem 12.3.1. Let $u = \psi(t)$ be continuous in D for which $\sup_{u \in D(t)} V(t, \psi(t), u) \to 0$ as $t \to t_0$. Assume also that, for every $t_1, t_0 < t_1 < t_0 + a$, (12.1.1) has a solution starting from $(t_1, \psi(t_1))$ and reaching the plane $t = t_0 + a$ in D. Then, there is a unique solution $u(t)$ of (12.1.1) with $u(t_0) = u_0 = \psi(t_0)$.

Proof. Denote by $u = \phi(t; t_1)$ a solution starting from $(t_1, \psi(t_1))$. Then, if we take $t_0 < t_1 < t_2$, we have

$$\frac{dV(t, \phi(t; t_1), \phi(t; t_2))}{dt} \leqslant 0 \quad \text{for} \quad t_2 \leqslant t \leqslant t_0 + a,$$

so that

$$V(t, \phi(t; t_1), \phi(t; t_2)) \leqslant V(t_2, \phi(t_2; t_1), \phi(t_2; t_2))$$
$$= V(t_2, \phi(t_2; t_1), \psi(t_2)) \to 0 \quad \text{as} \quad t_2 \to t_0.$$

Thus, $|\phi(t; t_1) - \phi(t; t_2)| \to 0$ if $t_1, t_2 \to t_0$. Let $\phi(t) = \lim_{t_1 \to t_0} \phi(t; t_1)$. Then, $(t, \phi(t)) \in D$ and $\phi(t)$ is a solution of (12.1.1), since, for arbitrary $\hat{t} > t_0$, we have

$$\phi(t; t_1) - \phi(\hat{t}; t_1) = \int_{\hat{t}}^{t} f(s, \phi(s; t_1)) \, ds, \quad t_0 < t_1 < t,$$

and hence, by letting $t_1 \to t_0$, we get

$$\phi(t) - \phi(\hat{t}) = \int_{\hat{t}}^{t} f(s, \phi(s)) \, ds.$$

This means that $\phi(t)$ is a solution of (12.1.1) in D. Uniqueness of solutions is obvious, since, if $\phi^*(t)$ is such a solution, then

$$V(t, \phi(t; t_1), \phi^*(t)) \leqslant V(t_1, \phi(t_1; t_1), \phi^*(t_1))$$
$$= V(t_1, \psi(t_1), \phi^*(t_1)) \to 0$$

as $t_1 \to t_0$ shows that

$$\phi(t; t_1) \to \phi^*(t).$$

This completes the proof.

Let H be a Hilbert space and $f(t, u)$ be a continuous function on $\Omega = [(t, u): t_0 < t \leqslant t_0 + a, \ |u - u_0| \leqslant c]$ taking values in H. Suppose that

(i) $f(t, u) \to 0$ as $(t, u) \to (t_0, u_0)$ in Ω;
(ii) $\operatorname{Re}(f(t, u) - f(t, v), u - v) \leqslant (t - t_0)^{-1} |u - v|^2$ on Ω.

Then, there is a unique solution $\phi(t)$ of (12.1.1) such that $\phi(t) \to u_0$ as $t \to t_0$. In fact, take

$$V(t, u, v) = [(t - t_0)^2]^{-1} |u - v|^2.$$

Then, for $(t, u), (t, v) \in \Omega$,

$$\frac{\partial V}{\partial t} + \frac{\partial V f(t, u)}{\partial u} + \frac{\partial V f(t, v)}{\partial v} \leqslant 0.$$

Let $\psi(t) = u_0$. By condition (i), we may assume that $|f(t, u)| \leqslant m$ if $t_0 < t \leqslant t_0 + a_1$, $a_1 \leqslant t_0 + a$, $|u - u_0| \leqslant c_1$, $c_1 \leqslant c$, and $ma_1 \leqslant c_1$. Set

$$\lambda(t) = \sup[|f(t, u)| : |u - u_0| \leqslant m(t - t_0)].$$

Then, $\lambda(t) \to 0$ as $t \to t_0$. Writing

$$\mu(t) = \int_{t_0}^{t} \lambda(s)\, ds,$$

we see that

$$D = [(t, u) : t_0 < t \leqslant t_0 + a_1, |u - u_0| \leqslant \mu(t)]$$

is contained in Ω and

$$\sup_{u \in D(t)} V(t, \psi(t), u) = [(t - t_0)^2]^{-1} |\mu(t)|^2 \to 0 \quad \text{as} \quad t \to t_0.$$

Since there exists a solution starting from $(t_1, \psi(t_1))$ for each t_1 such that $t_0 < t_1 < t_0 + a_1$, which reaches the plane $t = t_0 + a_1$, by Theorem 12.1.1, the desired uniqueness is a consequence of Theorem 12.3.2.

12.4. Continuous dependence and the method of averaging

Let us consider the differential equation

$$u' = f(t, u, \lambda), \tag{12.4.1}$$

where f is a function with values in a Banach space E, defined in a set $J \times H \times \Lambda$, where H is an open subset of E and Λ an arbitrary metric space.

Throughout this section, we shall assume that the following conditions hold:

(i) for each $\lambda \in \Lambda$, the mapping $(t, u) \to f(t, u, \lambda)$ is continuous in $J \times H$;

(ii) for some $\lambda_0 \in \Lambda$, there exists a solution $u_0(t)$ of (12.4.1) which is defined on J and has its values in H;

(iii) there is a neighborhood Γ_0 of λ_0 in Λ such that, for every $\lambda \in \Gamma_0$, (12.4.1) admits a solution $u(t)$ with $u(0) = u_0(0)$, which exists in some interval $[0, T(\lambda)) \subset J$.

We recall that condition (i) alone does not guarantee the existence of a solution of (12.4.1) unless E is finite dimensional.

Then, we can prove the following results using the arguments similar to the proofs of Lemma 3.20.1 and Theorem 3.20.1. We do not give the details.

LEMMA 12.4.1. Assume that

(i) $V \in C[J \times E, R_+]$, $V(t, 0) \equiv 0$, $V(t, u)$ is positive definite and satisfies a Lipschitz condition u for a constant $M > 0$;

(ii) $g \in C[J \times R_+, R]$, $g(t, 0) \equiv 0$, and, for any step function $v(t)$ on J with values in H and for every $t \in J$, $u \in H$,

$$D^+ U(t, u, \lambda_0) = \limsup_{h \to 0^+} h^{-1}[V(t + h, F(t, v(t), \lambda_0) - u$$
$$+ h\{f(t, v(t), \lambda_0) - f(t, u, \lambda_0)\}) - V(t, F(t, v(t), \lambda_0) - u)]$$
$$\leqslant g(t, V(t, v(t) - u)),$$

where

$$F(t, v(t), \lambda) = v(0) + \int_0^t f(s, v(s), \lambda) \, ds.$$

Then, given any compact interval $[0, T_0] \subset J$ and an $\epsilon > 0$, there is a $\delta = \delta(\epsilon) > 0$ such that, for any step function $v(t)$ in $[0, T_0]$, with $v(0) = u_0(0)$ and $|v(t) - u_0(t)| < \delta$ in $[0, T_0]$, there follows

$$\left| \int_0^t [f(s, v(s), \lambda_0) - f(s, u_0(s), \lambda_0)] \, ds \right| < \epsilon,$$

for every $t \in [0, T_0]$, $u_0(t)$ being any solution of

$$u' = f(t, u, \lambda_0),$$

defined for $t \in J$.

We assume that, for each $t \in J$ and $u \in H$,

$$\lim_{\lambda \to \lambda_0} \int_0^t f(s, u, \lambda) \, ds = \int_0^t f(s, u, \lambda_0) \, ds. \tag{12.4.2}$$

It then follows that, given any compact interval $[0, T_0]$ and any step function $v(t)$ in $[0, T_0]$ with values in H,

$$\lim_{\lambda \to \lambda_0} \int_0^t f(s, v(s), \lambda) \, ds = \int_0^t f(s, v(s), \lambda_0) \, ds,$$

uniformly in $[0, T_0]$. Hence, if the assumptions of Lemma 12.4.1 hold, there exists, for every $\epsilon > 0$, a constant $\delta = \delta(\epsilon) > 0$ such that, whenever $v(t)$ is a step function in $[0, T_0]$ with $v(0) = u_0(0)$ and $|v(t) - u_0(t)| < \delta$ in $[0, T_0]$, there is a neighborhood $\Gamma = \Gamma(\epsilon) \subset \Lambda$ of λ_0 for which $\lambda \in \Gamma$ implies

$$\left| \int_0^t [f(s, v(s), \lambda) - f(s, u_0(s), \lambda_0)] \, ds \right| < \epsilon,$$

$t \in [0, T_0]$. Thus, we have

$$|F(t, v(t), \lambda) - u_0(t)| < \epsilon, \qquad t \in [0, T_0],$$

which will be used in the main theorem that follows, as in Theorem 3.20.1.

THEOREM 12.4.1. Suppose that

(i) $V \in C[J \times E, R_+]$, $V(t, 0) \equiv 0$, $V(t, u)$ is positive definite and satisfies a Lipschitz condition in u for a constant $M > 0$;

(ii) $g \in C[J \times R_+, R]$, $g(t, 0) \equiv 0$, and $r(t) \equiv 0$ is the maximal solution of

$$y' = g(t, y)$$

passing through $(0, 0)$;

(iii) for any step function $v(t)$ on J, with values in H and for every $t \in J$, $u \in H$, $y \in \Lambda$,

$$D^+ U(t, u, \lambda) \leq g(t, V(t, v(t) - u));$$

(iv) the relation (12.4.2) holds.

Then, given any compact interval $[0, T_0] \subset J$ and any $\epsilon > 0$, there exists a neighborhood $\Gamma(\epsilon)$ of λ_0 such that, for every $\lambda \in \Gamma(\epsilon)$, (12.4.1) admits a unique solution $u(t)$ with $u(0) = u_0(0)$, which is defined on $[0, T_0]$ and satisfies

$$|u(t) - u_0(t)| < \epsilon, \qquad t \in [0, T_0].$$

12.5. Existence (*continued*)

Hereafter, we shall be concerned with the nonlinear evolution equation

$$u' = A(t)u + f(t, u), \qquad (12.5.1)$$

where $A(t)$ is a family of densely defined closed linear operators on a Banach space E and $f(t, u)$ is a function on $[t_0, t_0 + a] \times E$ taking values in E.

First of all, we shall summarize some of the known results for the linear equation

$$u' = A(t)u + F(t), \qquad (12.5.2)$$

where $F(t)$ is a function on $[t_0, t_0 + a]$ taking values in E. Usually $A(t)$ is unbounded.

Let us make the standing assumptions that there exists an evolution operator $U(t, s)$ associated with $A(t)$. This means that $\{U(t, s)\}$ is a family of bounded linear operators from E into E defined for $t_0 \leqslant s \leqslant t \leqslant t_0 + a$, strongly continuous in the two variables jointly and satisfying the conditions

$$U(t, s)\, U(s, r) = U(t, r), \qquad U(s, s) = I,$$

$$\left. \begin{array}{l} \dfrac{\partial U(t, s)}{\partial t} u = A(t)\, U(t, s) u, \\[6pt] \dfrac{\partial U(t, s)}{\partial s} u = U(t, s)\, A(s) u \end{array} \right\} \text{ for } u \text{ in a subset of } E, \text{ specified in each case.}$$

DEFINITION 12.5.1. A function $u(t)$ defined on $[t_0, t_0 + a]$ is said to be a *strict solution* of (12.5.2) with the initial value u_0, if $u(t)$ is strongly continuous on $[t_0, t_0 + a]$, $u(t_0) = u_0$, strongly continuously differentiable, and satisfies (12.5.2) on $(t_0, t_0 + a)$.

If $F(t)$ is continuous, any solution of (12.5.2) is of the form

$$u(t) = U(t, t_0) u_0 + \int_{t_0}^{t} U(t, s)\, F(s)\, ds. \tag{12.5.3}$$

DEFINITION 12.5.2. A function $u(t)$ is said to be a *mild solution* of (12.5.2) with the initial value u_0 if $u(t)$ is continuous on $[t_0, t_0 + a]$ and satisfies (12.5.3).

DEFINITION 12.5.3. The family $\{A(t)\}$ of operators is said to be *uniformly parabolic* if

(i) the spectrum of $A(t)$ is in a sector

$$S_\omega = [z : |\arg(z - \pi)| < \omega < \pi/2],$$
$$|(\lambda - A(t))^{-1}| \leqslant M/|\lambda|, \qquad \lambda \notin S_\omega,$$

and

$$|(A(t))^{-1}| \leqslant M,$$

where ω and M are independent of t;

(ii) for some $h = n^{-1}$ where n is a positive integer, the domain of $A(t)^h$ is independent of t, that is, $D[A(t)^h] = D$ and

$$|A(t)^h A(s)^{-h}| \leqslant M,$$
$$|A(t)^h A(s)^{-h} - I| \leqslant M\,|t - s|^k,$$

for $t, s \in [t_0, t_0 + a]$, $1 - h < k \leqslant 1$, M being independent of t.

DEFINITION 12.5.4. The family $\{A(t)\}$ of operators is said to be *hyperbolic* if $A(t)$, for each $t \in [t_0, t_0 + a]$, is the infinitesimal generator of a contraction semigroup, $D[A(t)]$ is independent of t, and $A(t) A(t_0)^{-1}$, which is a bounded linear operator, is strongly continuously differentiable.

It is known that, if the family $\{A(t)\}$ of operators is uniformly parabolic, there exists a unique evolution operator with the following properties:

(i) $U(t, s) E \subset D[A(t)]$ for $s < t$;

(ii) $U(t, s)u$ is Hölder continuous in t and s, for $u \in D = D[A(t_0)^h]$;

(iii) the mild solution is the strict solution of (12.5.2), if $F(t)$ is Hölder continuous on $[t_0, t_0 + a]$, where u_0 is an arbitrary element of E.

If, on the other hand, the family $\{A(t)\}$ is hyperbolic, there exists a unique evolution operator such that

$$U(t, s) D[A(s)] \subset D[A(t)],$$

and, if $F(t) \in D[A(t)]$ for every $t \in (t_0, t_0 + a)$, the mild solution is the strict solution of (12.5.2).

We shall now prove an existence theorem for mild solutions of the nonlinear evolution equation (12.5.1), which is parallel to Theorem 12.1.1.

THEOREM 12.5.1. Assume that

(i) $f \in C[[t_0, t_0 + a] \times E, E]$ and $f(t, u)$ maps bounded sets into bounded sets so that, for any $b > 0$, there exists a positive number M such that

$$|f(t, u)| \leqslant M$$

for $t \in [t_0, t_0 + a]$, $u \in S_b = [u \in E : |u - u_0| \leqslant b]$;

(ii) $V \in C[[t_0, t_0 + a] \times E, R_+]$, $V(t, u, v) > 0$ if $u \neq v$, $V(t, u, v) = 0$ if $u = v$, $V(t, u_n, v_n) \to 0$ implies $u_n - v_n \to 0$ uniformly in t;

(iii) $V(t, u, v)$ has continuous (bounded additive Fréchet) derivatives and

$$\frac{\partial V}{\partial t} + \frac{\partial V}{\partial u} f(t, u) + \frac{\partial V}{\partial v} f(t, v) \leqslant 0;$$

(iv) for any positive number M, $\partial V/\partial t$, $(\partial V/\partial u)x$, $(\partial V/\partial v)x$ are continuous in (u, v) uniformly $(t, u, v) \in [t_0, t_0 + a] \times S_b \times S_b$ and $|x| \leqslant M$;

(v) $(\partial V/\partial u) A(t)u + (\partial V/\partial v) A(t)v \leqslant 0$.

Then, the evolution equation (12.5.1) has a mild solution on $[t_0, t_0 + \alpha]$, where α is a positive number $\leq a$, under the conditions

(a) when $A(t)$ is parabolic, $u_0 \in D[A(t_0)]$ and $f(t, u)$ is Hölder continuous in t and u;

(b) when $A(t)$ is hyperbolic, $u_0 \in D[A(t_0)]$ and $A(t) f(t, u(t))$ is defined and strongly continuous whenever $u(t) \in D[A(t)]$.

Furthermore, in the latter case, the mild solution thus obtained is actually a strict solution if E is reflexive and if we assume, in addition, that $A(t) f(t, u)$ is bounded whenever u remains in a bounded set.

Proof. Choose a positive number α so small that

$$| U(t, t_0) u_0 | + M | U | (t - t_0) \leq b,$$

where $| U | = \sup | U(t, s)|$, $s, t \in [t_0, t_0 + \alpha]$. If $| t - t_0 | \leq \alpha$, let Δ be a subdivision of $[t_0, t_0 + \alpha]$, namely, $t_0 < t_1 < \cdots < t_n = t_0 + \alpha$. For $t_{k-1} \leq t \leq t_k$, we define

$$\phi_\Delta(t) = U(t, t_{k-1}) \phi_\Delta(t_{k-1}) + \int_{t_{k-1}}^{t} U(t, s) f(s, U(s, t_{k-1}) \phi_\Delta(t_{k-1})) \, ds,$$

$$\phi_\Delta(t_0) = u_0.$$

Then, we can write

$$\phi_\Delta(t) = U(t, t_0) u_0 + \int_{t_0}^{t} f_\Delta(t, s) \, ds,$$

where

$$f_\Delta(t, s) = U(t, s) f(s, U(s, t_{j-1}) \phi_\Delta(t_{j-1})), \qquad s \in [t_{j-1}, t_j].$$

Hence, $\phi_\Delta(t) \in S_b$.

Suppose that $A(t)$ is parabolic, $u_0 \in D[A(t_0)]$, and $f(t, u)$ is Hölder continuous in t and u. Then, we have $\phi_\Delta(t_{k-1}) \in D[A(t_{k-1})]$, and $U(t, t_{k-1}) \phi_\Delta(t_{k-1})$ is Hölder continuous in t for $t_{k-1} \leq t \leq t_k$. Hence, $\phi_\Delta(t)$ is differentiable in (t_{k-1}, t_k), and, as a consequence,

$$\phi_\Delta'(t) = A(t) U(t, t_{k-1}) \phi_\Delta(t_{k-1}) + f(t, U(t, t_{k-1}) \phi_\Delta(t_{k-1})). \qquad (12.5.4)$$

If, on the other hand, $A(t)$ is hyperbolic, $u_0 \in D[A(t_0)]$, $f(t, u) \in D[A(t)]$ whenever $u \in D[A(t)]$, and $A(t) f(t, U(t, t_{k-1}) \phi_\Delta(t_{k-1}))$ is strongly continuous, we can again deduce (12.5.4). Let Δ and Δ_1 be two subdivisions of $[t_0, t_0 + \alpha]$. Then, if t is not a subdivision point of either Δ or Δ_1,

12.5. EXISTENCE

we can take k and j such that $t_{k-1} < t < t_k$ and $t'_{j-1} < t < t'_j$ and obtain

$$\frac{dV}{dt}(t, \phi_\Delta(t), \phi_{\Delta_1}(t))$$

$$= \frac{\partial V}{\partial t} + \frac{\partial V}{\partial u}[A(t)\phi_\Delta(t) + f(t, U(t, t_{k-1})\phi_\Delta(t_{k-1}))]$$

$$+ \frac{\partial V}{\partial v}[A(t)\phi_{\Delta_1}(t) + f(t, U(t, t'_{j-1})\phi_{\Delta_1}(t'_{j-1}))]$$

$$= \left[\left(\frac{\partial V}{\partial t} - \frac{\partial \tilde{V}}{\partial t}\right) + \frac{\partial \tilde{V}}{\partial t}\right] + \left[\frac{\partial V}{\partial u}\{A(t)\phi_\Delta(t)\} + \frac{\partial V}{\partial v}\{A(t)\phi_{\Delta_1}(t)\}\right]$$

$$+ \left[\left(\frac{\partial V}{\partial u} - \frac{\partial \tilde{V}}{\partial u}\right) + \frac{\partial \tilde{V}}{\partial u}\right] f(t, U(t, t_{k-1})\phi_\Delta(t_{k-1}))$$

$$+ \left[\left(\frac{\partial V}{\partial v} - \frac{\partial \tilde{V}}{\partial v}\right) + \frac{\partial \tilde{V}}{\partial v}\right] f(t, U(t, t'_{j-1})\phi_{\Delta_1}(t'_{j-1})),$$

where

$$\frac{\partial V}{\partial t} = \frac{\partial V}{\partial t}(t, \phi_\Delta(t), \phi_{\Delta_1}(t)),$$

$$\frac{\partial \tilde{V}}{\partial t} = \frac{\partial V}{\partial t}(t, U(t, t_{k-1})\phi_\Delta(t_{k-1}), U(t, t'_{j-1})\phi_{\Delta_1}(t'_{j-1})),$$

with similar expressions for $\partial V/\partial u$, $\partial \tilde{V}/\partial u$, $\partial V/\partial v$, and $\partial \tilde{V}/\partial v$.

Since

$$|\phi_\Delta(t) - U(t, t_{k-1})\phi_\Delta(t_{k-1})|$$

$$= \left|\int_{t_{k-1}}^{t} U(t, s) f(s, U(s, t_{k-1})\phi_\Delta(t_{k-1}))\, ds\right|$$

$$\leq M \,|\, U \,|\,|\, \Delta \,|,$$

and, similarly,

$$|\phi_{\Delta_1}(t) - U(t, t'_{j-1})\phi_{\Delta_1}(t'_{j-1})| \leq M \,|\, U \,|\,|\, \Delta_1 \,|,$$

we see that $\{\phi_\Delta(t)\}$ is a strongly convergent family. Thus, there exists a $\phi(t)$ such that $\phi_\Delta(t) \to \phi(t)$. Furthermore, by assumption (ii) we find that this convergence is uniform. Hence,

$$f_\Delta(t, s) \to U(t, s) f(s, \phi(s)),$$

and consequently

$$\phi(t) = U(t, t_0)u_0 + \int_{t_0}^{t} U(t, s) f(s, \phi(s))\, ds,$$

which shows that $\phi(t)$ is a mild solution of (12.5.1).

When $A(t)$ is hyperbolic, we can prove that $\phi(t)$ thus obtained is a strict solution, under the assumption that $|A(t)f(t, u)|$ is bounded whenever u remains in a bounded set. This means that there is a positive integer L such that

$$|A(t)f(t, U(t, t_{k-1})\phi_\varDelta(t_{k-1}))| \leqslant L.$$

Then,

$$\phi_\varDelta'(t) = A(t)\, U(t, t_0)\, A(t_0)^{-1} A(t_0) u_0$$
$$+ \sum_{j=1}^{k-1} \int_{t_{j-1}}^{t_j} A(t)\, U(t, s)\, A(s)^{-1} A(s) f(s, U(s, t_{j-1})\phi_\varDelta(t_{j-1}))\, ds$$
$$+ \int_{t_{k-1}}^{t} A(t)\, U(t, s)\, A(s)^{-1} A(s) f(s, U(s, t_{k-1})\phi_\varDelta(t_{k-1}))\, ds$$
$$+ f(t, U(t, t_{k-1})\phi_\varDelta(t_{k-1}))$$

shows that

$$|\phi_\varDelta'(t)| \leqslant (|A(t_0)u_0| + L(t - t_0))A + M,$$

where $A = \sup |A(t)\, U(t, s)\, A(s)^{-1}|$. Thus, $\{A(t)\phi_\varDelta(t)\}$ is bounded. Since $A(t)$ is closed, this implies that $\phi(t) \in D[A(t)]$ when E is reflexive. It then follows that $\phi(t)$ is a strict solution of (12.5.1) because of the assumption that $A(t)f(t, \psi(t))$ is strongly continuous in t if $\psi(t) \in D[A(t)]$. The proof is therefore complete.

12.6. Approximate solutions and uniqueness

Suppose that $\{A(t)\}$ is a one-parameter family of closed linear operators defined for each $t \in J$. We shall assume that the domain $D[A(t)] = D$ of $A(t)$ is independent of t. Assume also that, for each $t \in J$, the resolvent set of $A(t)$ includes all sufficiently small positive real numbers α and that the domain of the resolvent $R(\alpha, A(t)) = [I - \alpha A(t)]^{-1}$ is dense in E. We observe that, since $A(t)$ is closed, $D[R(\alpha, A(t))] = E$, when α is in the resolvent set. Let us consider the differential inequality

$$|x' - A(t)x - f(t, x)| \leqslant w_1(t, |x|), \tag{12.6.1}$$

where $f \in C[J \times E, E]$, $w_1 \in C[J \times R_+, R_+]$.

DEFINITION 12.6.1. Let the function $x(t)$ be strongly continuous and defined for $t \in J$. Suppose further that $x(t)$ has strong derivative $x'(t)$

and that $x(t) \in D$. Assume that $x(t)$ satisfies the inequality (12.6.1) for $t \in J - S$, where S is at most a denumerable subset of J. Then we shall say that $x(t)$ is a solution of (12.6.1). If $w_1(t, u) = \epsilon$, $x(t)$ is said to be an ϵ-approximate solution of the differential equation

$$x' = A(t)x + f(t, x). \qquad (12.6.2)$$

THEOREM 12.6.1. Let $w_2(t, u)$ be a scalar function defined and continuous on $J \times R_+$. Suppose that $r(t)$ is the maximal solution of the scalar differential equation

$$u' = w(t, u), \qquad u(t_0) = u_0, \qquad (12.6.3)$$

where

$$w(t, u) = w_1(t, u) + w_2(t, u), \qquad (12.6.4)$$

existing to the right of t_0. Assume that, for each $t \in J$, $x \in E$,

$$\lim_{h \to 0^+} R(h, A(t))x = x, \qquad (12.6.5)$$

and

$$| R(h, A(t))x + hf(t, x)| \leqslant | x | + hw_2(t, | x |), \qquad (12.6.6)$$

for all sufficiently small $h > 0$ depending on t and x. Let $x(t)$ be any solution of (12.6.1) such that $| x(t_0)| \leqslant u_0$. Then

$$| x(t)| \leqslant r(t), \qquad t \geqslant t_0. \qquad (12.6.7)$$

Proof. Let $x(t)$ be any solution of (12.6.1) such that $| x(t_0)| \leqslant u_0$. Define $m(t) = | x(t)|$. For small $h > 0$, we have

$$\begin{aligned} m(t + h) &= | x(t + h)| - | R(h, A(t)) x(t) + hf(t, x(t))| \\ &\quad + | R(h, A(t)) x(t) + hf(t, x(t))| \\ &\leqslant | x(t + h) - R(h, A(t)) x(t) - hf(t, x(t))| \\ &\quad + | R(h, A(t)) x(t) + hf(t, x(t))|. \end{aligned} \qquad (12.6.8)$$

Since, for each $t \in J$, $A(t)$ is closed and $D[R(h, A(t))]$ is dense in E, it follows that $D[R(h, A(t))] = E$ and $R(h, A(t))(I - hA(t)) x = x$ for every $x \in D$. Hence,

$$R(h, A(t)) x(t) = x(t) + hA(t) x(t) + h[R(h, A(t)) A(t) x(t) - A(t) x(t)]. \qquad (12.6.9)$$

From (12.6.8) and (12.6.9), we obtain the inequality

$$\limsup_{h \to 0^+} h^{-1}[m(t + h) - m(t)] \leqslant w(t, m(t)),$$

because of the relations (12.6.1), (12.6.4), (12.6.5), and (12.6.6). Now, an application of Theorem 1.4.1 implies the stated result.

We can prove an analogous result for lower bounds.

THEOREM 12.6.2. Let $w_2(t, u)$ be a scalar function defined and continuous on $J \times R_+$. Suppose that $\rho(t)$ is the minimal solution of the scalar differential equation

$$u' = -w(t, u), \qquad \rho(t_0) = \rho_0 > 0,$$

existing on $[t_0, \infty)$. Assume that, for each $t \in J$ and $x \in E$,

$$\lim_{h \to 0^+} R(h, A(t))x = x$$

and

$$|R(h, A(t))x + hf(t, x)| \geq |x| - hw_2(t, |x|)$$

for all sufficiently small $h > 0$ depending on t and x. Let $x(t)$ be any solution of (12.6.1) such that $|x(t_0)| \geq \rho_0$. Then, for all t for which $\rho(t) \geq 0$, we have

$$|x(t)| \geq \rho(t).$$

Proof. Defining $m(t) = |x(t)|$ as before, it is easy to obtain the inequality

$$\liminf_{h \to 0^+} h^{-1}[m(t+h) - m(t)] \geq -w(t, m(t)).$$

This is enough to prove the stated result using an argument essentially similar to that of Theorem 1.4.1.

For various choices of w_1 and w_2, Theorems 12.6.1 and 12.6.2 extend many known results in ordinary differential equations to abstract differential equations. Suppose that $w_1 \equiv \epsilon$ and that $x(t)$ is an ϵ-approximate solution of (12.6.2). Let $w_2 = Ku$, $K > 0$. Then, Theorem 12.6.1 gives an estimate of the norm of ϵ-approximate solution, namely,

$$|x(t)| \leq |x(t_0)| e^{K(t-t_0)} + (\epsilon/K)(e^{K(t-t_0)} - 1), \qquad t \geq t_0,$$

whereas Theorem 12.6.2 yields a lower estimate,

$$|x(t)| \geq |x(t_0)| e^{-K(t-t_0)} + (\epsilon/K)(e^{-K(t-t_0)} - 1), \qquad t \geq t_0.$$

Again, suppose that $w_1 \equiv 0$ and that $x(t)$ is a solution of (12.6.2) existing on $[t_0, \infty)$. Let $w_2 = \lambda(t) g(u)$, where $g(u) > 0$ for $u > 0$ and

$\lambda \in C[J, R]$. Then, we obtain the following upper and lower bounds of the norm of a solution, namely,

$$G^{-1}\left[G(|x(t_0)|) - \int_{t_0}^{t} \lambda(s)\,ds\right] \leqslant |x(t)|$$
$$\leqslant G^{-1}\left[G(|x(t_0)|) + \int_{t_0}^{t} \lambda(s)\,ds\right],$$

where $G(u) = \int_{u_0}^{u} [g(s)]^{-1}\,ds$, $u_0 \geqslant 0$.

If we suppose that $w_1 = v(t)u$, $v(t) \geqslant 0$ is continuous on J, we have a variant of Theorem 12.6.1 which offers a sharper estimate.

THEOREM 12.6.3. Let the assumptions of Theorem 12.6.1 hold except that the condition (12.6.6) is replaced by

$$|R(h, A(t))x + hf(t, x)| \leqslant |x|(1 - \alpha h)$$
$$+ hw_2(t, |x| e^{\alpha(t-t_0)}) e^{-\alpha(t-t_0)}, \quad (12.6.10)$$

where $\alpha > 0$. Then (12.6.7) is replaced by

$$e^{\alpha(t-t_0)} |x(t)| \leqslant r(t), \qquad t \geqslant t_0.$$

Proof. Let $R(t)$ be the maximal solution of

$$R' = -\alpha R + w(t, Re^{\alpha(t-t_0)}) e^{-\alpha(t-t_0)},$$

such that $|x(t_0)| \leqslant R(t_0)$. Then, it is clear from (12.6.10) that Theorem 12.6.1 can be used to obtain the inequality

$$|x(t)| \leqslant R(t), \qquad t \geqslant t_0.$$

But $R(t) = r(t) e^{\alpha(t-t_0)}$, where $r(t)$ is the maximal solution of (12.6.3) such that $R(t_0) = u_0$. Verification is just the method of variation of parameters. Hence the result follows.

We next prove a uniqueness result analogous to Theorem 2.2.8.

THEOREM 12.6.4. Suppose that $x(t)$ and $y(t)$ are two solutions of the differential equation (12.6.2) with the initial condition $x(0) = y(0) = 0$. Let the condition

$$\lim_{t \to 0^+} \frac{|x(t) - y(t)|}{B(t)} = 0$$

be satisfied, where the function $B(t)$ is positive and continuous on

$0 < t < \infty$, with $B(0) = 0$. Let $g(t, u) \geqslant 0$ be continuous on $J \times R_+$. Suppose that the only solution $u(t)$ of

$$u' = g(t, u)$$

on $0 \leqslant t < \infty$ such that

$$\lim_{t \to 0^+} \frac{u(t)}{B(t)} = 0$$

is the trivial solution. Assume that, for each $t \in J$,

$$\lim_{h \to 0^+} R(h, A(t))x = x$$

for every $x \in E$ and that

$$| R(h, A(t))x - R(h, A(t))y + h[f(t, x) - f(t, y)]|$$
$$\leqslant | x - y | + hg(t, | x - y |)$$

for each $t \in (0, \infty)$, each $x, y \in E$, and for all sufficiently small $h > 0$, depending on t and x. Then, there exists at most one solution of (12.6.2) on J.

Proof. Suppose that there are two solutions $x(t)$ and $y(t)$ of (12.6.2) on J, with the initial condition $x(0) = y(0) = 0$. Let $m(t) = | x(t) - y(t)|$. Then, $m(0) = 0$. Now, using an argument similar to that of Theorem 12.6.1, we obtain

$$D^+ m(t) \leqslant g(t, m(t)).$$

From now on, we follow the proof of Theorem 2.2.8 with appropriate changes to complete the proof.

12.7. Chaplygin's method

By the one-parameter contraction semigroup of operators, we mean a one-parameter family $\{T(t)\}$, $t \geqslant 0$, of bounded operators acting from E to E, such that

(i) $T(t_1 + t_2) = T(t_1) T(t_2)$ for $t_1, t_2 \geqslant 0$;
(ii) $\lim_{h \to 0} T(h) x = x$ for $x \in E$;
(iii) $| T(t) | \leqslant 1$ for $t \in J$.

The infinitesimal generator A of $T(t)$ is defined by

$$Ax = \lim_{h \to 0} \frac{T(h) - I}{h} x$$

for every x, for which the limit exists. The limits mentioned previously, of course, are strong limits. The domain $D[A]$ of A is dense in E, A is closed, and, for $h > 0$, $|R(h, A)| \leqslant 1$. It is well known that, if A is closed and densely defined and if $|R(h, A)| \leqslant 1$, then there exists a unique contraction semigroup $\{T(t)\}$ such that A is its infinitesimal generator.

For $x \in D[A]$, the function $x(t) = T(t)x$ satisfies the equation

$$x'(t) = Ax(t), \quad x(0) = x, \quad t \geqslant 0.$$

Notice that, for $t, h \geqslant 0$,

$$|T(t+h)x| \leqslant |T(h) T(t)x| \leqslant |T(h)| |T(t)x|$$
$$\leqslant |T(t)x|.$$

Hence, it follows that

$$|x(t+h)| \leqslant |x(t)|,$$

that is, the norm of the solution $x(t)$ is a decreasing function.

We observe that $\lim_{h \to 0} R(h, A) x = x$ for every $x \in E$, if A is closed, $D(A)$ is dense in E, and $\lim_{h \to 0} \sup |R(h, A)| < \infty$. In view of this fact and on the basis of Theorem 12.6.1, we can prove the following

THEOREM 12.7.1. Assume that

(i) A is closed with dense domain such that $|R(h, A)| \leqslant 1$ for $h > 0$;

(ii) $g \in C[J \times R_+, R_+]$, $f \in C[J \times E, E]$, and $|f(t, x)| \leqslant g(t, |x|)$ for $t \in J$ and $x \in E$;

(iii) $r(t)$ is the maximal solution of $u' = g(t, u)$, $u(t_0) = u_0$ existing on J.

Then, if $x(t)$ is any solution of

$$x' = Ax + f(t, x) \tag{12.7.1}$$

such that $|x(t_0)| \leqslant u_0$ existing on J, we have

$$|x(t)| \leqslant r(t), \quad t \geqslant t_0.$$

We shall now prove a result that generates the Newtonian method of approximations in a version given by Chaplygin.

THEOREM 12.7.2. Suppose that

(i) A is an infinitesimal generator of contraction semigroup;

(ii) $f(t, x)$ is Fréchet differentiable in x to $f_x(t, x)$ and

$$|f_x(t, y) - f_x(t, z)| \leqslant g_1(t, |y - z|),$$

where $g_1 \in C[J \times R_+, R_+]$ and $g_1(t, u)$ is nondecreasing in u for each $t \in J$;

(iii) the sequence of functions $\{x_n(t)\}$ such that $|x_n(t)| \leqslant M$, $t \in [0, a]$, $n = 0, 1, 2,...$ satisfies

$$\begin{aligned}x'_{n+1}(t) &= Ax_{n+1}(t) + f_x(t, x_n(t))[x_{n+1}(t) - x_n(t)] \\ &\quad + f(t, x_n(t)), \quad x_n(0) = x_0, \quad t \in (0, a];\end{aligned} \quad (12.7.2)$$

(iv) $\sup_{t \in [0,a]} |f_x(t, 0)| < \infty$.

Then, $x_n(t)$ converges uniformly on $[0, a]$. Furthermore, if $x(t)$ is a solution of (12.7.1) such that $x(0) = x_0$, then there exists a well-defined sequence $\{w_n(t)\}$ such that

$$|x_n(t) - x(t)| \leqslant w_n(t), \quad (12.7.3)$$

where $|x_1(t) - x(t)| \leqslant w_1(t)$, $t \in [0, a]$, and

$$w_{n+1}(t) = \int_0^t \exp[K(t - s)] g(s, w_n(s)) \, ds, \quad (12.7.4)$$

K being given by

$$K = \sup_{t \in [0,a]} |f_x(t, 0)| + \sup_{t \in [0,a]} g(t, M).$$

If $w_n(t)$ is equibounded, then $x_n(t) \to x(t)$ as $n \to \infty$ uniformly on $[0, a]$.

Proof. Consider the sequence

$$z_n(t) = x_{n+1}(t) - x_n(t),$$

which satisfies the equation

$$\begin{aligned}z_n'(t) &= Az_n(t) + f_x(t, x_n(t)) z_n(t) + f(t, x_n(t)) \\ &\quad - f_x(t, x_{n-1}(t)) z_{n-1}(t) - f(t, x_{n-1}(t)) \\ &= Az_n(t) + R_n\,.\end{aligned}$$

Using assumption (ii), we obtain

$$|R_n| \leqslant K |z_n(t)| + g_1(t, |z_{n-1}(t)|) |z_{n-1}(t)|.$$

On the strength of Theorem 12.7.1, we get

$$|z_n(t)| \leq \int_0^t \exp[K(t-s)] g_1(s, |z_{n-1}(s)|) |z_{n-1}(s)| \, ds,$$

setting $g(t, u) = Ku + g_1(t, |z_{n-1}(t)|)|z_{n-1}(t)|$. Because of the monotonic character of g_1 and the fact $|z_n(t)| \leq 2M$, we have

$$|g_1(s, |z_{n-1}(s)|)| \leq g_1(s, 2M) \leq M_0, \quad t \in [0, a],$$

where M_0 is a suitable bound. Hence,

$$|z_n(t)| \leq \beta \int_0^t |z_{n-1}(s)| \, ds, \tag{12.7.5}$$

where $\beta = M_0 \exp(Ka)$. On the other hand, $|z_1(t)| \leq 2M$, and consequently, by (12.7.5),

$$|z_n(t)| \leq \frac{2M(\beta t)^{n-1}}{(n-1)!} \quad (n = 1, 2, 3, \ldots).$$

It then follows that $\{x_n(t)\}$ is uniformly convergent. If $x(t)$ is a solution of (12.7.1), under the assumption, $x(t)$ is uniquely determined. Also, notice that the inequality (12.7.3) is true for $n = 1$. Furthermore, $w_n(t)$ is well defined on $[0, a]$ because

$$w'_{n+1}(t) = Kw_{n+1}(t) + g_1(t, w_n(t)) w_n(t).$$

Suppose now that

$$|x_{n-1}(t) - x(t)| \leq w_{n-1}(t). \tag{12.7.6}$$

Observe that

$$[x_n(t) - x(t)]' = A[x_n(t) - x(t)] + P_n,$$

where

$$P_n = f_x(t, x_{n-1}(t))[x_n(t) - x(t)] + f_x(t, x_{n-1}(t))[x(t) - x_{n-1}(t)] + f(t, x_{n-1}(t)) - f(t, x(t)).$$

Then, assumption (ii) implies that

$$|P_n| \leq K|x_n(t) - x(t)| + g_1(t, |x_{n-1}(t) - x(t)|)|x_{n-1}(t) - x(t)|. \tag{12.7.7}$$

Since $g_1(t, u)$ is nondecreasing, (12.7.6) and (12.7.7) yield

$$|P_n| \leq K|x_n(t) - x(t)| + g_1(t, w_{n-1}(t)) w_{n-1}(t).$$

Theorem 12.7.1, with $g(t, u) = Ku + g_1(t, w_{n-1}(t)) w_{n-1}(t)$, shows that

$$|x_n(t) - x(t)| \leq \int_0^t \exp[K(t-s)] g_1(s, w_{n-1}(s)) w_{n-1}(s) \, ds$$
$$= w_n(t),$$

which proves (12.7.3). Suppose that $Q = \sup |w_n(t)|$ for $t \in [0, a]$, $n = 1, 2, \ldots$, is finite. Then,

$$w_n(t) \leq \int_0^t \exp[Ka] N w_{n-1}(s) \, ds,$$

where $N = \max_s g_1(s, a)$. Hence, $w_n(t) \to 0$ uniformly on $[0, a]$. The proof of the theorem is therefore complete.

Another version of Theorem 12.7.2, which depends on Theorem 1.4.4, will be given next.

THEOREM 12.7.3. Suppose that

(i) $f(t, x)$ is Fréchet differentiable in x to $f_x(t, x)$ and

$$|f_x(t, y) - f_x(t, z)| \leq g_1(t, |y - z|),$$

where $g_1 \in C[J \times R_+, R_+]$ and $g_1(t, u)$ is nondecreasing in u for each $t \in J$;

(ii) $z(t) = T(t) z_0$, where $\{T(t)\}$ is a contraction semigroup with generator A and $z_0 \in D[A]$;

(iii) $G, F \in C[[0, a], R_+]$, and, for $t \in [0, a]$,

$$|f_x(t, z(t))| \leq G(t),$$
$$|f(t, z(t))| \leq F(t);$$

(iv) $r(t)$ is the maximal solution of

$$u' = 3G(t)u + 3g_1(t, u)u + F(t), \quad u(0) = 0,$$

existing on $[0, a]$;

(v) $y(t)$ is a solution of the equation

$$y'(t) = Ay(t) + f_x(t, x(t))[y(t) - x(t)] + f(t, x(t)),$$

such that $y(0) = T(0) z_0 = z_0$.

Then, the inequality

$$|x(t) - z(t)| = |x(t) - T(t)z_0| \leqslant r(t), \quad t \in [0, a],$$

implies

$$|y(t) - z(t)| \leqslant r(t), \quad t \in [0, a].$$

Proof. We have, in view of the assumptions,

$$[y(t) - z(t)]' = A[y(t) - z(t)] + f_x(t, x(t))[y(t) - z(t)]$$
$$+ f_x(t, x(t))[z(t) - x(t)]$$
$$+ f(t, x(t)), |f_x(t, x(t))[y(t) - z(t)]|$$
$$\leqslant [G(t) + g_1(t, r(t))] |y(t) - x(t)|,$$

and

$$|f_x(t, x(t))[z(t) - x(t)] + f(t, x(t))|$$
$$\leqslant 2g_1(t, r(t)) r(t) + 2G(t) r(t) + F(t).$$

By Theorem 12.7.1, it follows that

$$|y(t) - z(t)| \leqslant R(t), \quad t \in [0, a], \qquad (12.7.8)$$

where $R(t)$ is the maximal solution of

$$u' = 2g_1(t, r(t)) r(t) + 2G(t) r(t) + F(t) + [g_1(t, r(t)) + G(t)]u,$$

existing on $[0, a]$. Setting

$$g(t, u, v) = 2g_1(t, v)v + 2G(t)v + F(t) + [g_1(t, v) + G(t)]u,$$

we see that all the assumptions of Theorem 1.4.4 are satisfied, and, as a result, $R(t) \equiv r(t)$ on $[0, a]$. The assertion of the theorem then follows from (12.7.8).

REMARK 12.7.1. It follows from the preceding theorem that, for a sequence of solutions $x_n(t)$ of

$$x'_{n+1}(t) = Ax_{n+1}(t) + f(t, x_{n-1}(t))[x_{n+1}(t) - x_n(t)] + f(t, x_n(t)),$$
$$x_n(0) = z_0 \in D[A],$$

the estimate

$$|x_n(t) - z(t)| \leqslant r(t), \quad t \in [0, a],$$

holds, provided $z(t) = x_1(t)$.

12.8. Asymptotic behavior

We shall now suppose that the norm in E is differentiable in the sense of Gateaux, namely,

$$\lim_{\lambda \to 0} \frac{|x + \lambda h| - |x|}{\lambda} = (\Gamma x, h),$$

where $\Gamma x = \mathrm{grad}\,|x|$ and $(1, x)$ is the value of the linear functional $1 \in E^*$, the conjugate space of E, at an element of E. It is easy to check that Γ acts from E into E^* and that

$$(\Gamma x, x) = |x|, \qquad \Gamma(\alpha x) = \Gamma x, \qquad \alpha > 0.$$

DEFINITION 12.8.1. Consider a function $\gamma(t)$ for which the estimate

$$(\Gamma x, A(t)x) \leqslant \gamma(t)\,|x| \tag{12.8.1}$$

holds for all $t \in J$ and $x \in D = D[A(t)]$. Introduce the notation

$$\Omega_\gamma = \limsup_{t \to \infty} t^{-1} \int_0^t \gamma(s)\,ds,$$

and

$$\Omega = \inf \Omega_\gamma,$$

where the inf is taken over all functions $\gamma(t)$. The number Ω is called the *central characteristic exponent*.

THEOREM 12.8.1. Assume that

(i) Ω is the central characteristic exponent;

(ii) $f(t, x)$ allows the estimate

$$(\Gamma x, f(t, x)) \leqslant \delta\,|x|, \qquad \delta > 0. \tag{12.8.2}$$

Then, given an $\epsilon > 0$, there exists a $\delta > 0$ such that any solution $x(t)$ of (12.6.2) admits an estimate

$$|x(t)| \leqslant |x(0)|\,C \exp[(\Omega + 2\epsilon)t], \qquad t \geqslant 0, \tag{12.8.3}$$

where the constant C depends on ϵ.

Proof. Let $m(t) = |x(t)|$. Then, using (12.8.1) and (12.8.2), we have

$$\begin{aligned} m'(t) &= (\Gamma x(t), x'(t)) \\ &= (\Gamma x(t), A(t)\,x(t)) + (\Gamma x(t), f(t, x(t))) \\ &\leqslant \gamma(t)\,|x(t)| + \delta\,|x(t)|, \end{aligned}$$

12.8. ASYMPTOTIC BEHAVIOR

so that
$$m'(t) \leqslant [\gamma(t) + \delta] m(t).$$

By Theorem 1.4.1, we get
$$|x(t)| \leqslant |x(0)| \exp\left[\int_0^t (\gamma(s) + \delta) \, ds\right], \qquad t \geqslant 0. \tag{12.8.4}$$

From the definition of Ω, given $\epsilon > 0$ there exists a function $\gamma(t)$ such that (12.8.1) is satisfied and, at the same time,
$$\Omega_\gamma < \Omega + \epsilon.$$
In other words,
$$\int_0^t \gamma(s) \, ds \leqslant C + (\Omega + \epsilon)t,$$
where C is a constant depending on ϵ. Then, taking $\delta \leqslant \epsilon$ and considering the last inequality, we obtain from (12.8.4) the desired inequality (12.8.3), and the theorem is proved.

As an application of Theorem 12.8.1, we prove the following theorem, which gives sufficient conditions for the asymptotic stability of the null solution of (12.6.2).

THEOREM 12.8.2. Suppose that

(i) the central characteristic exponent Ω is negative;

(ii) the function $f(t, x)$ satisfies
$$(\Gamma x, f(t, x)) \leqslant \beta \, |x|^{1+\alpha}, \qquad \alpha > 0. \tag{12.8.5}$$

Then, the trivial solution of (12.6.2) is asymptotically stable.

Proof. Choose a $\lambda > 0$ such that $\Omega_1 = \Omega + \lambda < 0$, and consider the function
$$x(t) = \exp(-\lambda t) \, y(t).$$
Then
$$y'(t) = [A(t) + \lambda I] y(t) + g(t, y(t)), \tag{12.8.6}$$
where
$$g(t, y) = \exp(\lambda t) f(t, \exp(-\lambda t) y).$$

It then follows from the properties of Γ and (12.8.5) that, setting $y = y(t)$,
$$\begin{aligned}(\Gamma y, g(t, y)) &= e^{\lambda t}(\Gamma y, f(t, e^{-\lambda t} y)) \\ &= e^{\lambda t}(\Gamma(e^{-\lambda t} y), f(t, e^{-\lambda t} y)) \\ &\leqslant e^{\lambda t} \beta \, |e^{-\lambda t} y|^{1+\alpha},\end{aligned}$$

and, consequently,
$$(\Gamma y, g(t, y)) \leq \beta e^{-\lambda \alpha t} |y|^{1+\alpha}. \tag{12.8.7}$$

Furthermore, using (12.8.1), we derive
$$\begin{aligned}(\Gamma y, [A(t) + \lambda I] y) &= (\Gamma y, A(t) y) + \lambda (\Gamma y, y) \\ &= (\Gamma y, A(t) y) + \lambda |y| \\ &\leq [\gamma(t) + \lambda] |y|.\end{aligned}$$

Hence, the central characteristic exponent of the operator $A(t) + \lambda I$ is equal to $\Omega_1 = \Omega + \lambda$, and, therefore, choosing an $\epsilon > 0$ such that $\Omega_1 + 2\epsilon < 0$, we can find a function $\gamma(t)$ that satisfies (12.8.1) and $\Omega < \Omega_1 + \epsilon < 0$, because of the definition of Ω. Let $\delta > 0$ be such that $\delta < \epsilon$. Take the initial time $t_0 > 0$ so large that, for $t \geq t_0$ and small $|y|$, we get, by (12.8.7),
$$(\Gamma y, g(t, y)) \leq \delta |y|.$$

Thus, the operator $g(t, y)$ verifies the hypotheses of Theorem 12.8.1, and hence
$$|y(t)| \leq |y(t_0)| C \exp[(\Omega_1 + 2\epsilon)t], \quad t \geq t_0.$$

Since $\Omega_1 + 2\epsilon < 0$, the null solution of (12.8.6) is asymptotically stable, and, as a result, the trivial solution of (12.6.2) is also asymptotically stable. The theorem is proved.

Another set of conditions for the asymptotic stability is given by the following

THEOREM 12.8.3. Assume that

(i) $g \in C[J \times R_+ , R]$, and the solutions $u(t)$ of the scalar differential equation
$$u' = g(t, u), \quad u(t_0) = u_0 \geq 0, \tag{12.8.8}$$
are bounded on $[t_0 , \infty]$;

(ii) for each $t \in J$, $x \in E$,
$$\lim_{h \to 0^+} R(h, A(t)) x = x$$
and
$$|R(h, A(t)) x + h f(t, x)| \leq |x| (1 - \alpha h) + h g(t, |x| e^{\alpha(t-t_0)}) e^{-\alpha(t-t_0)},$$

where $\alpha > 0$, for all sufficiently small $h > 0$ depending on t and x.

Then, the trivial solution of (12.6.2) is asymptotically stable.

Proof. Following the proof of Theorem 12.6.3, we obtain

$$|x(t)| \leq r(t) \exp[-\alpha(t - t_0)], \qquad t \geq t_0, \qquad (12.8.9)$$

where $r(t)$ is the maximal solution of (12.8.8) and $x(t)$ is any solution of (12.6.2). By assumption, $r(t)$ is bounded on $[t_0, \infty]$. Hence, the asymptotic stability of the trivial solution of (12.6.2) is immediate from the estimate (12.8.9). The proof is complete.

12.9. Lyapunov function and comparison theorems

We shall continue to consider the differential equation (12.6.2) under the same assumptions on the family of operators $\{A(t)\}$ as in Sect. 12.6. Let us prove the following comparison theorems.

THEOREM 12.9.1. Assume that

(i) $V \in C[J \times E, R_+]$ and

$$|V(t, x_1) - V(t, x_2)| \leq c(t) |x_1 - x_2|, \qquad (12.9.1)$$

for $t \in J$, $x_1, x_2 \in E$, $c(t) \geq 0$ being a continuous function on J;

(ii) $g \in C[J \times R_+, R]$, $r(t)$ is the maximal solution of the scalar differential equation

$$u' = g(t, u), \qquad u(t_0) = u_0 \geq 0, \qquad t_0 \geq 0, \qquad (12.9.2)$$

existing on J, and, for $t \in J$, $x \in E$,

$$D^+ V(t, x) = \limsup_{h \to 0^+} h^{-1}[V(t + h, R(h, A(t))x + hf(t, x)) - V(t, x)]$$

$$\leq g(t, V(t, x)); \qquad (12.9.3)$$

(iii) for each $t \in J$, $\lim_{h \to 0^+} R(h, A(t)) x = x$, $x \in E$, and $x(t)$ is any solution of (12.6.2) existing on $[t_0, \infty)$ such that

$$V(t_0, x(t_0)) \leq u_0.$$

Under these assumptions, we have

$$V(t, x(t)) \leq r(t), \qquad t \geq t_0. \qquad (12.9.4)$$

Proof. Let $x(t)$ be any solution of (12.6.2) existing on $[t_0, \infty)$, satisfying $V(t_0, x(t_0)) \leq u_0$. Consider the function

$$m(t) = V(t, x(t)),$$

so that $m(t_0) \leqslant u_0$. Furthermore, for small $h > 0$,

$$m(t+h) - m(t) \leqslant c(t)[|\, x(t+h) - R(h, A(t))\, x(t) - hf(t, x(t))|]$$
$$+ V(t+h, R(h, A(t))\, x(t) + hf(t, x(t))) - V(t, x(t)),$$
(12.9.5)

because of (12.9.1). Since, for every $x \in D[A(t)]$,

$$R(h, A(t))[I - hA(t)]x = x,$$

it follows that

$$R(h, A(t))x + hf(t, x) = x + h[A(t)x + f(t, x)]$$
$$+ h[R(h, A(t))\, A(t)x - A(t)x],$$

which, together with (12.9.5), implies that

$$m(t+h) - m(t) \leqslant c(t)[|\, x(t+h) - x(t) - h[A(t)\, x(t) + f(t, x(t))]|]$$
$$+ c(t)\, h[|\, R(h, A(t))\, A(t)\, x(t) - A(t)\, x(t)|]$$
$$+ V(t+h, R(h, A(t))\, x(t) + hf(t, x(t))) - V(t, x(t)).$$

Using the relations (12.6.2), (12.9.3), and assumption (iii), we obtain the inequality

$$D^+ m(t) \leqslant g(t, m(t)).$$

An application of Theorem 1.4.1 now yields the stated inequality (12.9.4), and the proof is complete.

THEOREM 12.9.2. *Let the assumptions of Theorem 12.9.1 be satisfied except that the condition (12.9.3) be replaced by*

$$p(t)\, D^+ V(t, x) + V(t, x)\, D^+ p(t) \leqslant g(t, V(t, x)\, p(t)),$$
(12.9.6)

where $p(t) > 0$ is continuous on J. Then, whenever

$$p(t_0)\, V(t_0, x(t_0)) \leqslant u_0,$$

the inequality (12.9.4) takes the form

$$p(t)\, V(t, x(t)) \leqslant r(t), \quad t \geqslant t_0.$$

Proof. Define $L(t, x) = p(t)\, V(t, x)$. Then, using (12.9.6), we have

$$L(t+h, R(h, A(t))x + hf(t, x)) - L(t, x)$$
$$\leqslant p(t+h)[V(t, x) + hp(t)^{-1}\{g(t, L(t, x)) - D^+ p(t)\, V(t, x) + \epsilon\}] - p(t)\, V(t, x),$$

where $\epsilon \to 0$ as $h \to 0$; a rearrangement of the right-hand side gives

$$L(t+h, R(h, A(t))x + hf(t,x)) - L(t,x)$$
$$\leqslant V(t,x)[p(t+h) - p(t)]$$
$$+ hp(t+h)\,p(t)^{-1}[g(t, L(t,x)) + \epsilon - D^+p(t)\,V(t,x)].$$

It then follows that

$$D^+L(t,x) = \limsup_{h \to 0^+} h^{-1}[L(t+h, R(h, A(t))x + hf(t,x)) - L(t,x)]$$
$$\leqslant g(t, L(t,x)),$$

which implies that Theorem 12.9.2 can be reduced to Theorem 12.9.1 with $L(t,x)$ in place of $V(t,x)$. Hence we have the proof.

12.10. Stability and boundedness

Let M be a nonempty subset of E containing $\{0\}$, and let $d(x, M)$ denote the distance between an element $x \in E$ and the set M. Denote the sets $[x: d(x, M) < \eta]$ and $[x: d(x, M) \leqslant \eta]$ by $S(M, \eta)$ and $\bar{S}(M, \eta)$, respectively. Suppose that $x(t)$ is any solution of (12.6.2) existing in the future. Then, we may formulate the various definitions of stability and boundedness with respect to the set M and the differential system (12.6.2) corresponding to the definitions (S_1) to (S_{10}) and (B_1) to (B_{10}) given in Chapter 3. As an example, (S_1) would run as follows.

DEFINITION 12.10.1. The set M, with respect to the system (12.6.2), is said to be (S_1) *equistable* if, for any $\epsilon > 0$, $t_0 \in J$, there exists a $\delta = \delta(t_0, \epsilon)$ that is continuous in t_0 for each ϵ, such that

$$x(t) \subset S(M, \epsilon), \quad t \geqslant t_0,$$

provided that $x(t_0) \in \bar{S}(M, \delta)$.

The following theorem gives sufficient conditions for stability.

THEOREM 12.10.1. Assume that

(i) $g \in C[J \times R_+, R]$ and $g(t, 0) \equiv 0$;

(ii) $V \in C[J \times S(M, \rho), R_+]$, $V(t, x)$ is locally Lipschitzian in x, and, for $(t, x) \in J \times S(M, \rho)$,

$$b(d(x, M)) \leq V(t, x) \leq a(t, d(x, M)),$$

where $b \in \mathscr{K}$, $a \in C[J \times [0, \rho), R_+]$, $a \in \mathscr{K}$ for each $t \in J$;

(iii) for $(t, x) \in J \times S(M, \rho)$,

$$D^+ V(t, x) \leq g(t, V(t, x));$$

(iv) $\lim_{h \to 0^+} R(h, A(t)) x = x$ for $t \in J$ and $x \in E$.

Then, the equistability of the null solution of (10.2.1) implies the equistability of the set M with respect to the system (12.6.2).

Proof. Let $0 < \epsilon < \rho$, $t_0 \in J$ be given. Assume that the trivial solution of (10.2.1) is equistable. Then, given $b(\epsilon) > 0$, $t_0 \in J$, there exists a $\delta = \delta(t_0, \epsilon)$ that is continuous in t_0 for each ϵ such that

$$u(t, t_0, u_0) < b(\epsilon), \quad t \geq t_0, \tag{12.10.1}$$

provided $u_0 \leq \delta_1$, where $u(t, t_0, u_0)$ is any solution of (10.2.1). Choose $u_0 = V(t_0, x(t_0))$. Because of the hypothesis on $a(t, u)$, there exists a $\delta_1 = \delta_1(t_0, \epsilon)$ satisfying the inequalities

$$d(x(t_0), M) \leq \delta_1 \quad \text{and} \quad a(t_0, d(x(t_0), M)) \leq \delta$$

at the same time. We claim that, if $x(t_0) \in \bar{S}(M, \delta_1)$, $x(t) \subset S(M, \epsilon)$, $t \geq t_0$. Suppose that this is not true. Then, there would exist a solution $x(t)$ with $x(t_0) \in \bar{S}(M, \delta_1)$ and a $t_1 > t_0$ such that

$$d(x(t_1), M) = \epsilon \quad \text{and} \quad d(x(t), M) \leq \epsilon, \quad t \in [t_0, t_1],$$

so that

$$b(\epsilon) \leq V(t_1, x(t_1)). \tag{12.10.2}$$

Since this implies that $x(t) \in S(M, \rho)$, $t \in [t_0, t_1]$, the choice $u_0 = V(t_0, x(t_0))$ and condition (iii) yield, by Theorem 12.9.1, the estimate

$$V(t, x(t)) \leq r(t, t_0, u_0), \quad t \in [t_0, t_1], \tag{12.10.3}$$

where $r(t, t_0, u_0)$ is the maximal solution of (10.2.1). It is easy to see that the relations (12.10.1), (12.10.2), and (12.10.3) lead us to the following contradiction:

$$b(\epsilon) \leq V(t_1, x(t_1)) \leq r(t_1, t_0, u_0) < b(\epsilon).$$

Hence, the stated result is true, proving the theorem.

12.10. STABILITY AND BOUNDEDNESS

THEOREM 12.10.2. Let the assumptions of Theorem 12.10.1 hold except that the function $a(t, u)$ in condition ii is independent of t, that is, $a(t, u) \equiv a(u)$, where $a \in \mathcal{K}$. Then, one of the stability conditions of the trivial solution of (10.2.1) implies the corresponding one of the stability conditions of the set M with respect to the system (12.6.2).

On the basis of the proof of Theorem 12.10.1 and that of the parallel theorems in Chapter 3, the proof of Theorem 12.10.2 may be constructed easily. We therefore omit its proof.

As an application of Theorem 12.9.2, we shall give a result that offers sufficient conditions for equi-asymptotic stability.

THEOREM 12.10.3. Let the hypotheses of Theorem 12.10.1 hold except that assumption (iii) is replaced by

$$p(t) D^+V(t, x) + V(t, x) D^+p(t) \leqslant g(t, V(t, x) p(t)),$$

where $p(t) > 0$ is continuous on J and $p(t) \to \infty$ as $t \to \infty$. Then, the equistability of the trivial solution of (10.2.1) assures the equi-asymptotic stability of the set M with respect to the system (12.6.2).

Proof. The proof of this theorem is analogous to the proof of Theorem 3.4.7, if we introduce the necessary changes.

Finally, we state a theorem giving conditions for various boundedness notions, the proof of which may be constructed on the basis of the proof of Theorem 12.10.1 and the corresponding boundedness results of Chapter 3.

THEOREM 12.10.4. Assume that

(i) $V \in C[J \times E, R_+]$, $V(t, x)$ is locally Lipschitzian in x, and, for $(t, x) \in J \times E$,

$$b(d(x, M)) \leqslant V(t, x) \leqslant a(d(x, M)),$$

where $a, b \in \mathcal{K}$ and $b(u) \to \infty$ as $u \to \infty$;

(ii) $g \in C[J \times R_+, R]$, and, for $(t, x) \in J \times E$,

$$D^+V(t, x) \leqslant g(t, V(t, x));$$

(iii) $\lim_{h \to 0+} R(h, A(t)) x = x$ for $t \in J$ and $x \in E$.

Then, one of the boundedness conditions with respect to the scalar differential equation (10.2.1) implies the corresponding one of the boundedness conditions of the set M with respect to the system (12.6.2).

12.11. Notes

The existence theorems 12.1.1 and 12.1.2 of Sect. 12.1 are due to Murakami [3]. See also Browder [2] and T. Kato [4]. Theorem 12.2.1 is taken from the work of Mlak [5]. Theorems 12.3.1, 12.3.2, and the example that follows are due to Murakami [3].

The results of Sect. 12.4 are due to Antosiewicz [1]. Section 12.5 consists of the work of Murakami [3]. The results of Sect. 12.6 are taken from Lakshmikantham [4].

For the work contained in Sect. 12.7, see Mlak [8, 13]. Theorems 12.8.1 and 12.8.2 are due to Mamedov [2], whereas Theorem 12.8.3 is adopted from the work of Lakshmikantham [4]. The results of Sects. 12.9 and 12.10 are due to Lakshmikantham [3].

For the existence of solutions of functional equations using Lyapunov functions, see Hartman [1], where, applying analogous arguments, existence theorem for an initial value problem for ordinary differential equations in Hilbert spaces is discussed. Concerning abstract differential inequalities, see Cohen and Lees [1] and Edmunds [1]. Differential equations on cones in Banach spaces are treated in Coffman [1–3], Chandra and Fleishman [1, 2], and Szarski [8]. For an excellent treatment on linear differential equations and function spaces, see Massera and Schaffer [1].

For related work, see Agmon and Nirenberg [1], Browder [4], Foias *et al.* [3], T. Kato [1–4], Kato and Tanabe [1], Kisynski [2], Krasnoselskii [2], Krasnoselskii, *et al.* [1, 2], Krein and Prozorovskaya [1], Lees [1], Lions [1], Lyubic [1], Minty [1, 2], Mlak [5, 10, 12, 14, 15, 18], Ramamohan Rao and Tsokos [1], Sobolevski [1, 2], Taam and Welch [1], and Tanabe [1, 2].

COMPLEX DIFFERENTIAL EQUATIONS

Chapter 13

13.0. Introduction

An important extension of the initial value problem of ordinary differential equations is to the case where the time variable t and the space variables x may be complex. We treat in the present chapter various problems connected with this type of differential systems. We consider the existence and uniqueness problems, obtain error estimates of approximate solutions, discuss the singularity-free regions of solutions, and derive *a priori* bounds. Introducing Lyapunov-like functions, we prove comparison theorems, which may be used for stability criteria. All our considerations depend crucially on the technique of reducing the study of the behavior of solutions of complex differential systems to that of certain ordinary differential equations.

13.1. Existence, approximate solutions, and uniqueness

Let C^n denote complex euclidean n space, and we shall use C to represent C^1. For any element $y \in C^n$, we mean by $\|y\|$ one of the usual norms $\|y\| = \sum_{i=1}^{n} |y^i|$, $\|y\| = (y \cdot \bar{y})^{1/2}$, and $\|y\| = \max |y_i|$, $i = 1, 2, \ldots, n$. Let f be an analytic complex valued vector function defined on the domain

$$D = [(z, y) \in C^{n+1} : 0 \leqslant |z| < a, \|y\| < b, a, b > 0].$$

We consider the initial value problem of complex differential system given by

$$\frac{dy}{dz} = f(z, y), \quad y(0) = y_0, \quad y_0 \in C^n. \qquad (13.1.1)$$

The existence and uniqueness of the solutions of (13.1.1) may be inferred from the method of successive approximations. We merely state this well-known result.

THEOREM 13.1.1. Suppose that the function f is regular-analytic and bounded in D. Let $\sup_{(z,y) \in D} \| f(z, y) \| = M$ and $\alpha = \min(a, b/M)$. Then, there exists a unique regular-analytic function $y(z)$ defined on $0 \leqslant |z| < \alpha$, which satisfies (13.1.1).

We now consider the following complex differential inequality,

$$\left\| \frac{dy}{dz} - f(z, y) \right\| \leqslant \epsilon, \tag{13.1.2}$$

where f is regular-analytic in D and $\epsilon > 0$ is a real number.

DEFINITION 13.1.1. Any complex valued vector function $y(z, \epsilon)$ is said to be an *ϵ-approximate solution* of (13.1.1) if the following conditions are satisfied:

(i) $y(z, \epsilon)$ is regular-analytic in $0 \leqslant |z| < a$;

(ii) $y(z, \epsilon)$ satisfies the complex differential inequality (13.1.2) for $0 \leqslant |z| < a$.

If $\epsilon = 0$ in (13.1.2), it is to be understood that $y(z)$ is a solution of (13.1.1).

The following result gives the upper and lower bounds of the norm of the difference between any two approximate solutions of (13.1.1).

THEOREM 13.1.2. Let $g \in C[[0, a) \times R_+, R_+]$, $f(z, y)$ be regular-analytic in D, and

$$\| f(z, y_1) - f(z, y_2) \| \leqslant g(|z|, \| y_1 - y_2 \|). \tag{13.1.3}$$

Suppose further that $y(z, \epsilon_1)$ and $y(z, \epsilon_2)$ are ϵ_1- and ϵ_2-approximate solutions of (13.1.1) such that

$$0 < \rho(0) \leqslant \| y_1(0, \epsilon_1) - y_2(0, \epsilon_2) \| \leqslant r(0). \tag{13.1.4}$$

Assume that $r(t)$ and $\rho(t)$ are maximal and minimal solutions of ordinary differential equations

$$\begin{aligned} r' &= g(t, r) + \epsilon_1 + \epsilon_2, \\ \rho' &= -[g(t, \rho) + \epsilon_1 + \epsilon_2], \end{aligned} \tag{13.1.5}$$

respectively, existing on $0 \leqslant t < a$. Then, on each ray $z = te^{i\theta}$ and $0 \leqslant |z| < a$, we have

$$\rho(t) \leqslant \| y(z, \epsilon_1) - y(z, \epsilon_2) \| \leqslant r(t). \tag{13.1.6}$$

13.1. EXISTENCE, APPROXIMATIONS, UNIQUENESS

Proof. Let $y(z, \epsilon_1)$ and $y(z, \epsilon_2)$ be the approximate solutions of (13.1.1) satisfying (13.1.4). Let $z = te^{i\theta}$ and $\arg z = \text{const}$. Define

$$m(t) = \| y(te^{i\theta}, \epsilon_1) - y(te^{i\theta}, \epsilon_2)\|.$$

For small $h > 0$, we obtain

$$| m(t + h) - m(t)| \leq \|[y((t + h)e^{i\theta}, \epsilon_1) - y(te^{i\theta}, \epsilon_1)$$
$$- y((t + h) e^{i\theta}, \epsilon_2) - y(te^{i\theta}, \epsilon_2)]\|. \quad (13.1.7)$$

Also, we have

$$\left\| \frac{dy(te^{i\theta}, \epsilon_1)}{dt} - \frac{dy(te^{i\theta}, \epsilon_2)}{dt} \right\|$$
$$= \left\| \frac{dy(z, \epsilon_1)e^{i\theta}}{dz} - \frac{dy(z, \epsilon_2)e^{i\theta}}{dz} \right\|$$
$$= \left\| \frac{dy(z, \epsilon_1)}{dz} - \frac{dy(z, \epsilon_2)}{dz} \right\|. \quad (13.1.8)$$

Hence, using (13.1.7) and (13.1.8), we get

$$| m_+'(t)| \leq \left\| \frac{dy(z, \epsilon_1)}{dz} - \frac{dy(z, \epsilon_2)}{dz} \right\|,$$

where $m_+'(t)$ is the right-hand derivative of $m(t)$ with respect to t. Moreover,

$$\left\| \frac{dy(z, \epsilon_1)}{dz} - \frac{dy(z, \epsilon_2)}{dz} \right\|$$
$$\leq \left\| \frac{dy(z, \epsilon_1)}{dz} - f(z, y(z, \epsilon_1)) \right\|$$
$$+ \left\| \frac{dy(z, \epsilon_2)}{dz} - f(z, y(z, \epsilon_2)) \right\|$$
$$+ \| f(z, y(z, \epsilon_1)) - f(z, y(z, \epsilon_2))\|.$$

It therefore follows, in view of (13.1.2) and (13.1.3), that

$$| m_+'(t)| \leq g(t, m(t)) + \epsilon_1 + \epsilon_2,$$

and, consequently,

$$-[g(t, m(t)) + \epsilon_1 + \epsilon_2] \leq m_+'(t) \leq g(t, m(t)) + \epsilon_1 + \epsilon_2.$$

It is now easy to prove the stated estimate (13.1.6) by repeating the proof of Theorem 1.2.1 with appropriate changes.

COROLLARY 13.1.1. The function $g(t, u) = ku$, $k > 0$, is admissible in Theorem 13.1.2.

Indeed, this choice of g yields

$$\rho(0)e^{-kt} + \frac{\epsilon_1 + \epsilon_2}{k}(e^{-kt} - 1) \leqslant \| y(te^{i\theta}, \epsilon_1) - y(te^{i\theta}, \epsilon_2)\|$$
$$\leqslant r(0)e^{kt} + \frac{\epsilon_1 + \epsilon_2}{k}(e^{kt} - 1),$$

which corresponds to a well-known inequality in ordinary differential equations.

We now prove a general uniqueness theorem for complex differential equations which extends a similar result in ordinary differential equations, namely, Theorem 2.2.8.

Suppose that $y_1(z)$ and $y_2(z)$ are any two solutions of (13.1.1). As before, let $\arg z = \text{const}$, $|z| = t$, and

$$m(t) = \| y_1(z) - y_2(z)\|. \tag{13.1.9}$$

Let the function $B(t) > 0$ be continuous on $0 < t < a$ and $B(0^+) = 0$. Suppose that

$$\lim_{t \to 0^+} \frac{m(t)}{B(t)} = 0. \tag{13.1.10}$$

THEOREM 13.1.3. Let $y_1(z)$ and $y_2(z)$ be any two solutions of (13.1.1) satisfying (13.1.10), where $B(t)$ is the function just defined. Let $g(t, u) \geqslant 0$ be continuous on $0 < t < a$ and $u \geqslant 0$ and $g(t, 0) \equiv 0$. Assume that the only solution $u(t)$ of

$$u' = g(t, u) \tag{13.1.11}$$

such that

$$\lim_{t \to 0^+} \frac{u(t)}{B(t)} = 0 \tag{13.1.12}$$

is the trivial solution. Suppose further that

$$\| f(z, y_1) - f(z, y_2)\| \leqslant g(|z|, \| y_1 - y_2 \|)$$

for $0 < |z| < a$, $\| y_1 \|, \| y_2 \| < b$. Then, there exists at most one solution of (13.1.1) on $0 \leqslant |z| < a$.

Proof. Suppose that there exist two solutions $y_1(z)$ and $y_2(z)$ of (13.1.1) and that (13.1.10) holds. Proceeding as in the proof of Theorem 13.1.2 with necessary modifications, it is easy to obtain the differential inequality

$$m_+'(t) \leqslant g(t, m(t)), \qquad 0 < t < a,$$

where $m(t)$ is the function defined by (13.1.9). The rest of the proof follows closely the proof of Theorem 2.2.8. Hence the proof is complete.

REMARK 13.1.1. If $B(t) = t$, Theorem 13.1.3 reduces to an extension of Kamke's general uniqueness theorem 2.2.2 for complex differential equations.

Let us consider the following example. Let

$$y' = 2y\,z, \qquad y(0) = 0.$$

This has a unique solution $y(z) \equiv 0$ if we consider the solutions in the classical sense, whereas, for each complex K, $y(z) = Kz^2$ represents a solution in the extended sense.

In the first case, taking $B(t) = t^2$, we see that the condition (13.1.10) is verified. Moreover, since the differential equation (13.1.11) takes the form $u' = 2u/t$, all the assumptions of the uniqueness theorem are fulfilled. On the other hand, when we consider the latter case, even though the choice of $B(t) = t^{2-\epsilon}$, $\epsilon > 0$ being an arbitrarily small number, satisfies the condition (13.1.11), the assumption (13.1.12) is violated, since the solutions of (13.1.11) are $u(t) = Lt^2$, where L is an arbitrary nonnegative constant.

13.2. Singularity-free regions and growth estimates

Let us first of all be concerned with the solutions of the complex equation

$$\frac{d^2y}{dz^2} + F(z, y) = 0, \qquad (13.2.1)$$

where $F(z, y)$ is an entire function of y and regular-analytic in z for $|z| < a$. In addition, we assume that

$$|F(z, y)| \leqslant G(|z|, |y|) \qquad (13.2.2)$$

for all y, $|z| < a$, where $G \in C[[0, a) \times R_+, R_+]$ and $G(t, u)$ is nondecreasing in u for each t.

DEFINITION 13.2.1. A region S is said to be *singularity-free* for a solution $y(z)$ if $y(z)$ is regular-analytic in S.

THEOREM 13.2.1. Assume that $F(z, y)$ is subjected to the conditions stated previously. If there exists a function $u(t)$, which is defined and continuous in $[0, a)$ such that

$$u(0) = \alpha \geqslant 0, \qquad u'(0) = \beta \geqslant 0, \qquad \alpha^2 + \beta^2 > 0,$$

and that

$$u''(t) \geqslant G(t, u(t)), \qquad 0 \leqslant t < a, \tag{13.2.3}$$

then any solution of (13.2.1) for which

$$|y(0)| = \alpha, \qquad \left|\frac{dy(0)}{dz}\right| = \beta,$$

is regular for $|z| < a$.

Proof. Suppose that $\arg z = \theta = \text{const.}$ Then, from (13.2.1), we have the integral equation

$$y(z) = y(0) + z\,\frac{dy(0)}{dz} - \int_0^z (z - \xi) F(\xi, y(\xi))\, d\xi,$$

where the integration is carried out along the ray $\theta = \text{const.}$ Then,

$$|y(z)| \leqslant \alpha + \beta t + \int_0^t (t - s)\, G(s, y(se^{i\theta}))\, ds, \tag{13.2.4}$$

where $s = |\xi|$, $t = |z|$. Also, since $u(t)$ satisfies (13.2.3), it follows that

$$u(t) = \alpha + \beta t + \int_0^t (t - s)\, u''(s)\, ds$$

$$\geqslant \alpha + \beta t + \int_0^t (t - s)\, G(s, u(s))\, ds, \tag{13.2.5}$$

from which, using (13.2.4),

$$u(t) - |y(z)| \geqslant \int_0^t (t - s)[G(s, u(s)) - G(s, |y(se^{i\theta})|)]\, ds.$$

Setting $m(t) = |y(te^{i\theta})|$ for fixed θ, it is easily checked, as in Theorem 13.1.2, that

$$|m_+'(t)| \leqslant \left|\frac{dy(z)}{dz}\right|.$$

Hence, in particular,

$$|m_+'(0)| \leqslant \left|\frac{dy(0)}{dz}\right| = \beta < \beta + \delta \qquad \text{for every} \qquad \delta > 0.$$

We consider the function $u(t, \delta)$ that satisfies (13.2.3) and

$$u(0, \delta) = \alpha, \qquad u'(0, \delta) = \beta + \delta.$$

13.2. SINGULARITY-FREE REGIONS

It is easy to show that $m(t) < u(t, \delta)$ for $0 \leqslant t < a$. Supposing the contrary, there exists a $t_1 > 0$ such that

$$m(t_1) = u(t_1, \delta) \tag{13.2.6}$$

and

$$m(t) \leqslant u(t, \delta), \qquad t \leqslant t_1. \tag{13.2.7}$$

Notice that $u(t) < u(t, \delta)$, $0 \leqslant t < a$. Using this fact, the monotony of $G(t, u)$ in u, the inequalities (13.2.5) and (13.2.7) lead to

$$u(t_1) - |y(t_1 e^{i\theta})| \geqslant \int_0^{t_1} (t_1 - s)[G(s, u(s)) - G(s, u(s, \delta))]\, ds$$
$$\geqslant 0.$$

In other words,

$$u(t_1, \delta) > u(t_1) \geqslant |y(t_1 e^{i\theta})| = m(t_1),$$

which contradicts (13.2.7). Thus, we have

$$m(t) = |y(te^{i\theta})| < u(t, \delta), \qquad 0 \leqslant |z| < a.$$

Hence, letting $\delta \to 0$, we arrive at the inequality

$$|y(z)| \leqslant u(t), \qquad 0 < |z| < a. \tag{13.2.8}$$

Since $u(t)$ is assumed to exist on $[0, a)$, the stated result follows from (13.2.8), and the proof is complete.

COROLLARY 13.2.1. The complex equation

$$\frac{d^2 y}{dz^2} + p(z) y^n = 0, \qquad n > 1, \tag{13.2.9}$$

has solutions, which are regular in $|z| < 1$ if

$$|p(z)| \leqslant \alpha(1 - |z|)^{\epsilon - 2}, \qquad \epsilon > 0. \tag{13.2.10}$$

Proof. In view of Theorem 13.2.1, it is sufficient to show that there exists a solution of

$$u'' = \frac{\alpha u^n}{(1-t)^{2-\epsilon}}, \tag{13.2.11}$$

which is continuous on $[0, 1)$. We consider whether (13.2.11) can have a solution of the form $\beta(1-t)^{-\mu}$, μ, $\beta > 0$. If such a solution exists, we must have

$$\frac{\mu(\mu+1)}{(1-t)^{\mu+2}} = \frac{\alpha\beta^{n-1}}{(1-t)^{2+n\mu-\epsilon}},$$

that is to say,

$$\mu(\mu+1) = \alpha\beta^{n-1} \quad \text{and} \quad \mu = n\mu - \epsilon.$$

Since $n > 1$, we obtain

$$\mu = \frac{\epsilon}{n-1} > 0, \quad \frac{\epsilon}{n-1}\left(\frac{\epsilon}{n-1}+1\right) = \alpha\beta^{n-1}.$$

If β is determined by the last equation, $u(t) = \beta(1-t)^{-\mu}$ will be a solution of (13.2.11) for which $u(0)$ and $u'(0)$ are positive. The proof is complete.

We shall next consider the complex differential system (13.1.1), where f is regular in z, $0 \leqslant |z| < a$ and entire in $y \in C^n$. The following theorem gives an upper bound of the norm of solutions of (13.1.1) along each ray $z = te^{i\theta}$.

THEOREM 13.2.2. Assume that

(i) $f \in C^n$, $f(z, y)$ is regular-analytic in z, $0 \leqslant |z| < a$, entire in $y \in C^n$, and, for each fixed θ, $0 \leqslant \theta < 2\pi$, and $z = te^{i\theta}$,

$$\|f(z, y)\| \leqslant g(|z|, \|y\|), \tag{13.2.12}$$

where $g \in C[[0, a) \times R_+, R_+]$;

(ii) $r(t)$ is the maximal solution of the scalar differential equation

$$u' = g(t, u), \quad u(0) = u_0 > 0, \tag{13.2.13}$$

whose maximal interval of existence is $[0, b(\theta))$, $b(\theta) \leqslant a$.

Then, every solution $y(z)$ of (13.1.1), such that $0 < \|y(0)\| \leqslant u_0$, is regular-analytic in a region that contains the set

$$E = [z : z = te^{i\theta}, 0 \leqslant t < b(\theta), 0 \leqslant \theta < 2\pi].$$

Furthermore,

$$\|y(te^{i\theta})\| \leqslant r(t)$$

on each ray $z = te^{i\theta}$, $0 \leqslant t < b(\theta)$.

Proof. Let $z = te^{i\theta}$, and fix θ, $0 \leq \theta < 2\pi$. Define

$$m(t) = \|y(te^{i\theta})\|,$$

where $y(z)$ is any solution of (13.1.1) such that $0 < \|y(0)\| \leq u_0$. Then, as in Theorem 13.1.2, we derive the differential inequality

$$D^+m(t) \leq \left\|\frac{dy(te^{i\theta})}{dt}\right\| = \left\|\frac{dy(te^{i\theta})e^{i\theta}}{dz}\right\|$$

$$= \left\|\frac{dy(z)}{dz}\right\| = \|f(z, y)\|$$

$$\leq g(|z|, \|y(z)\|)$$

$$= g(t, m(t)).$$

Moreover, $m(0) \leq u_0$. Hence, by Theorem 1.4.1, we have

$$m(t) = \|y(te^{i\theta})\| \leq r(t) \quad \text{on} \quad [0, b(\theta)).$$

Therefore, $y(z)$ is regular for $z = te^{i\theta}$, $0 \leq t < b(\theta)$, $0 \leq \theta < 2\pi$, and the proof is complete.

We state a corollary that generalizes Theorem 13.2.1.

COROLLARY 13.2.2. *If, in addition to the hypotheses of Theorem 13.2.2, we assume that the solutions of (13.2.13) are unique and there exist continuous functions $u(t)$ on $[0, \alpha(\theta))$, $\alpha(\theta) > 0$, $0 \leq \theta < 2\pi$, such that*

$$u'(t) \geq g(t, u(t)), \quad u(0) = u_0,$$

on $[0, \alpha(\theta))$, then every solution of (13.1.1) for which $\|y(0)\| \leq u_0$ is regular in a region containing the set

$$E = [z : z = te^{i\theta}, 0 \leq t < \alpha(\theta), 0 \leq \theta < 2\pi].$$

COROLLARY 13.2.3. *Let hypothesis (i) of Theorem 13.2.2 hold. Suppose that $\rho(t)$ is the minimal solution of $u' = -g(t, u)$, $u(0) = u_0 > 0$ existing on $[0, b(\theta))$. Then, every solution $y(z)$ of (13.1.1) such that $\|y(0)\| \geq u_0$ satisfies*

$$\|y(te^{i\theta})\| \geq \rho(t), \quad 0 \leq t < b(\theta).$$

THEOREM 13.2.3. *Let the assumptions of Theorem 13.2.2 hold, except that the condition (13.2.12) is replaced by a weaker condition,*

$$\|y + hf(z, y)\| \leq \|y\| + hg(|z|, \|y\|) + 0(h),$$

for sufficiently small $h > 0$. Then, the conclusion of the Theorem 13.2.2 remains valid.

It is easy to see that the weaker condition of Theorem 13.2.3 is sufficient to get the inequality

$$D^+m(t) \leqslant g(t, m(t)),$$

as in the proof of Theorem 13.2.2. In the present case, since $g(t, u)$ need no longer be nonnegative, the growth estimates are sharper.

13.3. Componentwise bounds

We shall first prove a result that offers componentwise bounds for solutions of the complex differential system (13.1.1).

THEOREM 13.3.1. Assume that

(i) $f \in C^n$, $f(z, y)$ is regular-analytic in z, $0 \leqslant |z| < a$, entire in $y \in C^n$, and, for each fixed θ, $0 \leqslant \theta < 2\pi$, and $z = te^{i\theta}$,

$$|f(z, y)| \leqslant g(|z|, |y|), \qquad (13.3.1)$$

where $g \in C[[0, a) \times R_+^n, R_+^n]$, $g(t, u)$ is quasi-monotone nondecreasing in u for each $t \in [0, a)$;

(ii) the maximal solution $r(t, 0, u_0)$ of the ordinary differential system

$$u' = g(t, u), \qquad u(0) = u_0 \geqslant 0,$$

exists on $[0, a)$.

Then, every solution $y(z)$ of (13.1.1) such that $|y(0)| \leqslant u_0$ satisfies

$$|y(te^{i\theta})| \leqslant r(t, 0, u_0), \qquad t \geqslant 0,$$

where $|z| = t$.

Proof. Let $z = te^{i\theta}$, and we fix θ, $0 \leqslant \theta < 2\pi$. Define the vector function

$$m(t) = |y(te^{i\theta})|,$$

where $y(z)$ is any solution of (13.1.1) such that $|y(0)| \leqslant u_0$. Proceeding as in the proof of Theorem 13.1.2 with obvious modifications, it is easy to obtain the differential inequality

$$D^+m(t) \leqslant g(t, m(t)).$$

Corollary 1.7.1 now assures the stated componentwise bounds.

Analogous to Theorem 13.2.3, we can state a theorem for componentwise bounds which yields sharper bounds in some situations.

THEOREM 13.3.2. Let the condition (13.3.1) in Theorem 13.3.1 be replaced by

$$|y + hf(z, y)| \leq |y| + hg(|z|, |y|) + 0(h)$$

for all small $h > 0$, where $g \in C[[0, a) \times R_+^n, R^n]$, and $g(t, u)$ is quasi-monotone nondecreasing in u for each $t \in [0, a)$, other assumptions remaining the same. Then, the conclusion of Theorem 13.3.1 is true.

Instead of the complex differential inequality (13.1.2), we shall consider the system of inequalities

$$\left| \frac{dy}{dz} - f(z, y) \right| \leq \epsilon, \qquad (13.3.2)$$

where ϵ is a positive vector. Definition 13.1.1 has to be slightly modified in an obvious way. Corresponding to Theorem 13.1.2, we have the following

THEOREM 13.3.3. Let $g \in C[[0, a) \times R_+^n, R_+^n]$, $g(t, u)$ be quasi-monotone nondecreasing in u for each $t \in [0, a)$, and $r(t)$ be the maximal solution of the system

$$u' = g(t, u) + \epsilon, \qquad u(0) = u_0 \geq 0$$

existing on $[0, a)$. Suppose further that $f(z, y)$ is regular-analytic in D and

$$|f(z, y_1) - f(z, y_2)| \leq g(|z|, |y_1 - y_2|).$$

If $y_1(z, \epsilon_1)$ and $y_2(z, \epsilon_2)$ are ϵ_1- and ϵ_2-approximate solutions of (13.1.1) such that

$$|y_1(0, \epsilon_1) - y_2(0, \epsilon_2)| \leq u_0,$$

then we have, on each ray $z = te^{i\theta}$ and $0 \leq |z| < a$,

$$|y_1(z, \epsilon_1) - y_2(z, \epsilon_2)| \leq r(t), \qquad t \geq 0, \qquad |z| = t.$$

When $\epsilon_1 = 0$, $\epsilon_2 = 0$, $y_1(z)$ and $y_2(z)$ are any two solutions of (13.1.1), and Theorem 13.3.3 gives componentwise upper bounds for the difference of these solutions.

13.4. Lyapunov-like functions and comparison theorems

Let D denote the region of the complex plane $|z| \geq a$ and $\alpha \leq \arg z \leq \beta$, where a, α, β are real numbers. We consider the complex differential system

$$y' = f(z, y), \quad y(z_0) = y_0, \quad z_0 \in D, \qquad (13.4.1)$$

where y and f are n vectors, and the function $f(z, y)$ is regular-analytic in z on D and entire in $y \in C^n$.

As before, we shall denote $|z|$ by t and $\arg z$ by θ. Moreover, let $|\xi| = (|\xi_1|, |\xi_2|, ..., |\xi_n|)$.

Let a function $V(z, y)$, where V and y are n vectors be defined, regular-analytic in z on D and entire in $y \in C^n$. The ith component of V will be denoted by $V_i(z, y)$ or synonymously by $V_i(z, y_1, y_2, ..., y_n)$ whenever necessary. We define

$$V^*(z, y) = \frac{\partial V}{\partial z} + \frac{\partial V}{\partial y} \cdot f(z, y). \qquad (13.4.2)$$

With respect to the functions defined previously, we state the following comparison theorems.

THEOREM 13.4.1. Assume that

(i) $g \in C[J \times R_+^n, R_+^n]$, $g(t, u)$ is quasi-monotone nondecreasing in u for each t, and

$$|V^*(z, y)| \leq g(|z|, |V_1(z, y)|, ..., |V_n(z, y)|); \qquad (13.4.3)$$

(ii) $r(t) = r(t, t_0, u_0)$ is the maximal solution of the ordinary differential system

$$u' = g(t, u), \quad u(t_0) = u_0, \qquad (13.4.4)$$

existing on $[t_0, \infty]$;

(iii) $y(z)$ is any solution of (13.4.1) such that

$$|V(z_0, y(z_0))| \leq r(t_0), \quad |z_0| = t_0. \qquad (13.4.5)$$

Then we have, for $z \in D$,

$$|V(z, y(z))| \leq r(t), \quad t \geq t_0, \qquad (13.4.6)$$

where $|z| = t$.

13.4. LYAPUNOV-LIKE FUNCTIONS

THEOREM 13.4.2. Let assumptions (i) and (ii) of Theorem 13.4.1 hold except that the condition (13.4.3) is replaced by

$$\left| V^*(z, y) \, p(z) + V(z, y) \frac{dp(z)}{dz} \right|$$
$$\leqslant g(|z|, |V_1(z, y) \, p(z)|, \ldots, |V_n(z, y) \, p(z)|), \qquad (13.4.7)$$

where $p(z)$ is regular-analytic in z on D. If $y(z)$ is any solution of (13.4.1) satisfying

$$|V(z_0, y(z_0)) \, p(z_0)| \leqslant r(t_0), \qquad |z_0| = t_0, \qquad (13.4.8)$$

then

$$|V(z, y(z)) \, p(z)| \leqslant r(t), \qquad z \in D, \qquad |z| = t, \qquad (13.4.9)$$

for all $t \geqslant t_0$.

We shall prove below Theorem 13.4.2, since Theorem 13.4.1 can be deduced from Theorem 13.4.2 by taking $p(z) = 1$. We have stated Theorem 13.4.1 separately, as it is a basic comparison theorem by itself.

Proof of Theorem 13.4.2. Define

$$L(z, y(z)) = V(z, y(z)) \, p(z),$$

where $y(z)$ is any solution of (13.4.1) verifying (13.4.8). For each fixed θ, set

$$m(t) = |L(te^{i\theta}, y(te^{i\theta}))|.$$

Then, if $h > 0$ is sufficiently small,

$$m(t + h) - m(t) \leqslant |L((t+h)e^{i\theta}, y((t+h)e^{i\theta})) - L(te^{i\theta}, y(te^{i\theta}))|.$$

We can easily verify that

$$\left| \frac{dm(t)}{dt} \right| \leqslant \left| \frac{dL(te^{i\theta}, y(te^{i\theta}))}{dt} \right|.$$

Also,

$$\left| \frac{dL(te^{i\theta}, y(te^{i\theta}))}{dt} \right| = \left| \frac{dL(z, y(z))e^{i\theta}}{dz} \right|$$
$$= \left| \frac{dL(z, y(z))}{dz} \right|.$$

It therefore follows from the foregoing considerations that

$$D^+m(t) \leq \left|\frac{dL(z, y(z))}{dz}\right|$$

$$\leq \left|V^*(z, y(z))\, p(z) + V(z, y(z))\, \frac{dp(z)}{dz}\right|$$

$$\leq g(t, m_1(t), \ldots, m_n(t)).$$

Now a straightforward application of Corollary 1.7.1 yields the desired inequality (13.4.9).

13.5. Notes

The results of Sect. 13.1 are due to Deo and Lakshmikantham [1]. Theorem 13.2.1 and Corollary 13.2.1 are taken from the work of Das [1]. See also Das [4]. Theorem 13.2.2 is due to Wend [2], whereas Theorem 13.2.3 is new. The results of Sects. 13.3 and 13.4 are adapted from the work of Kayande and Lakshmikantham [1]. For further results, see Deo and Lakshmikantham [2] and Kayande and Lakshmikantham [1], where stability and boundedness criteria are discussed.

Bibliography

AGMON, S., AND NIRENBERG, L.

[1] Properties of solutions of ordinary differential equations in Banach spaces, *Comm. Pure Appl. Math.* **16** (1963), 121–239.

ALEXIEWICZ, A., AND ORLICZ, W.

[1] Some remarks on the existence and uniqueness of solutions of the hyperbolic equation $\partial^2 z/\partial x\, \partial y = f(x, y, z, \partial z/\partial x, \partial z/\partial y)$, *Studia Math.* **15** (1956), 201–215.

ANTOSIEWICZ, H. A.

[1] Continuous parameter dependence and the method of averaging, *Proc. Int. Symp. Nonlinear Oscillations, 2nd,* Izd. Akad. Nauk. Ukrain, SSR, Kiev, 1963, pp. 51–58.

ARONSON, D. G.

[1] On the initial value problem for parabolic systems of differential equations, *Bull. Amer. Math. Soc.* **65** (1959), 310–318.

[2] Uniqueness of solutions of the initial value problem for parabolic systems of differential equations, *J. Math. Mech.* **11** (1962), 403–420.

[3] Uniqueness of positive weak solutions of second order parabolic equations, *Ann. Polon. Math.* **16** (1965), 285–303.

ARONSON, D. G., AND BESALA, P.

[1] Uniqueness of solutions of the Cauchy problem for parabolic equations, *J. Math. Anal. Appl.* **13** (1966), 516–526.

AZIZ, A. K.

[1] A functional integral equation with applications to hyperbolic partial differential equations, *Duke Math. J.* **32** (1965), 579–592.

[2] Periodic solutions of hyperbolic partial differential equations, *Proc. Amer. Math. Soc.* **17** (1966), 557–566.

AZIZ, A. K., AND DIAZ, J.

[1] On a mixed boundary value problem for linear hyperbolic partial differential equations in two independent variables, *Arch. Rational Mech. Anal.* **10** (1962), 1–28.

Aziz, A. K., and Maloney, J. P.
   [1] An application by Tychonoff's fixed point theorem to hyperbolic partial differential equations, *Math. Ann.* **162** (1965), 77–82.

Barrar, R. B.
   [1] Some estimates for solutions of parabolic equations, *J. Math. Anal. Appl.* **3** (1961), 373–397.

Bellman, R.
   [1] On the boundedness of solutions of nonlinear differential and difference equations, *Trans. Amer. Math. Soc.* **62** (1947), 357–386.
   [2] On the existence and boundedness of solutions of nonlinear partial differential equations of parabolic type, *Trans. Amer. Math. Soc.* **64** (1948), 21–44.
   [3] On the existence and boundedness of solutions of nonlinear differential–difference equations, *Ann. of Math.* **50** (1949), 347–355.
   [4] The stability theory of differential–difference equations, RAND Corp. Paper No. P-381, 1953.
   [5] Terminal control, time lags, and dynamic programming, *Proc. Nat. Acad. Sci. U.S.A.* **43** (1957), 927–930.
   [6] Asymptotic series for the solutions of linear differential–difference equations, *Rend. Circ. Mat. Palermo* [2] **7** (1958), 1–9.
   [7] On the computational solution of differential–difference equations, *J. Math. Anal. Appl.* **2** (1961), 108–110.

Bellman, R., and Cooke, K. L.
   [1] Stability theory and adjoint operators for linear differential–difference equations, *Trans. Amer. Math. Soc.* **92** (1959), 470–500.
   [2] Asymptotic behavior of solutions of differential–difference equations, *Mem. Amer. Math. Soc.* **35** (1959).
   [3] On the limit of solutions of differential–difference equations as the retardation approaches zero, *Proc. Nat. Acad. Sci. U.S.A.* **45** (1959), 1026–1028.
   [4] "Differential–Difference Equations." Academic Press, New York, 1963.

Bellman, R., and Danskin, J. M.
   [1] The stability theory of differential–difference equations, *Proc. Symp. Nonlinear Circuit Anal., New York, 1953*, pp. 107–123.
   [2] A survey of mathematical theory of time lag, retarded control, and hereditary processes, RAND Corp. Rep. No. R-256, 1956.

Bellman, R., Danskin, J. M., and Glicksberg, I.
   [1] A bibliography of the theory and application of differential–difference equations, RAND Corp. Res. Memo. No. RM-688, 1952.

Besala, P. (see Aronson, D. G.)
   [1] On solutions of nonlinear parabolic equations defined in unbounded domains, *Bull. Acad. Polon. Sci. Sér. Sci. Math. Astronom. Phys.* **9** (1961), 531–535.
   [2] On solution of Fourier's first problem for a system of nonlinear parabolic equations in an unbounded domain, *Ann. Pol. Math.* **13** (1963), 247–265.

[3] Concerning solutions of an exterior boundary-value problem for a system of non-linear parabolic equations, *Ann. Pol. Math.* **14** (1964), 289–301.

[4] Limitations of solutions of non-linear parabolic equations in unbounded domains, *Ann. Pol. Math.* **17** (1965), 25–47.

BHATIA, N. P.

[1] Stability and Lyapunov functions in dynamical systems, *Cont. Differential Equations* **3** (1964), 175–188.

[2] On exponential stability of linear differential systems, *J. SIAM Control. Ser. A.* **2** (1965), 181–191.

[3] Weak attractors in dynamical systems, *Bol. Soc. Mat. Mexicana* **11** (1966), 56–64.

[4] On asymptotic stability in dynamical systems, *Math. Systems Theory* **1** (1967), 113–127.

BHATIA, N. P., AND SZEGO, G. P.

[1] Weak attractors in R^n, *Math. Systems Theory* **1** (1967), 129–133.

[2] "Dynamical Systems: Stability Theory and Applications" (Lecture Notes in Math., No. 35). Springer, New York, 1967.

BIELECKI, A.

[1] Une remarque sur l'application de la methode de Banach–Cacciopoli–Tikhonov dans la théorie de l'équation $s = f(x, y, z, p, q)$, *Bull. Acad. Polon. Sci. Cl. III* **4** (1956), 265–268.

BOGDANOWICZ, W. M., AND WELCH, J. N.

[1] On a linear operator connected with almost periodic solutions of linear differential equations in Banach spaces, *Math. Ann.* **172** (1967), 327–335.

BROWDER, F. E.

[1] Non-linear parabolic boundary value problems of arbitrary order, *Bull. Amer. Math. Soc.* **69** (1963), 858–861.

[2] Nonlinear equations of evolution, *Ann. of. Math.* **80** (1964), 485–523.

[3] Strongly non-linear parabolic boundary value problems, *Amer. J. Math.* **85** (1964), 339–357.

[4] Nonlinear monotone operators and convex sets in Banach spaces, *Bull. Amer. Math. Soc.* **71** (1965), 780–785.

BRZYCHCZY, S.

[1] Some theorems on second order partial differential inequalities of parabolic type, *Ann. Polon. Math.* **15** (1964), 143–151.

[2] Extensions of Chaplighin's method to the system of non-linear parabolic equations in an unbounded domain, *Bull. Acad. Polon. Sci. Sér. Sci. Math. Astronom. Phys.* **13** (1965), 27–30.

BULLOCK, R. M.

[1] A uniqueness theorem for a delay-differential system, *SIAM Rev.* **9** (1967), 737–740.

CALDERON. A. P.
- [1] Uniqueness in the Cauchy problem for partial differential equations, *Amer. J. Math.* **80** (1958), 16–36.

CAMERON, R. H.
- [1] Nonlinear Volterra functional equations and nonlinear parabolic differential systems, *J. Analyse Math.* (1956/57), 136–182.

CESARI, L.
- [1] Existence in the large of periodic solutions of hyperbolic partial differential equations, *Arch. Rational Mech. Anal.* **3** (1965), 170–190.
- [2] Periodic solutions of hyperbolic partial differential equations, *Intern. Symp. Nonlinear Eqs. and Nonlinear Mechanics*, Academic Press, New York, 1963. pp. 36–57.

CHANDRA, J., AND FLEISHMAN, B. A.
- [1] Positive solution of nonlinear equations, *Proc. U.S.–Japan Sem. Differential and Functional Eqs. Minneapolis*, Benjamin, New York, 1967, pp. 435–442.
- [2] Bounds and maximal solutions of nonlinear functional equations, *Bull. Amer. Math. Soc.* **74** (1968), 512–516.

CHANG, H.
- [1] On the stability of systems of differential equations with time lag, *Acta Math. Sinica.* **10** (1960), 202–211.

CHU, S. C.
- [1] On a mixed boundary value problem for linear third order hyperbolic partial differential equations, *Rend. Circ. Mat. Palermo* [2] **13** (1964), 199–208.

CILIBERTO, C.
- [1] Sul problema di Darboux per l'equazione $s = f(x, y, z, p, q)$, *Rend. Acad. Sci. Fis. Mat. Napoli* **22** (1956), 221–225.
- [2] Sulle equazioni quasi-lineari di tipo parabolico in due variabili, *Ricerche Mat.* **5** (1956b), 97–125.
- [3] Sull'approxximazione delle soluzioni del problema di Darboux per l'equazione $s = f(x, y, z, p, q)$, *Ricerche Mat.* **10** (1961), 106–138.
- [4] Teoremi di confronto e di unicita per le soluzioni del problema di Darboux, *Ricerche Mat.* **10** (1961a), 214–243.
- [5] Sul problema di Darboux, *Atti Acad. Naz. Lincei. Rend. Cl. Sci. Fis. Mat. Natur.* **30** (1961b), 460–466.

COFFMAN, C. V.
- [1] Linear differential equations on cones in Banach spaces, *Pacific J. Math.* **12** (1962), 69–75.
- [2] Nonlinear differential equations on cones in Banach spaces, *Pacific J. Math.* **14** (1964), 9–16.
- [3] Asymptotic behavior of solutions of ordinary difference equations, *Trans. Amer. Math. Soc.* **110** (1964), 22–51.

COHEN, P., AND LEES, M.
  [1] Asymptotic decay of solutions of differential inequalities, *Pacific J. Math.* **11** (1961), 1235–1249.

CONLAN, J.
  [1] The Cauchy problem and the mixed boundary value problem for a nonlinear hyperbolic partial differential equation in two independent variables, *Arch. Rational Mech. Anal.* **3** (1959), 355–380.
  [2] An existence theorem for the equation $u_{xyz} = f$, *Arch. Rational Mech. Anal.* **9** (1962), 64–76.

CONTI, R.
  [1] Sul problema di Darboux per l'equazione $z_{xy} = f(x, y, z, z_x, z_y)$, *Ann. Univ. Ferrara Sez. VII* (N.S.) **2** (1953), 129–140.

COOKE, K. L. (see Bellman, R.)
  [1] The asymptotic behavior of the solutions of linear and nonlinear differential–difference equations, *Trans. Amer. Math. Soc.* **75** (1953), 80–105.
  [2] The rate of increase of real continuous solutions of algebraic differential–difference equations of the first order, *Pacific J. Math.* **4** (1954), 483–501.
  [3] Forced periodic solutions of a stable nonlinear differential–difference equation, *Ann. of Math.* **61** (1955), 381–387.
  [4] A symbolic method for finding integrals of linear difference and differential equations, *Math. Mag.* **31** (1958), 121–126.
  [5] Functional–differential equations: some models and perturbation problems, *Proc. Intern. Symp. Differential Eqs. and Dynamical Systems, Puerto Rico, 1965*, Academic Press, New York, 1967, pp. 167–183.
  [6] Functional differential equations close to differential equations, *Bull. Amer. Math. Soc.* **72** (1966), 285–288.
  [7] Functional differential equations with asymptotically vanishing lag, *Rend. Circ. Mat. Palermo* **16** (1967), 39–56.
  [8] Asymptotic theory for the delay-differential equation $u'(t) = -au(t - r(u(t)))$, *J. Math. Anal. Appl.* **19** (1967), 160–173.

CORDUNEANU, C.
  [1] Théorèmes d'existence globale pour les systèmes différentielles à argument retarde, *Studii Cere. Mat. Iasi* **12** (1961), 251–258.
  [2] Sur la stabilité des systèmes perturbes à argument retarde, *An. Sti. Univ. "Al. I. Cuza" Iasi* **11** (1965), 99–105.

DAS, K. M.
  [1] Singularity-free regions for solutions of second order non-linear differential equations, *J. Math. Mech.* **13** (1964), 73–84.
  [2] Comparison and monotonicity theorems for second order non-linear differential equations, *Acta Math. Acad. Sci. Hungar.* **15** (1964), 449–456.
  [3] Properties of solutions of certain non-linear differential equations, *J. Math. Anal. Appl.* **8** (1964), 445–451.
  [4] Bounds for solutions of second order complex differential equations, *Proc. Amer. Math. Soc.* **18** (1967), 220–225.

DEO, S. G., AND LAKSHMIKANTHAM, V.
- [1] On complex differential inequalities, *Bol. Soc. Mat. Sao Paulo* **18** (1966), 3–8.
- [2] Complex differential inequalities and extension of Lyapunov method, *Proc. Nat. Acad. Sci. India Sect. A* **36** (1966), 217–242.
- [3] Conditional stability and functional differential equations. *Proc. Symp. Anal., Annamalai Univ., India, Math. Student* (Special Issue) (1966).
- [4] Functional differential inequalities, *Bull. Calcutta Math. Soc.* **58** (1966), 89–95.

DIAZ, J. B. (see Aziz, A. K.)
- [1] On an analogue of the Euler–Cauchy polygon method for the numerical solution of $u_{xy} = f(x, y, u, u_x, u_y)$, *Arch. Rational Mech. Anal.* **1** (1958), 357–390.
- [2] On existence, uniqueness, and numerical evaluation of solutions of ordinary and hyperbolic differential equations, *Ann. Mat. Pura Appl.* **52** (1960), 163–181.

DRIVER, R. D.
- [1] Delay-differential equation and an application to a two-body problem of classical electrodynamics, Tech. Rep., Univ. of Minnesota, Minneapolis, 1960.
- [2] Existence theory for a delay-differential system, *Contr. Differential Equations* **1** (1961), 317–336.
- [3] Existence and stability of solutions of a delay-differential system, *Arch. Rational Mech. Anal.* **10** (1962), 401–426.
- [4] Existence and continuous dependence of solutions of a neutral functional–differential equation, *Arch. Rational Mech. Anal.* **19** (1965), 149–166.
- [5] On Ryabov's asymptotic characterization of solutions of quasi-linear differential equations with small delays. *SIAM Rev.* **10** (1968), 329–341.

EDMUNDS, D. E.
- [1] Differential inequalities, *Proc. London Math. Soc.* **15** (1965), 361–372.

EIDELMAN, S. D.
- [1] Bounds for solutions of parabolic systems and some applications, *Mat. Sb.* **33** (1953), 359–382.
- [2] On the connection between the fundamental matrix-solution of parabolic and elliptic systems, *Mat. Sb.* **35** (1954), 57–72.
- [3] On the fundamental solution of parabolic systems, *Mat. Sb.* **38** (1956), 51–92.
- [4] The fundamental matrix of general parabolic systems, *Dokl. Akad. Nauk SSSR* (N.S.), **120** (1958), 980–983.
- [5] On the fundamental solution of parabolic systems, *Mat. Sb.* **95** (1961), 73–136.

EL'SGOL'TS, L. E.
- [1] An approximate method of solution of differential equations with a retarded argument, *Uspehi Mat. Nauk* **6** (1951), 1–30.
- [2] On approximate integration of differential equations with retarded argument, *Prikl. Mat. Meh.* **15** (1951), 771–772.
- [3] Some questions in the theory of differential–difference equations, *Uspehi Mat. Nauk* **7** (1952), 147–148.

[4] Stability of solutions of differential–difference equations, *Uspehi Mat. Nauk* **9** (1954), 95–112.

[5] On the integration of linear partial differential equations with retarded argument, *Moskov Gos. Univ. Ucen. Zap. Mat.* **181** (1956), 57–58.

[6] On the theory of stability of differential equations with perturbed arguments, *Vestnik Moskov. Univ. Ser. Mat. Meh. Astronom Fiz. Him.* **5** (1959), 65–72.

[7] Qualitative methods in mathematical analysis (trans. from Russian), *Amer. Math. Soc. Providence, 1964.*

El'sgol'ts, L. E., and Myshkis, A. D.

[1] Some results and problems in the theory of differential equations, *Russian Math. Surveys* **22** (1967), 19–57.

El'sgol'ts, L. E., Myshkis, A. D., and Shimanov, S. N.

[1] Stability and oscillations of systems with time lag, *Int. Union Theoret. Appl. Mech. Symp. Nonlinear Vibrations*, Ukr. SSR Acad. Sci. Publ., Kiev, 1961.

El'sgol'ts, L. E., Zverkin, A. M., Kamenskii, G. A., and Norkin, S. B.

[1] Differential equations with a perturbed argument, *Russian Math. Surveys* **17** (1962), 61–146.

[2] Differential equations with retarded arguments, *Uspehi Mat. Nauk* **17** (1962), 77–164.

Fodcuk, V. I.

[1] Existence and properties of an integral manifold for nonlinear systems of equations with retarded argument and variable coefficients. Approximative methods of solving differential equations, *Izv. Akad. Nauk SSSR* (1963), 129–140.

Foias, C., Gussi, G., and Poenaru, V.

[1] Une méthode directe dans l'étude des équations aux derivées partielles hyperboliques quasilinéaires en deux variables, *Math. Nachr.* **15** (1956), 89–116.

[2] Verallgemeinerte Losungen fur parabolische quasilineare systems, *Math. Nachr.* **17** (1958), 1–8.

[3] Sur les solutions generalisées de certaines équations linéaires et quasi-linéaires dans l'espace de Banach, *Rev. Math. Pure Appl.* **3** (1958), 283–304.

Franklin, J.

[1] On the existence of solutions of functional differential equations, *Proc. Amer. Math. Soc.* **5** (1954), 363–366.

Friedman, A.

[1] Classes of solutions of linear systems of partial differential equations of parabolic type, *Duke Math. J.* **24** (1957), 443–442.

[2] On the regularity of solutions of nonlinear elliptic and parabolic systems of partial differential equations, *J. Math. Mech.* **7** (1958), 43–60.

[3] Interior estimates for parabolic systems of partial differential equations, *J. Math. Mech.* **7** (1958), 393–418.

[4] Boundary estimates for second order parabolic equations and their applications, *J. Math. Mech.* **7** (1958), 771–792.

[5] On quasi-linear parabolic equations of the second order, *J. Math. Mech.* **7** (1958), 793–809.

[6] Remarks on the maximum principle for parabolic equations and its applications, *Pacific J. Math.* **8** (1958), 201–211.

[7] Convergence of solutions of parabolic equations to a steady state, *J. Math. Mech.* **8** (1959), 57–76.

[8] Asymptotic behavior of solutions of parabolic equations, *J. Math. Mech.* **8** (1959), 387–392.

[9] Free boundary problems for parabolic equations. I. Melting of Solids, *J. Math. Mech.* **8** (1959), 499–518.

[10] Parabolic equations of the second order, *Trans. Amer. Math. Soc.* **93** (1959), 509–530.

[12] On the uniqueness of the Cauchy problem for parabolic equations, *Amer. J. Math.* **81** (1959b), 503–511.

[13] Free boundary problems for parabolic equations. II. Evaporations or condensation of a liquid drop, *J. Math. Mech.* **9** (1960), 19–66.

[14] Free boundary problems for parabolic equations. III. Dissolution of a gas bubble in liquid, *J. Math. Mech.* **9** (1960), 327–345.

[15] On quasi-linear parabolic equations of the second order. II, *J. Math. Mech.* **9** (1960a), 539–556.

[16] Remarks on Stefan-type free boundary problems for parabolic equations, *J. Math. Mech.* **9** (1960), 885–904.

[17] Mildly nonlinear parabolic equations with applications to flow of gases through porous media, *Arch. Rational Mech. Anal.* **5** (1960), 238–248.

[18] Asymptotic behavior of solutions of parabolic equations of any order, *Acta Math.* **106** (1961), 1–43.

[19] A strong maximum principle for weakly subparabolic functions, *Pacific J. Math* **11** (1961), 175–184.

[21] "Generalized Functions and Partial Differential Equations." Prentice-Hall, Englewood Cliffs, New Jersey, 1963.

[22] "Partial Differential Equations of Parabolic Type." Prentice-Hall, Englewood Cliffs, New Jersey, 1964.

[23] Asymptotic behavior of solutions of parabolic differential equations and of integral equations, *Proc. Intern. Symp. Differential Eqs. and Dynamical Systems, Puerto Rico, 1965*, pp. 409–426.

[24] Remarks on nonlinear parabolic equations, *Proc. Symp. Appl. Math.* **17** (1965), 3–23.

GIULIANO, L.

[1] Sull'unicita della soluzione per una classe di equazioni differenziali alle derivate parziali, paraboliche, non lineari, *Atti Accad. Naz. Lincei Rend. Cl. Sci. Fis. Mat. Natur.* **12** (1952), 260–265.

GUGLIELMINO, F.

[1] Sulla risoluzione del problema de Darboux per l'equazione $s = f(x, y, z)$, *Boll. Un. Mat. Ital.* **13** (1958), 308–318.

[2] Sul problema di Darboux, *Richerche Mat.* **8** (1959), 180–196.

[3] Sull'esistenza delle soluzioni dei problemi realtive alle equazioni non lineari di tipo iperbolico in due variabili, *Matematiche* **14** (1959), 67–80.

[4] Sul problema di Goursat, *Ricerche Mat.* **8** (1960), 91–105.

HAHN, W.

[1] "Stability of Motion." Springer, New York, 1967.

HALANAY, A.

[1] Some qualitative questions in the theory of differential equations with a delayed argument, *Rev. Math. Pures Appl.* **11** (1957), 127–144.

[2] Solutions périodiques des systèmes linéaires à argument retarde, *C. R. Acad. Sci. Paris Sér. A–B* (1959), 2708–2709.

[3] The method of averaging in equations with delay, *Rev. Math. Pures Appl.* **4** (1959), 467–483.

[4] Periodic and almost periodic solutions of systems of differential equations with retarded arguments, *Rev. Math. Pures Appl.* **4** (1959), 685–692.

[5] Stability criterions for systems of differential equations with a delayed argument, *Rev. Math. Pures Appl.* **5** (1960), 367–374.

[6] Integral stability for systems of differential equations with a delayed argument, *Rev. Math. Pures Appl.* **5** (1960), 541–548.

[7] Sur les systèmes d'équations différentielles linéaires à argument retarde, *C. R. Acad. Sci. Paris Sér. A–B* **250** (1960), 797–798.

[8] Solutions périodiques des systèmes généraux à retardement, *C. R. Acad. Sci. Paris Sér. A–B* **250** (1960), 3557–3559.

[9] Almost periodic solutions of systems of differential equations with a delayed argument, *Rev. Math. Pures Appl.* **5** (1960), 75–79.

[10] Differential inequalities with time-lag and an application to a problem of the stability theory for systems with time-lag, *Com. Acad. R. P. Romaine* **11** (1961), 1305–1310.

[11] Periodic solutions of linear systems with delay, *Rev. Math. Pures Appl.* **6** (1961), 141–158.

[12] Stability theory of linear periodic systems with delay, *Rev. Math. Pures Appl.* **6** (1961), 633–653.

[13] Perturbations singulières des systèmes à retardement, *C. R. Acad. Sci. Paris Ser. A–B*, **253** (1961), 1659–1650.

[14] Singular perturbations of systems with time lag, *Rev. Math. Pures Appl.* **7** (1962), 627–631.

[15] Asymptotic stability and small perturbations of periodic systems of differential equations with retarded arguments, *Uspehi Mat. Nauk* **17** (1962), 231–233.

[16] Periodic and almost-periodic solutions of some singularly perturbed systems with time lag, *Rev. Math. Pures Appl.* **8** (1963), 397–403.

[17] Almost-periodic solutions of linear systems with time lag, *Rev. Math. Pures Appl.* **9** (1964), 71–79.

[18] Systèmes à retardement. Résultats et problèmes, *3rd Conf. Nonlinear Vibrations, East Berlin, 1964*, Akademie-Verlag, Berlin, 1965.

[19] Periodic invariant manifold for a class of systems with time lag, *Rev. Roumaine Math. Pures Appl.* **10** (1965), 251–259.

[20] Invariant manifolds for systems with time lag, *Proc. Intern. Symp. Differential Eqs. and Dynamical Systems, Puerto Rico,* 1965, pp. 199–213.

[21] An invariant surface for some singularly perturbed systems with time lag, *J. Differential Equations* **2** (1966), 33–46.

[22] "Differential Equations." Academic Press, New York, 1966.

HALE, J. K.

[1] Asymptotic behavior of the solutions of differential–difference equations, RIAS Tech. Rep. No. 61-10, 1961.

[2] Functional-differential equations with parameters, *Contr. Differential Equations* **1** (1962), 401–410.

[3] A class of functional-differential equations, *Contr. Differential Equations* **1** (1962), 411–423.

[4] Linear functional-differential equations with constant coefficients, *Contr. Differential Equations* **2** (1963), 291–317.

[5] A stability theorem for functional-differential equations, *Proc. Nat. Acad. Sci. U.S.A.* **50** (1963), 942–946.

[6] Periodic and almost periodic solutions of functional-differential equations, *Arch. Rational Mech. Anal.* **15** (1964), 289–304.

[7] Geometric theory of functional-differential equations, *Proc. Intern. Symp. Differential Eqs. and Dynamical Systems, Puerto Rico, 1965,* pp. 247–266.

[8] Sufficient conditions for stability and instability of autonomous functional-differential equations, *J. Differential Equations* **1** (1965), 452–482.

[9] Averaging methods for differential equations with retarded arguments and a small parameter, *J. Differential Equations* **2** (1966), 57–73.

[10] Linear asymptotically autonomous functional differential equations, *Rendi. cinc. Matem. Palermo* **15** (1966), 331–351.

[11] Periodic solutions of a class of hyperbolic equations containing a small parameter, *Arch. Rational Mech. Anal.* **23** (1967), 380–398.

HALE, J. K., AND PERELLO, C.

[1] The neighborhood of a singular point of functional–differential equations, *Contr. Differential Equations* **3** (1964), 351–375.

HARTMAN, P.

[1] Generalized Lyapunov functions and functional equations, *Ann. Mat. Pura Appl.* **69** (1965), 305–320.

[2] On homotopic harmonic maps, *Canad. J. Math.* **19** (1967), 673–687.

[3] On hyperbolic partial differential equations, *Amer. J. Math.* **74** (1952), 834–864.

HASTINGS, S. P.

[1] Variation of parameters for non-linear differential difference equations, *Proc. Amer. Math. Soc.* **19** (1968), 1211–1216.

[2] Backward existence and uniqueness for retarded functional differential equations, *J. Differential Eqs.* **5** (1969), 441–451.

ILIN, A. M., KALASHNIKOV, A. S., AND OLEINIK, O. A.

[1] Linear second order parabolic equations, *Uspehi Math. Nauk* **17** (1962), 3–146.

Ito, S.
[1] Fundamental solutions of parabolic differential equations and boundary value problems, *Japan J. Math.* **27** (1957), 55–102.

Jones, G. S.
[1] The existence of periodic solutions of $f'(x) = f(x-1)(1+f(x))$, *J. Math. Anal. Appl.* **5** (1962), 435–450.
[2] Asymptotic fixed point theorems and periodic systems of functional-differential equations, *Contr. Differential Equations* **2** (1963), 385–405.
[3] Periodic motions in Banach space and applications to functional-differential equations, *Contr. Differential Equations* **3** (1964), 75–106.
[4] Hereditary dependence in dynamics, *Proc. Intern. Symp. Differential Eqs. and Dynamical Systems, Puerto Rico, 1965*, pp. 189–198.
[5] Hereditary dependence in the theory of differential equations, I, II, Univ. of Maryland Reps 1965.

Kakutani, S., and Markus, L.
[1] On the nonlinear difference–differential equation $y'(t) = [A + By(t-\gamma)] y(t)$, *Contr. Theory Nonlinear Oscillations, Princeton* **4** (1958), 411–418.

Kamenskii, G. A. (see El'sgol'ts, L. E., Zverkin, A. M., and Norkin, S. B.)
[1] Asymptotic behavior of solutions of linear differential equations of the second order with lagging argument, *Moskov. Gos. Univ. Učen. Zap. Mat.* **165** (1954), 195–204.
[2] Existence and uniqueness of solutions of differential–difference equations of neutral type, *Moskov. Gos. Univ. Učen. Zap. Mat.* **181** (1956), 83–89.
[3] On the general theory of equations with perturbed argument, *Dokl. Akad. Nauk SSSR* **120** (1958), 697–700.
[4] Existence, uniqueness, and continuous dependence on initial values of the solutions of systems of differential equations with deviating argument of neutral type, *Mat. Sb.*(N.S.) **55** (1961), 363–378.
[5] Differential equations with a perturbed argument, *Russian Math. Surveys* **17** (1962), 61–416.

Kaplan, S.
[1] On the growth of solutions of quasi-linear parabolic equations, *Comm. Pure Appl. Math.* **16** (1963), 304–330.

Karasik, G. J.
[1] On conditions for the existence of periodic solutions of difference equations, *Izv. Vyss. Učebn. Zaved.* **4** (1959), 70–79.
[2] On the existence of periodic solutions of a system of differential equations with retarded argument, *Sibirsk Mat. Z.* **2** (1961), 551–555.

Kato, T.
[1] Asymptotic behaviors in functional–differential equations, *Tohoku Math. J.* **18** (1966), 174–215.

KATO, T.

[1] Integration of the equation of evolution in Banach space, *J. Math. Soc. Japan* **5** (1953), 208–234.

[2] On linear differential equations in Banach spaces, *Comm. Pure Appl. Math.* **9** (1956), 479–486.

[3] Abstract evolution equations of parabolic type in Banach and Hilbert spaces, *Nagoya Math. J.* **19** (1961), 93–125.

[4] Nonlinear evolution equations in Banach spaces, Mimeographed notes, Univ. of California, Berkeley (1964).

KATO, T., AND TANABE, H.

[1] On the abstract evolution equation, *Osaka J. Math.* **14** (1962), 107–133.

KAYANDE, A. A., AND LAKSHMIKANTHAM, V.

[1] Complex differential systems and extension of Lyapunov's method. *J. Math. Anal. Appl.* **13** (1966), 337–347.

[2] General dynamical systems and conditional stability, *Proc. Cambridge Philos. Soc.* **53** (1967), 199–207.

[3] Lyapunov functions defined over sets, to appear.

KISYNSKI, J.

[1] Sur l'existence et l'unicité des solutions des problèmes classiques rélatifs à l'équation $s = F(x, y, z, p, q)$, *Ann. Univ. Mariae Curie-Sklodowska. Sect. A* **11** (1957), 73–112.

[2] Sur les équations différentielles dans les espaces de Banach, *Bull. Acad. Polon. Sci.* **7** (1959), 381–385.

[3] Sur a convergence des approximations successives pour l'équation $\partial^2 z/\partial x\, \partial y = f(x, y, z, \partial z/\partial x, \partial z/\partial y)$, *Ann. of. Math.* **7** (1960), 233–240.

KOMATSU, H.

[1] Abstract analyticity in time and unique continuation property of solutions of a parabolic equation, *J. Fac. Sci. Univ. Tokyo* **9** (1961), 1–11.

KRASNOSELSKII, M. A.

[1] An alternative principle for the existence of periodic solutions for differential equations with retarded argument, *Dokl. Akad. Nauk SSR* **152** (1963), 801–804.

[2] "Positive Solutions of Operator Equations," Noordhoff, (1964).

KRASNOSELSKII, M. A., KREIN, S. G., AND SOBOLEVSKI, P. E.

[1] On differential equations with unbounded operators in Banach spaces, *Dokl. Akad. Nauk SSSR* **111** (1956), 19–22.

[2] On differential equations with unbounded operators in a Hilbert space, *Dokl. Akad. Nauk SSSR* **112** (1957), 990–993.

KRASOVSKII, N. N.

[1] On the application of Lyapunov's second method to equations with time lags, *Prikl. Mat. Meh.* **20** (1956), 315–327.

[2] On the asymptotic stability of systems with after effects, *Prikl. Mat. Meh.* **20** (1956), 513–518.

[3] On periodical solution of differential equations involving a time lag, *Dokl. Akad. Nauk SSSR* **114** (1957), 252–255.

[4] The stability of quasilinear systems with after effects, *Dokl. Akad. Nauk SSSR* **119** (1958), 435–438.

[5] Some problems in the theory of stability of motion, *Goz. Izd. Fiz.-Mat. Lit.*, Moscow, 1959. English transl. Stanford Univ. Press, Stanford, California, 1963.

KREIN, S. G., AND PROZOROVSKAYA, D. I.

[1] Analytic semigroups and incorrect problems for evolutionary equations. *Dokl. Akad. Nauk. SSSR. (N. S.)* **133** (1960), 277–280.

KRZYZANSKI, M.

[1] Sur les solutions des équations du type parabolique déterminées dans une région illimitée, *Bull. Amer. Math. Soc.* **47** (1941), 911–915.

[2] Sur les solutions de l'équation linéaire du type parabolique déterminées par les conditions initiales, *Ann. Soc. Polon. Math.* **18** (1945), 145–156.

[3] Note complémentaire, *Ann. Soc. Polon. Math.* **20** (1947), 7–9.

[4] Sur les solutions de l'équation linéaire du type parabolique déterminées par les conditions initiales, *Ann. Soc. Polon. Math.* **20** (1948), 7–9.

[5] Sur l'allure asymptotique des solutions d'équation du type parabolique, *Bull. Acad. Polon. Sci. Cl. III* **4** (1956), 247–251.

[6] Évaluations des solutions de l'équation aux dérivées partielles du type parabolique, déterminées dans un domaine non borné, *Ann. Polon. Math.* **4** (1957), 93–97.

[7] Recherches concernant l'allure des solutions de l'équation dy type parabolique lorsque la variable du temps tend vers l'infini, *Atti Accad. Naz. Lincei. Rend. Cl. Sci. Fis. Mat. Natur.* **23** (1957a), 28–32.

[8] Certaines inégalités relatives aux solutions de l'équation parabolique linéaire normal, *Bull. Acad. Polon. Sci. Sér. Sci. Math. Astronom. Phys.* **7** (1959), 131–135.

[9] Sur l'unicité des solutions des second et troisième problèmes de Fourier relatifs à l'équation linéaire normale du ype parabolique, *Ann. Polon. Math.* **7** (1960), 201–208.

[10] Une propriété de solution de l'équation linéaire du type parabolique à coefficients non bornés, *Ann. Polon. Math.* **12** (1962), 209–212.

LADYZHENSKAJA, O. A.

[1] On the uniqueness of solutions of the Cauchy problem for linear parabolic equations, *Mat. Sb.* **27** (1950), 175–184.

LAKSHMIKANTHAM, V.

[1] Lyapunov function and a basic inequality in delay–differential equations, *Arch. Rational Mech. Anal.* **10** (1962), 305–310.

[2] On the uniqueness and boundedness of solutions of hyperbolic differential equations, *Proc. Cambridge Philos. Soc.* **58** (1962), 583–587.

[3] Differential equations in Banach spaces and extensions of Lyapunov's method, *Proc. Cambridge Philos. Soc.* **59** (1963), 373–381.

[4] Properties of solutions of abstract differential inequalities, *Proc. London Math. Soc.* **14** (1964), 74–82.

[5] Parabolic differential equations and Lyapunov-like functions, *J. Math. Anal. Appl.* **9** (1964), 234–251.

[6] Functional-differential systems and extensions of Lyapunov's method, *J. Math. Anal. Appl.* **8** (1964), 392–405.

[7] Some results in functional differential equations, *Proc. Nat. Acad. Sci. India* **34** (1964), 299–306.

LAKSHMIKANTHAM, V., AND LEELA, S.

[1] Parabolic differential equations and conditional stability, *Rend. Circ. Mat. Palermo* **15** (1966), 179–192.

[2] Conditional stability of conditional invariant sets and functional differential equations, *Rend. Circ. Mat. Palermo* **15** (1966), 257–272.

[3] Almost periodic systems and differential inequalities, *J. Indian Inst. Tech.* **1** (1967), 286–301.

LAKSHMIKANTHAM, V., AND SHENDGE, G. R.

[1] Functional differential inequalities, *An. Acad. Brasil. Ci.* **39** (1967), 7–14.

LAX, P. D., AND MILGRAM, A.

[1] Parabolic equations, "Contributions to the Theory of Partial Differential Equations." Pp. 167–190, Princeton Univ. Press, Princeton, New Jersey, 1954.

LEES, M. (see Cohen, P.)

[1] Energy inequalities for the solution of differential equations, *Trans. Amer. Math. Soc.* **94** (1960), 58–73.

LEES, M., AND PROTTER, M. H.

[1] Unique continuation for parabolic differential equations and inequalities, *Duke Math. J.* **28** (1961), 369–382.

LEVIN, J. J., AND NOHEL, J. A.

[1] On a nonlinear delay equations, *J. Math. Anal. Appl.* **8** (1964), 31–44.

LIBERMAN, L. H.

[1] Stability of solutions of differential-operator equations with retarded times under perturbations bounded in the mean, *Sibirsk. Mat. Z.* **4** (1963), 138–144.

LIONS, J. L.

[1] "Équations Différentielles. Opérationelles et Problèmes aux Limites." Springer, New York, 1961.

LIONS, J. L., AND MALGRANGE, B.

[1] Sur l'unicité rétrograde dans les problèmes mixtes paraboliques, *Math. Scand.* **8** (1960), 277–286.

LYUBIC, YU. I.

[1] Conditions for the uniqueness of the solution to Cauchy's abstract problem, *Dokl. Akad. Nauk. SSSR* (N.S.) **130** (1960), 969–972.

MAMEDOV, JA. D.
[1] One-sided estimates in examination of solutions to the Cauchy limit problem for differential equations with unbounded operators, *Sibirsk Mat. Z.* **7** (1966), 313–317.
[2] One-sided estimates in the conditions for asymptotic stability of solutions of differential equations involving unbounded operators, *Dokl. Akad. Nauk SSR*, **166** (1966), 533–535.

MAPLE, C. G., AND PETERSON, L. G.
[1] Stability of solutions of nonlinear diffusion problems, *J. Math. Anal. Appl.* **14** (1966), 221–241.

MASSERA, J. L., AND SCHAFFER, J. J.
[1] "Linear Differential Equations and Functions Spaces." Academic Press, New York, 1966.

McNABB, A.
[1] Notes on criteria for the stability of steady state solutions of parabolic equations, *J. Math. Anal. Appl.* **4** (1962), 193–201.

MILICER-GRUZEWSKA, H.
[1] Le théorème d'unicité de la solution du système parabolique des équations linéaires avec les coefficients hölderiens, *Bull. Acad. Polon. Sci. Sér. Sci. Math. Astronom. Phys.* **7** (1959), 593–599.
[2] Le second théorème d'unicité de solution d'un système parabolique d'équations linéaires avec les coefficients hölderiens, *Bull. Acad. Polon. Sci. Sér. Sci. Math. Astronom. Phys.* **7** (1959), 719–720.

MILLER, R. K.
[1] Asymptotic behavior of nonlinear delay–differential equations, *J. Differential Equations* **1** (1965), 293–305.

MINORSKY, N.
[1] "On Certain Applications of Difference–Differential Equations." Stanford Univ. Press, Stanford, California, 1948.

MINTY, G. J.
[1] Monotone (non-linear) operators in Hilbert space, *Duke Math. J.* **29** (1962), 341–346.
[2] On a "monotonicity" method for the solution of non-linear equations in Banach spaces, *Proc. Nat. Acad. Sci. U.S.A.* **50** (1963), 1038–1041.

MITRYAKOV, A. P.
[1] On periodic solutions of the nonlinear hyperbolic equation, *Trudy Inst. Mat. Meh. Akad. Nauk Uzbek. SSR* **7** (1949), 137–149.

MIZOHATA, S.
[1] Le problème de Cauchy pour les équations paraboliques, *J. Math. Soc. Japan* **8** (1956), 269–299.

[2] Hypoellipticité des équations parabolique, *Bull. Soc. Math. France* **85** (1957), 15–49.

[3] Le problème de Cauchy le passe pour quelques équations paraboliques, *Proc. Japan Acad.* **34** (1958), 693–696.

MLAK, W.

[1] The epidermic effect for partial differential inequalities of the first order, *Ann. Polon. Math.* **3** (1956), 157–164.

[2] Remarks on the stability problem for parabolic equations, *Ann. Polon. Math.* **3** (1957), 343–348.

[3] Differential inequalities of parabolic type, *Ann. Polon. Math.* **3** (1957), 349–354.

[4] Differential inequalities in linear spaces, *Ann. Polon. Math.* **5** (1958), 95–101.

[5] A note on non-local existence of solutions of ordinary differential equations, *Ann. Polon. Math.* **4** (1958), 344–347.

[6] Limitation of solutions of parabolic equations, *Ann. Polon. Math.* **5**(1958/59), 237–245.

[7] The first boundary-value problem for a non-linear parabolic equation, *Ann. Polon. Math.* **5** (1958/59), 257–262.

[8] Limitations and dependence on parameter of solutions of non-stationary differential operator equations, *Ann. Polon. Math.* **6** (1959), 305–322.

[9] On a linear differential inequality of parabolic type, *Bull. Acad. Polon. Sci. Sér. Sci. Math. Astronom. Phys.* **7** (1959), 653–656.

[10] Differential inequalities with unbounded operators in Banach spaces, *Ann. Math. Pol.* **9** (1960), 101–111.

[11] Parabolic differential inequalities and Chaplighin's method, *Ann. Polon. Math.* **8** (1960), 139–153.

[12] Integration of differential equation with unbounded operators in abstract (L)-spaces, *Bull. Acad. Polon. Sci. Sér. Sci. Math. Astronom. Phys.* **8** (1960), 163–168.

[13] Note on abstract differential inequalities and Chaplighin method, *Ann. Polon. Math.* **10** (1961), 253–271.

[14] A note on approximation of solutions of abstract differential equations, *Ann. Polon. Math.* **10** (1961), 273–278.

[15] Note on abstract linear differential inequalities, *Rev. Math. Pures Appl. Acad. Rep. Pop. Roumaine* **6** (1961), 655–657.

[16] Estimates of solutions of hyperbolic systems of differential equations in two independent variables, *Ann. Polon. Math.* **12** (1962), 191–197.

[17] An example of the equation $u_t = u_{xx} + f(x, t, u)$ with distinct maximum and minimum solutions of a mixed problem, *Ann. Polon. Math.* **13** (1963), 101–103.

[18] Integration of linear differential inequalities with distributions, *Contr. to Differential Equations* **2** (1963), 265–268.

MURAKAMI, H.

[1] On non-linear partial differential equations of parabolic types, *Proc. Japan Acad.* **3** (1957), 530–535, 616–627.

[2] Semi-linear partial differential equations of parabolic type, *Funkcial. Ekvac.* **3** (1960), 1–50.

[3] On non-linear ordinary and evolution equations, *Funkcial. Ekvac.* **9** (1966), 151–162.

MYSHKIS, A. D. (see El'sgol'ts, L. E., and Shimanov, S. N.)
[1] Hystero-differential equations, *Uspehi Mat. Nauk* **4** (1949), 190–193.
[2] General theory of differential equations with a retarded argument, *Uspehi Mat. Nauk* **4** (1949), 99–141.
[3] Supplementary bibliographical material to the paper "General theory of differential equations with retarded argument," *Uspehi Mat. Nauk.* **5** (1950), 148–154.
[4] Linear homogeneous differential equations of the first order with retarded argument, *Uspehi Mat. Nauk* **5** (1950), 160–162.
[5] On the solutions of linear homogeneous differential equations of the first order of unstable type with a retarded argument, *Dokl. Akad. Nauk SSSR* **70** (1950), 953–956.
[6] "Linear Differential Equations with Delay in the Argument." Gostehizdat, 1951.
[7] A differential functional inequality, *Uspehi Mat. Nauk* **15** (1960), 157–161.

NAGUMO, M., AND SIMODA, S.
[1] Note sur l'inégalité différentielle concernant les équations du type parabolique, *Proc. Japan Acad.* **27** (1951), 536–539.

NARASIMHAN, R.
[1] On the asymptotic stability of solutions of parabolic differential equations, *J. Rational. Mech. Anal.* **3** (1954), 303–313.

NICKEL, K.
[1] Fehlerabschatzungen bei parabolischen Differentialgleichungen, *Math. Z.* **71** (1959), 268–282.
[2] Parabolic equations with applications to boundary layer theory. "Partial Differential Equations and Continuum Mechanics," Univ. of Wisconsin Press, Madison, Wisconsin, 1961.
[3] Gestaltaussagen über Losungen parabolischer Differentialgleichungen, *J. Reine Angew. Math.* **211** (1962), 78–94.

NIRENBERG, L. (see Agmon, S.)
[1] A strong maximum principle for parabolic equations, *Comm. Pure Appl. Math.* **6** (1953), 167–177.
[2] Existence theorems in partial differential equations, New York Univ. Notes.

NOHEL, J. A. (see Ergen, W. K., and Lipkin, H. J.; Levin, J. J.)
[1] A class of nonlinear delay differential equations, *J. Math. Phys.* **38** (1960), 295–311.

OĞUZTÖRELI, N. N.
[1] "Time-lag Control System." Academic Press, New York, 1966.
[2] On the existence and uniqueness of solutions of an ordinary differential equation in the case of Banach space, *Bull. Acad. Polon. Sci. Sér. Sci. Math. Astronom. Phys.* **8** (1960), 667–673.

ONUCHIC, N.
[1] On the asymptotic behavior of the solutions of functional-differential equations, *Proc. Intern. Symp. Differential Eqs. and Dynamical Systems, Puerto Rico, 1965*, pp. 223–233.
[2] On the uniform stability of a perturbed linear functional-differential equation, *Proc. Amer. Math. Soc.* **19**, (1968), 528–532.

PALCZEWSKI, B.
[1] On the uniqueness of solutions and the convergence of successive approximations in the Darboux problem under the conditions of the Krasnosielski and Krein type, *Ann. Polon. Math.* **14** (1964), 183–190.

PALCZEWSKI, B., AND PAWELSKI, W.
[1] Some remarks on the uniqueness of solutions of the Darboux problem with conditions of the Krasnosielski–Krein type," *Ann. Polon. Math.* **14** (1964), 97–100.

PELCZAR, A.
[1] On the existence and uniqueness of solutions of the Darboux problem for the equation $z_{xy} = f(x, y, z, z_x, z_y)$, *Bull. Acad. Polon. Sci. Sér. Sci. Math. Astronom Phys.* **12** (1964), 703–707.
[2] On the existence and the uniqueness of solutions of certain initial-boundary problems for the equation $z_{xy} = f(x, y, z, z_x, z_y)$, *Bull. Acad. Polon. Sci. Sér. Sci. Math. Astronom. Phys.* **12** (1964), 709–714.
[3] On a modification of the method of Euler polygons for the ordinary differential equation, *Ann. Polon. Math.* **15** (1964), 195–202.
[4] On the convergence to zero of oscillating solutions of some partial differential equations, *Zeszyty Nauk. Uniw. Jagiello. Prace Mat.* **11** (1966), 49–51.
[5] On the method of successive approximations for some operator equations with applications to partial differential hyperbolic equations, *Zeszyty Nauk. Uniw. Jagiello. Prace Mat.* **11** (1966), 59–68.

PERELLO, C. (see Hale, J. K.)
[1] A note on periodic solutions of nonlinear differential equations with time lags, *Proc. Intern. Symp. Differential Eqs. and Dynamical Systems, Puerto Rico, 1965*, pp. 185–187.

PHILIPS, R. S.
[1] Dissipative operators and hyperbolic systems of partial differential equations, *Trans. Amer. Math. Soc.* **90** (1959), 193–254.

PICONE, M.
[1] "Appunti di Analisi Superiore." Napoli, 1941.

PINI, B.
[1] Un problema di valori al contorno, generalizzato per l'equazione a derivate parziali lineare paraboliche del secondo ordine, *Riv. Mat. Parma* **3** (1952), 153–187.
[2] Maggioranti e minoranti delle solzioni delle equazioni paraboliche, *Ann. Mat. Pura Appl.* **37** (1954), 249–264.

[3] Sur primo problema di valori al contorno per l'equazione parabolica non lineare del secondo ordine, *Rend. Sem. Mat. Univ. Padova* **27** (1957), 149–161.

PINNEY, E.
[1] "Differential–Difference Equations." Univ. of California Press, Berkeley, California, 1958.

PLIS, A.
[1] The problem of uniqueness for the solution of a system of partial differential equations, *Bull. Acad. Polon. Sci. Cl. III* **2** (1954), 55–57.
[2] On the uniqueness of the non-negative solution of the homogeneous Cauchy problem for a system of partial differential equations, *Ann. Polon. Math.* **2** (1955), 314–318.
[3] On the estimation of the existence domain for solutions of a nonlinear partial differential equation of the first order, *Bull. Acad. Polon. Sci. Cl. III* **4** (1956), 3, 125.
[4] Generalization of the Cauchy problem for a system of partial differential equations, *Bull. Acad. Polon. Sci. Cl. III* **4** (1956), 741–744.
[5] On the uniqueness of the non-negative solution of the homogeneous mixed problem for a system of partial differential equation, *Ann. Polon. Math.* **7** (1960), 255–258.
[6] Remark on a partial differential inequality of the first order, *Ann. Polon. Math.* **16** (1965), 371–375.

POGORZELSKI, W.
[1] Étude de la solution fondamentale de l'équation parabolique, *Ricerche Mat.* **5** (1956), 25–57.
[2] Propriétés des intégrales généralisées de Poisson Weierstrass et problème de Cauchy pour un système parabolique, *Ann. Ec. Norm.* **76** (1959), 125–149.

PRODI, G.
[1] Questioni di stabilita per equazioni non lineari alle derivate parziali di tipo parabolico, *Atti Accad. Naz. Lincei. Rend. Cl. Sci. Fis. Mat. Natur.* **10** (1951), 365–370.
[2] Teoremi di esistenza per equazioni alle derivate parziali non lineari di tipo parabolico. I. II, *Rend. Inst. Lombardo* (1953), 1–47.
[3] Soluzioni periodiche di equazioni alle derivate parziali di tipo iperbolico non-lineari, *Ann. Mat. Pura Appl.* **42** (1956), 25–49.

PROTTER, M. H. (see Lees, M.)
[1] Asymptotic behavior and uniqueness theorems for hyperbolic equations and inequalities, Univ. of California, Berkeley, Tech. Rep. No. 9, March, 1960.
[2] Properties of solutions of parabolic equations and inequalities, *Canad. J. Math.* **13** (1961), 331–345.

RAMAMOHAN RAO, M., AND TSOKOS, C. P.
[1] On the stability and boundedness of differential systems in Banach spaces. *Proc. Cambridge Philos. Soc.* **65** (1969), 507–512.

RAMAMOHAN RAO, M., AND VISWANATHAM, B.
[1] On the asymptotic behaviour of solutions of partial differential equations of hyperbolic type. *Proc. Nat. Acad. Sci. (India)*, **33** (1963), 279–284.

RAZUMIKHIN, B. S.
[1] Stability of motion, *Prikl. Mat. Meh.* **20** (1956), 266–270.
[2] Stability of systems with lag, *Prikl. Mat. Meh.* **20** (1956), 500–512.
[3] Estimates of solutions of a system of differential equations of the perturbed motion with variable coefficients, *Prikl. Mat. Meh.* **21** (1957), 119–120.
[4] Stability in the first approximation of systems with lag, *Prikl. Mat. Meh.* **22** (1958), 155–166.
[5] On the application of Liapunov's method to stability problems, *Prikl. Mat. Meh.* **22** (1958), 338–349.
[6] Application of Lyapunov's method to problems in the stability of systems with a delay, *Avtomat. i Telemeh.* **21** (1960), 740–748.

REKLISHKII, Z. I.
[1] Stability of solutions of certain linear differential equations with a lagging argument in Banach space, *Dokl. Akad. Nauk SSSR* **111** (1956), 29–32.
[2] Necessary and sufficient conditions for stability of solutions of certain systems of linear differential equations with lagging argument in Banach space, *Proc. 3rd All Union Math. Congr.* **2** (1956), 128–129.
[3] Tests for boundedness of solutions of linear differential equations with variable lag of the argument, *Dokl. Akad. Nauk SSSR* **118** (1958), 447–449.
[4] Tests for boundedness of solutions of linear differential equations with several lags of the argument, *Dokl. Akad. Nauk SSSR* **125** (1959), 46–47.
[5] Tests for boundedness of solutions of differential equations with continuous lag of the argument in a Banach space, *Dokl. Akad. Nauk SSSR* **127** (1959), 971–974.

SANTORO, P.
[1] Sul problema di Darboux per l'equazione $s = f(x, y, z, p, q)$ e il fenomeno di Peano, *Rend. Accad. Naz. XL* **10** (1959), 3–17.

SARKOVA, N. V.
[1] Approximate solution of differential equations with lagging argument, (Coll. of papers on oscillating functions.) *Učen. Zap. Perm. Univ.* **13** (1959), 63–73.

SATO, T.
[1] Sur l'équation aux dérivées partielles $\Delta z = f(x, t, z, p, q)$, *Compositio. Math.* **12** (1954), 157–177; and II, *Ibid.* **14** (1959), 151–171.

SEIFERT, G.
[1] A condition for almost periodicity with some applications to functional–differential equations, *J. Differential Equations* **1** (1965), 393–408.

SERRIN, J.
[1] Local behavior of solutions of quasilinear equations, *Acta Math.* **111** (1964), 247–302.

SHANAHAN, J. P.
   [1] On uniqueness questions for hyperbolic differential equations, *Pacific J. Math.* **10** (1960), 677–688.

SHIMANOV, S. N. (see El'sgol'ts, L. E., and Myshkis, A. D.)
   [1] Almost periodic solutions of non-homogeneous linear differential equations with lag, *Izv. Vyss. Učebn. Zaved. Mat.* **4** (1958), 270–274.
   [2] Almost periodic oscillations in nonlinear systems with lag. *Dokl. Akad. Nauk SSSR* **125** (1959), 1203–1206.
   [3] On the theory of oscillations of quasilinear systems with delay, *Prikl. Mat. Meh.* **23** (1959), 836–844.
   [4] On the instability of the motion of systems with retardation, *Prikl. Mat. Meh.* **24** (1960), 55–63.
   [5] On stability in the critical case of a zero root with a time lag, *Prikl. Mat. Meh.* **24** (1960), 447–457.
   [6] Almost periodic oscillations of quasilinear systems with time lag in degenerate case, *Dokl. Akad. Nauk SSSR* **133** (1960), 36–39.
   [7] Oscillations of quasilinear autonomous systems with time lag, *Izv. Vysš. Učebn. Zaved. Radiofizika*, **3** (1960), 456–466.
   [8] On stability of quasiharmonic systems with retardation, *Prikl. Mat. Meh.* **25** (1961), 992–1002.
   [9] On the theory of linear differential equations with periodic coefficients and time lag, *Prikl. Mat. Meh.* **27** (1963), 450–458.
   [10] On the theory of linear differential equations with retardations, Differencial'nye Uravnenija **1** (1965), 102–116.

SLOBODETSKI, L. N.
   [1] On the fundamental solution and the Cauchy problem for parabolic systems, *Math. Sb.* **46** (1958), 229–258.

SMIRNOVA, G. N.
   [1] On uniqueness classes for the solution of Cauchy's problem for parabolic equations, *Dokl. Akad. Nauk SSSR* **153** (1963), 1269–1272.

SOBOLEVSKI, P. E. (see Krasnoselskii, M. A., and Krein, S. G.)
   [1] Approximate methods of solving differential equations in Banach space, *Dokl. Akad. Nauk SSSR* **115** (1957), 240–243.
   [2] On equations of parabolic type in Banach space, *Trudy Moscov. Mat. Obšč.* **10** (1961), 298–350.

SOLOV'EV, P. V.
   [1] Some remarks on periodic solutions of the nonlinear equation of hyperbolic type, *Izv. Akad. Nauk SSSR Ser. Mat.* **2** (1939), 150–164.

STOKES, A. P.
   [1] A Floquet theory for functional-differential equations, *Proc. Nat. Acad. Sci. U.S.A.* **48** (1962), 1330–1334.
   [2] On the stability of a limit cycle of an autonomous functional-differential equations, *Contr. Differential Equations* **3** (1964), 121–140.

SUGIYAMA, S.

[1] On the existence and uniqueness theorems of difference–differential equations, *Kōdai Math. Sem. Rep.* **12** (1960), 179–190.

[2] A note on difference–differential equations with a parameter, *Mem. School Sci. Engrg. Waseda Univ. Tokyo* **25** (1961), 23–36.

[3] Existence theorems on difference–differential equations, *Proc. Japan Acad.* **38** (1962), 145–149.

[4] Dependence properties of solutions on retardation and initial values in the theory of difference–differential equations, *Kōdai Math. Sem. Rep.* **15** (1963), 67–78.

[5] On the theory of difference–differential equations, I, *Bull. Sci. Engrg. Res. Lab. Waseda Univ.* **26** (1964), 97–111.

[6] On the theory of difference–differential equations, II, *Bull. Sci. Engrg. Res. Lab. Waseda Univ.* **27** (1964), 74–84.

[7] On the theory of difference–differential equations, III. Applications to Lyapunov functions, *Mem. School Sci. Engrg. Waseda Univ.* **28** (1964), 73–85.

[8] On a certain functional differential inequality, *Kōdai Math. Sem. Rep.* **17** (1965), 273–280.

[9] On a certain functional differential equation, *Proc. Japan Acad.* **41** (1965), 360–362.

SZARSKI, J.

[1] Sur certains systèmes d'inégalités différentielles aux dérivées partielles du premier ordre, *Ann. Soc. Pol. Math.* **21** (1948), 7–26.

[2] Sur certaines inégalités entre les intégrales des équations différentielles aux dérivées partielles du premier ordre, *Ann. Soc. Pol. Math.* **22** (1949), 1–34.

[3] Systèmes d'inégalités différentielles aux dérivées partielles du premier ordre, et leurs applications, *Ann. Polon. Math.* **1** (1954), 149–165.

[4] Sur la limitation et l'unicité des solutions d'un systèmè non-linéaire d'équations paraboliques aux dérivées partielles du second ordre, *Ann. Pol. Math.* **2** (1955), 237–249.

[5] Sur la limitation et l'unicité des solutions des problèmes de Fourier pour un système non-linéaire d'équations paraboliques, *Ann. Pol. Math.* **6** (1959), 211–216.

[6] "Characteristics and Cauchy Problem for Non-linear Partial Differential Equations of the First Order." Univ. of Kansas, Lawrence, Kansas, 1959.

[7] Sur un système non-linéaire d'inégalités différentielles paraboliques, *Ann. Pol. Math.* **15** (1964), 15–22.

[8] "Differential Inequalities." PWN, Polish Sci. Publ., Warsaw, 1965.

SZMYDT, Z.

[1] Sur un nouveau type de problèmes pour un système d'équations différentielles hyperboliques du second ordre a deux variables indépandantes, *Bull. Acad. Polon. Sci. Cl. III* **4** (1956), 67–72.

[2] Sur une généralisation des problèmes classiques concernant un système d'équations différentielles hyperboliques du second ordre a deux variables indépendantes, *Bull. Acad. Polon. Sci. Cl. III* **4** (1956), 579–584.

[3] Sur l'existence de solutions de certains nouveaux problèmes pour un système d'équations différentielles hyperboliques du second ordre a deux variables indépendantes, *Ann. Polon. Math.* **4** (1957), 40–60.

[4] Sur l'existence d'une solution unique de certains problèmes pour un système d'équations différentielles hyperboliques du second ordre a deux variables indépendantes, *Ann. Polon. Math.* **4** (1958), 165–182.

[5] Sur l'existence de solutions de certains problèmes aux limites relatifs à un système d'équations différentielles hyperboliques, *Bull. Acad. Polon. Sci. Sér. Sci. Math. Astronom. Phys.* **6** (1958), 31–36.

TAAM, C. T., AND WELCH, J. N.

[1] Compact solutions of non-linear differential equations in Banach spaces, *Michigan Math. J.* **13** (1966), 271–284.

TANABE, H. (see Kato, T.)

[1] On the equations of evolution in a Banach space, *Math. J. Osaka* **12** (1960), 363–376.

[2] Convergence to a stationary state of the solution of some kind of differential equations in a Banach space, *Proc. Japan Acad.* **37** (1961), 127–130.

TICHONOV, A. N.

[1] Functional equations of Volterra type and their applications to some problems of mathematical physics, *Byull. Moskov. Univ., Sect. A* **1** (8) (1938), 1–25.

TURSKI, S.

[1] On certain generalization of theorems on the existence and uniqueness of integrals of hyperbolic equations of type $\partial^2 z/\partial x\, \partial y = f(x, y, z, \partial z/\partial x, \partial z/\partial y)$, *Dodatek do Roczn. Pol. Tow. Mat. Krakow*, 1935.

VEJVODA, O.

[1] The stability of a system of differential equations in the complex domain, *Czechoslovak Math. J.* **7** (1957), 137–159.

VISHIK, M. I.

[1] Solvability of boundary problems for quasi-linear parabolic equations of higher order, *Math. Sb.* **59** (1962), 289–335.

VOLKOV, D. M.

[1] An analogue of the second method of Liapunov for nonlinear boundary value problems of hyperbolic equations, *Leningrad Gos. Univ. Učen. Zap. Ser. Math. Nauk* **33** (1958), 90–96.

WALTER, W.

[1] Uber die Euler–Poisson–Darboux–Gleichung, *Math. Z.* **67** (1967), 361–376.

[2] Uber die Differentialgleichung $u_{xy} = f(x, y, u, u_x, u_y)$. I, *Math. Z.* **71** (1959), 308–324.

[3] Uber die Differentialgleichung $u_{xy} = f(x, y, u, u_x, u_y)$. II, *Math. Z.* **71** (1959), 436–453.

[4] Uber die Differentialgleichung $u_{xy} = f(x, y, u, u_x, u_y)$. III, *Math. Z.* **73** (1960), 268–279.

[5] Eindeutigkeitssatze für gewohnliche, parabolische und hyperbolische Differentialgleichungen, *Math. Z.* **74** (1960), 191–208.

[6] Fehlerabschatzungen bei hyperbolischen Differentialgleichungen, *Arch. Rational Mech. Anal.* **7** (1961), 249–272.

[7] Fehlerabschatzungen und Eindeutigkeitssatze für gewohnliche und partielle Differentialgleichungen, *Z. Angew. Math. Mech.* **42** (1962), T49–T62.

[8] "Differential und Integral Ungleichungen." Springer, Berlin, 1964.

WAZEWSKI, T.

[1] Sur l'appréciation du domaine d'existence des intégrales de l'équation aux dérivées partielles du premier ordre, *Ann. Soc. Polon. Math.* **14** (1936), 149–177.

[2] Sur le problème de Cauchy relatif à un système d'équations aux dérivées partielles, *Ann. Soc. Polon. Math.* **15** (1936), 101.

[3] Ueber die Bedingungen des Existenz der Integrale partieller Differentialgleichungen erster Ordnung, *Math. Z.* **43** (1948), 522–532.

[4] Une généralisation des théorèmes sur les accroissements finis au cas des espaces de Banach et applications à la généralisation du théorème de l'Hospital, *Ann. Soc. Polon. Math.* **24** (1953), 132–147.

WEND, D. V. V.

[1] On the zeros solutions of some linear complex differential equations, *Pacific J. Math.* **10** (1960), 713–722.

[2] Singularity-free regions for solutions of non-linear complex differential equations, *Proc. Amer. Math. Soc.* **18** (1967), 592–597.

WESTPHAL, H.

[1] Zur Abschatzung der Losungen nichtlinearer parabolischer Differentialgleichungen, *Math. Z.* **51** (1949), 690–695.

WONG, J. S. W.

[1] On the convergence of successive approximation in the Darboux problem, *Ann. Polon. Math.* **17** (1966), 329–336.

[2] Remarks on the uniqueness theorem of solutions of the Darboux problem, *Canad. Math. Bull.*, **8** (1965), 791–796.

WRIGHT, E. M.

[1] The linear difference-differential equation with asymptotically constant coefficients, *Amer. J. Math.* **70** (1948), 221–238.

[2] The linear difference-differential equation with constant coefficients, *Proc. Roy. Soc. Edinburgh Sect. A* **62** (1949), 387–393.

[3] The stability of solutions of nonlinear difference–differential equations, *Proc. Roy. Soc. Edinburgh Sect. A* **63** (1950), 18–26.

[4] A nonlinear difference–differential equation, *J. Reine Angew. Math.* **194** (1955), 66–87.

YOSHIZAWA, T.

[1] Extreme stability and almost periodic solutions of functional–differential equations, *Arch. Rational, Mech. Anal.* **17** (1964), 148–170.

[2] Asymptotic stability of solution of an almost periodic system of functional–differential equations, *Rend. Circ. Mat. Palermo* [2] **13** (1964), 1–13.

[3] The stability theory by Liapunov's second method, *Math. Soc. Japan, Tokyo* (1967).

ZERAGIA, P. K.

[1] Solution of a basic boundary problem for a certain system of differential equations of parabolic type, *Dokl. Akad. Nauk GSSR* **26** (1951), 257–264.

[2] Boundary problems for certain nonlinear equations of parabolic type, *Trudy Tbil. Math. Inst. Akad. Nauk GSSR* **24** (1957), 195–222.

ZVERKIN, A. M. (see Els'gol'ts, L. E., Kamenskii, G. A. and Norkin, S.B.)

[1] Dependence of stability of solutions of linear differential equations with lagging argument upon the choice of initial moment, *Vestnik Moscov. Univ. Ser. Mat. Meh. Astronom. Fiz. Him.* **5** (1959), 15–20.

[2] On the theory of linear equations with a lagging argument and periodic coefficients, *Dokl. Akad. Nauk SSSR* **128** (1959), 882–885.

[3] General solution of a linear differential equation with perturbed argument, *Naucn. Dokl. Vyss. Skoly, Fis. Mat. Nauik* **1** (1960), 30–37.

Author Index

A

Agmon, S., 272, 289
Alexiewicz, A., 233, 289
Antosiewicz, H. A., 272, 289
Aronson, D. G., 219, 289
Aziz, A. K., 233, 289, 290

B

Barrar, R. B., 219, 290
Bellman, R., 42, 219, 290
Besala, P., 219, 290
Bhatia, N. P., 291
Bielecki, A., 233, 291
Bogdanowicz, W. M., 291
Browder, F. E., 272, 291
Brzychaczy, S., 219, 291
Bullock, R. M., 42, 291

C

Calderon, A. P., 292
Cameron, R. H., 219, 292
Cesari, L., 233, 292
Chandra, J., 272, 292
Chang, H., 292
Chu, S. C., 233, 292
Ciliberto, C., 219, 233, 292
Coffman, C. V., 272, 292
Cohen, P., 272, 293
Conlan, J., 233, 293
Conti, R., 233, 293
Cooke, K. L., 42, 290, 293
Corduneanu, C., 80, 293

D

Danskin, J. M., 42, 290
Das, K. M., 288, 293
Deo, S. G., 288, 294
Diaz, J. B., 233, 289, 294
Driver, R. D., 42, 80, 110, 294

E

Edmunds, D. E., 272, 294
Eidelman, S. D., 219, 294
El'sgol'ts, L. E., 42, 110, 294, 295

F

Fleishman, B. A., 272, 292
Fodcuk, V. I., 42, 295
Foias, C., 219, 272, 295
Franklin, J., 42, 295
Friedman, A., 219, 295

G

Giuliano, L., 296
Glicksberg, I., 290
Guglielmino, F., 233, 296
Gussi, G., 295

H

Hahn, W., 42, 297
Halanay, A., 42, 80, 110, 297
Hale, J. K., 42, 80, 233, 297
Hartman, P., 272, 297
Hastings, S. P., 42, 297

I

Ilin, A. M., 219, 298
Ito, S., 219, 299

J

Jones, G. S., 42, 299

K

Kakutani, S., 42, 299
Kalashnikov, A. S., 219, 298
Kamenskii, G. A., 42, 295, 299
Kaplan, S., 219, 299
Karasik, G. J., 42, 299
Kato, J., 80, 299
Kato, T., 272, 300
Kayande, A. A., 288, 300
Kisynski, J., 233, 272, 300
Komatsu, H., 219, 300
Krasnosel'skii, M. A., 272, 300
Krasovskii, N. N., 42, 80, 110, 300
Krein, S. G., 272, 300, 301
Krzyzanski, M., 219, 301

L

Ladyzhenskaja, O. A., 219, 301
Lakshmikantham, V., 42, 80, 110, 219, 233, 272, 288, 294, 300, 301, 302
Lax, P. D., 219, 302
Leela, S., 80, 110, 219, 302
Lees, M., 219, 272, 302
Levin, J. J., 302
Liberman, L. H., 80, 302
Lions, J. L., 219, 272, 302
Lyubic, Yu. I., 272, 302

M

McNabb, A., 219, 303
Malgrange, B., 219, 302
Maloney, J. P., 233, 290
Mamedov, Ja. D., 272, 303
Maple, C. G., 219, 303
Markus, L., 42, 299
Massera, J. L., 272, 303
Milgram, A., 219, 302

Milicer-Gruzewska, H., 219, 303
Miller, R. K., 80, 303
Minorsky, N., 303
Minty, G. J., 272, 303
Mitryakov, A. P., 303
Mizohata, S., 219, 303
Mlak, W., 219, 272, 304
Murakami, H., 219, 272, 304
Myshkis, A. D., 42, 295, 305

N

Nagumo, M., 219, 305
Narasimhan, R., 219, 305
Nickel, K., 219, 305
Nirenberg, L., 219, 272, 289, 305
Nohel, J. A., 302, 305
Norkin, S. G., 42, 295

O

Ogustoreli, N. N., 42, 110, 305
Olenik, O. A., 219, 298
Onuchic, N., 42, 306
Orlicz, W., 233, 289

P

Palczewski, B., 233, 306
Pawelski, W., 233, 306
Pelczar, A., 233, 306
Perello, C., 42, 298, 306
Peterson, L. G., 219, 303
Phillips, R. S., 233, 306
Picone, M., 219, 306
Pini, B., 219, 306
Pinney, E., 42, 307
Plis, A., 148, 307
Poenaru, V., 295
Pogorzelski, W., 219, 307
Prodi, G., 219, 307
Protter, M. H., 219, 233, 302, 307
Prozorovskaya, D. I., 272, 301

R

Ramamohan Rao, M., 272, 307, 308
Razumikhin, B. S., 80, 110, 308
Reklishii, Z. I., 80, 308

S

Santoro, P., 233, 308
Sarkova, N. V., 308
Sato, T., 308
Schaffer, J. J., 272, 303
Seifert, G., 80, 308
Serrin, J., 219, 308
Shanahan, J. P., 233, 309
Shendge, G. R., 42, 302
Shimanov, S. N., 42, 295, 309
Simoda, S., 219
Slobodetski, L. N., 219, 309
Smirnova, G. N., 219, 309
Sobolevski, P. E., 272, 300
Solov'ev, P. V., 309
Stokes, A. P., 42, 309
Sugiyama, S., 42, 80, 310
Szarski, J., 148, 219, 233, 272, 310
Szego, G. P., 291
Szmydt, Z., 233, 310

T

Taam, C. T., 272, 311
Tanabe, H., 272, 300, 311
Tichonov, A. N., 311
Tsokos, C. P., 272, 307

Turski, S., 311

V

Vejvoda, O., 311
Vishik, M. I., 219, 311
Viswanatham, B., 308
Volkov, D. M., 233, 311

W

Walter, W., 219, 233, 311
Wazewski, T., 312
Welch, J. N., 272, 291, 311
Wend, D. V. V., 288, 312
Westphal, H., 219, 312
Wong, J. S. W., 233, 312
Wright, E. M., 42, 312

Y

Yoshizawa, T., 42, 80, 313

Z

Zeragia, P. K., 219, 313
Zverkin, A. M., 42, 313

Subject Index

A

Almost periodic solutions, 77
Approximate solutions, 9, 38, 134, 170, 231, 255, 275
Asymptotic behaviour, 24, 105
Asymptotic equivalence, 25
Asymptotic in the sense of Wintner, 32
Autonomous systems, 58

B

Boundary value problem
    Dirichlet type, 164
    exterior, 213
    first, 163
    first Fourier, 164
    mixed, 164
    Newmann type, 164
    second, 164
Boundedness, 60, 190, 196, 200, 269
Bounds, 13, 127, 163, 229, 284.

C

Chaplygin's method, 259
Characteristic exponent, 264
Comparison theorems, 37, 81, 118, 155, 267, 287
Continuous dependence, 18, 247
Converse theorems, 49
    for extreme stability, 67
    for generalized exponential stability, 49, 50

D

Demi-continuity, 240
Differential inequalities
    functional, 34
    hyperbolic, 221
    parabolic, 149, 181, 205
    partial (of the first order), 113, 136
Diffusion equation, 160
Directional derivative, 151

E

Egress points, 29
Ellipticity, 152
Error estimates, 38, 170, 229, 231
Estimate of Time lag, 39, 100
Existence theorems
    for abstract differential equations, 237, 241
    for evolution equations, 249
    for functional differential equations, 5

G

Gateaux derivative, 264
Growth damping factor, 207

H

Homotopic, 241

I

Instability, 61
Invariant sets, 58

SUBJECT INDEX

L

$\mathscr{L}$-condition, 205
Lyapunov functions, 81, 267
Lyapunov functionals, 43
Lyapunov-like functions, 144, 186, 287

M

Maximal and minimal solutions, 36
   existence, 36
Maximum and minimum principles, 159
Mayer's transformation, 143
Method of averaging, 247
Mild solution, 250
Monotonicity condition, 240

N

Nonlinear boundary conditions, 154

P

Parabolicity, 152
Perturbed systems, 23, 62, 97, 197

R

Retarded arguments, 3

S

Several Lyapunov functions, 110
Singularity-free regions, 279

Stability, 21, 43, 87, 146, 175, 190, 267
   asymptotic, 22, 45, 89, 177, 192, 271
   conditional, 200
   eventual, 101
   extreme, 66
   generalized exponential, 49
   perfect, 74
   of steady state solutions, 174
   strong, 74
Strict solution, 250
Systems with repulsive forces, 32

T

Topological principle, 29

U

Under and over functions, 35, 163, 222
Uniqueness criteria,
   for abstract differential equations, 243, 257
   for complex differential equations, 278
   for functional differential equations, 11, 39, 72
   for hyperbolic differential equations, 223, 225
   for nonlinear evolution equations, 257
   for parabolic differential equations, 174, 210, 214
   for partial differential equations of first order, 136

Mathematics in Science and Engineering

A Series of Monographs and Textbooks

Edited by RICHARD BELLMAN, *University of Southern California*

1. T. Y. Thomas. Concepts from Tensor Analysis and Differential Geometry. Second Edition. 1965

2. T. Y. Thomas. Plastic Flow and Fracture in Solids. 1961

3. R. Aris. The Optimal Design of Chemical Reactors: A Study in Dynamic Programming. 1961

4. J. LaSalle and S. Lefschetz. Stability by by Liapunov's Direct Method with Applications. 1961

5. G. Leitmann (ed.). Optimization Techniques: With Applications to Aerospace Systems. 1962

6. R. Bellman and K. L. Cooke. Differential-Difference Equations. 1963

7. F. A. Haight. Mathematical Theories of Traffic Flow. 1963

8. F. V. Atkinson. Discrete and Continuous Boundary Problems. 1964

9. A. Jeffrey and T. Taniuti. Non-Linear Wave Propagation: With Applications to Physics and Magnetohydrodynamics. 1964

10. J. T. Tou. Optimum Design of Digital Control Systems. 1963.

11. H. Flanders. Differential Forms: With Applications to the Physical Sciences. 1963

12. S. M. Roberts. Dynamic Programming in Chemical Engineering and Process Control. 1964

13. S. Lefschetz. Stability of Nonlinear Control Systems. 1965

14. D. N. Chorafas. Systems and Simulation. 1965

15. A. A. Pervozvanskii. Random Processes in Nonlinear Control Systems. 1965

16. M. C. Pease, III. Methods of Matrix Algebra. 1965

17. V. E. Benes. Mathematical Theory of Connecting Networks and Telephone Traffic. 1965

18. W. F. Ames. Nonlinear Partial Differential Equations in Engineering. 1965

19. J. Aczel. Lectures on Functional Equations and Their Applications. 1966

20. R. E. Murphy. Adaptive Processes in Economic Systems. 1965

21. S. E. Dreyfus. Dynamic Programming and the Calculus of Variations. 1965

22. A. A. Fel'dbaum. Optimal Control Systems. 1965

23. A. Halanay. Differential Equations: Stability, Oscillations, Time Lags. 1966

24. M. N. Oguztoreli. Time-Lag Control Systems. 1966

25. D. Sworder. Optimal Adaptive Control Systems. 1966

26. M. Ash. Optimal Shutdown Control of Nuclear Reactors. 1966

27. D. N. Chorafas. Control System Functions and Programming Approaches (In Two Volumes). 1966

28. N. P. Erugin. Linear Systems of Ordinary Differential Equations. 1966

29. S. Marcus. Algebraic Linguistics; Analytical Models. 1967

30. A. M. Liapunov. Stability of Motion. 1966

31. G. Leitmann (ed.). Topics in Optimization. 1967

32. M. Aoki. Optimization of Stochastic Systems. 1967

33. H. J. Kushner. Stochastic Stability and control. 1967

34. M. Urabe. Nonlinear Autonomous Oscillations. 1967

35. F. Calogero. Variable Phase Approach to Potential Scattering. 1967

36. A. Kaufmann. Graphs, Dynamic Programming, and Finite Games. 1967

37. A. Kaufmann and R. Cruon. Dynamic Programming: Sequential Scientific Management. 1967

38. J. H. Ahlberg, E. N. Nilson, and J. L. Walsh. The Theory of Splines and Their Applications. 1967

39. Y. Sawaragi, Y. Sunahara, and T. Nakamizo. Statistical Decision Theory in Adaptive Control Systems. 1967

40. R. Bellman. Introduction to the Mathematical Theory of Control Processes Volume I. 1967 (Volumes II and III in preparation)

41. E. S. Lee. Quasilinearization and Invariant Imbedding. 1968

42. W. Ames. Nonlinear Ordinary Differential Equations in Transport Processes. 1968

43. W. Miller, Jr. Lie Theory and Special Functions. 1968

44. P. B. Bailey, L. F. Shampine, and P. E. Waltman. Nonlinear Two Point Boundary Value Problems. 1968.

45. Iu. P. Petrov. Variational Methods in Optimum Control Theory. 1968

46. O. A. Ladyzhenskaya and N. N. Ural'tseva. Linear and Quasilinear Elliptic Equations. 1968

47. A. Kaufmann and R. Faure. Introduction to Operations Research. 1968

48. C. A. Swanson. Comparison and Oscillation Theory of Linear Differential Equations. 1968

49. R. Hermann. Differential Geometry and the Calculus of Variations. 1968

50. N. K. Jaiswal. Priority Queues. 1968

51. H. Nikaido. Convex Structures and Economic Theory. 1968

52. K. S. Fu. Sequential Methods in Pattern Recognition and Machine Learning. 1968

53. Y. L. Luke. The Special Functions and Their Approximations (In Two Volumes). 1969

54. R. P. Gilbert. Function Theoretic Methods in Partial Differential Equations. 1969

55. V. Lakshmikantham and S. Leela. Differential and Integral Inequalities (In Two Volumes.) 1969

56. S. H. Hermes and J. P. LaSalle. Functional Analysis and Time Optimal Control. 1969.

57. M. Iri. Network Flow, Transportation, and Scheduling: Theory and Algorithms. 1969

58. A. Blaquiere, F. Gerard, and G. Leitmann. Quantitative and Qualitative Games. 1969

59. P. Falb and J. De Jong. Successive Approximation Methods in Control and Oscillation Theory. 1969

In preparation

R. Bellman. Methods of Nonlinear Analysis, Volume I

R. Bellman, K. L. Cooke, and J. A. Lockett. Algorithms, Graphs, and Computers

A. H. Jazwinski. Stochastic Processes and Filtering Theory

S. R. McReynolds and P. Dyer. The Computation and Theory of Optimal Control

J. M. Mendel and K. S. Fu. Adaptive, Learning, and Pattern Recognition Systems: Theory and Applications

G. Rosen. Formulations of Classical and Quantum Dynamical Theory

E. J. Beltrami. Methods of Nonlinear Analysis and Optimization

H. H. Happ. The Theory of Network Diakoptics

QA
371
L23
v.2

APR 19 1971